METHODS OF STRUCTURAL SAFETY

H. O. MADSEN

Det norske Veritas,
P.O. Box 300,
1322 Hovik, Oslo,
Norway

S. KRENK

Risø National Laboratory,
4000 Roskilde,
Denmark

N.C. LIND

University of Waterloo,
Waterloo, Ontario N2L 3G1,
Canada

Prentice-Hall, Inc., Englewood Cliffs, NJ 07632

Library of Congress Cataloging in Publication Data

Madsen, H. O.
    Methods of structural safety
    (Prentice-Hall international series in civil
engineering and engineering mechanics)
    Bibliography: p.
    Includes index.
    1. Structural stability.   2. Structural failures.
I. Krenk, S.   II. Lind, N.C. (Niels Christian),
1930–   .   III. Title.   IV. Series.
TA656.M33 1986   624.1'71   85-9272
ISBN 0-13-579475-7

Editorial/production supervision: *Mary Carnis*
Cover design: *Debra Watson*
Manufacturing buyer: *Rhett Conklin*

Printed in the United States of America

10 9 8 7 6 5 4 3 2 1

Prentice-Hall International, Inc., *London*
Prentice-Hall of Australia Pty. Limited, *Sydney*
Editora Prentice-Hall do Brasil, Ltda., *Rio de Janeiro*
Prentice-Hall Canada Inc., *Toronto*
Prentice-Hall Hispanoamericana, S.A., *Mexico*
Prentice-Hall of India Private Limited, *New Delhi*
Prentice-Hall of Japan, Inc., *Tokyo*
Prentice-Hall of Southeast Asia Pte. Ltd., *Singapore*
Whitehall Books Limited, *Wellington, New Zealand*

CONTENTS

PREFACE

Traditionally, structural design relies on deterministic analysis. Suitable dimensions, material properties, and loads are assumed, and an analysis is then performed to provide a more or less detailed description of the structure. However, fluctuations of the loads, variability of the material properties, and uncertainties regarding the analytical models all contribute to a generally small probability that the structure does not perform as intended. In response to this problem, methods have been developed to deal with the statistical nature of loads and material properties, and more recently, a general framework for comparing and combining these statistical effects has emerged. The methods are rapidly finding application to structural design and reassessment of the safety of existing structures.

The present book provides an introduction to the general problem of consistent evaluation of the safety of structures and gives an up-to-date presentation of the associated methods of analysis. It contains an extensive account of safety index methods, in which failure is described in terms of one or more limit states. In the mathematical formulation the limit states are described as surfaces in a parameter space, and both the parameters and the shape of the surfaces are subject to statistical variation. Although a safety index can make no claim of representing an absolute measure of the failure probability, it does provide an efficient means of comparing the reliability of different structures. It also enables an evaluation of the effect of redundancy of complex structures, and an associated sensitivity analysis can identify critical load or design parameters.

Some of the factors affecting the safety of structures can be described in more detail by specific stochastic models, and several of these are presented. The necessary background is provided in a chapter on stochastic processes and extreme-value theory, and these ideas are further developed in the analysis of response of structures. Statistical properties of material behavior are also treated, notably in connection with size effects, fatigue damage and crack propagation theories. Stochastic models are also able to capture essential features of some important types of loads, and accounts are given of the natural loads from wind, waves and earthquakes, and man-made loads in buildings and from traffic.

The reader is assumed to be familiar with elementary probability theory. In addition, some knowledge of structural analysis is necessary to appreciate the specific problems and examples. References are provided for background material and further reading. However, the literature relating to various aspects of the subject is vast, and only a selection could be included.

It is a pleasure to thank Det norske Veritas for its positive attitude and practical support during the writing of the book.

H. O. Madsen
S. Krenk
N. C. Lind

1

INTRODUCTION

1.1 HISTORICAL BACKGROUND

Early history. The development of the theory of structural reliability has a history of some 50 to 60 years. The first phase, from about 1920 to 1960, appears in retrospect as a very slow beginning. During this period one or two dozen pioneers, scattered over many countries, worked independently on various elements of the theory of reliability applied to problems of structural strength. They questioned established thought and developed the basic concepts of random structural events, departing radically from the classical notions of structural engineering.

Until about 1960 this pioneering work was largely ignored. The main body of the structural engineering profession was occupied with other developments. Insofar as theory is concerned, most attention was given to the development of linear elastic, and later plastic, analysis of structural systems and to the mathematical demands of particular forms, such as shells. The challenge of ordinary stress analysis was enough to absorb the attention of serious engineers; why bother them with tedious questions about safety factors? Questions of safety margin appeared intractable because some of the concepts necessary for rational discourse were not available.

It should not be believed, of course, that the structural theory of this and of earlier periods was naively deterministic. Strength and load were considered uncertain, but it was believed or postulated that an absolute upper limit to any load, and a lower limit to any strength, could be established. Safety factors were then applied to separate these limits "sufficiently" for every member of every structure. Why was a safety factor applied if absolute limits of load and strength existed? The answer might begin: "For example, because of fluctuating workmanship . . ." Workmanship, then, was thought of as a random factor, but varying within limits that made it possible conceptually to counter its influence with a fixed safety factor. All other uncertainties were dealt with in a similar recursive way, using the concepts of absolute limits to uncertain quantities. The notion was that the only missing ingredient was fact. With enough data about the world (went the creed), the applicable limits could be established in principle and in fact if it were justified economically. Safety factors were established only by means of "engineering judgment" as a matter of expediency, not necessity.

Thus over the first two thirds of this century, the structural uncertainty problem was thought in principle to have a scientific "fix." However, structural engineers of that period did not usually think it would be economically justifiable to collect information on the uncertainty or the dispersion of the random variables per se. Structural design codes seemed "engraved in stone" — they changed slowly and were apparently written by omniscient authority. Any new type of structure, for example, an airframe or a transmission tower structure, would develop its own set of safety factors, first from judgment and later from accumulated experience. Rapidly the process would converge on a workable set of values, and their genesis would be forgotten. Very little of the history of the early codes of this century is now available.

The relative dispersion of the random quantities involved was known early to be an important factor influencing the necessary safety margin. Mayer (1926) in Germany and later the Swiss E. Basler (1960) proposed safety measures akin to the Cornell (1967) safety index. Their work was, however, ignored and had no influence on design practice. Except as it may be necessary for the exposition, it is not the purpose of this book to give an account of the early history of the subject. Following is a list of references of early works that are outstanding, are noteworthy, or have been influential. Forssell (1924) stated the principle of optimality: that design should minimize the total expected cost of a structure, being the sum of certain initial costs and the expected cost of failure. Mayer (1926) suggested design based on the mean and variance of the random variables. Plum (1950), considering reinforced concrete slabs in buildings, established the apparent contradiction between economically optimum safety levels and the much lower rates of failures observed in actual practice. Johnson (1953) gave the first comprehensive presentation of the theory of structural reliability and of economical design, including statistical theories of strength developed by Weibull (1939). Freudenthal (1947) presented the fundamental problems of structural safety of a member under random variable load; his presentation was the first to evoke a measure of acceptance among structural engineers.

First transition period. During the period from 1967 to 1974 there was a rapid growth of academic interest in structural reliability theory and a growing acceptance by engineers of probability-based structural design. The classical structural reliability theory became well developed and widely known through a few influential publications, such as Freudenthal (1947), Johnson (1953), Pugsley (1966) and Ferry-Borges and Castanheta (1971). Yet there was little acceptance professionally, for several reasons. First, there seemed little need to change paradigm; deterministic design served very well. Structural failures were few and, when they occurred, they could be attributed to human error as a matter of routine. Moreover, probabilistic design seemed cumbersome; the theory was intractable mathematically and numerically. Finally, few data were available, certainly not enough to define the important "tails" of the distributions of load and strength. Indeed, among those who felt that deterministic design was not possible, there was disillusionment even with rigorous reliability theory. First, all rational analysis utilizes mathematical models that imperfectly reflect reality. The error in modeling is unknown. Second, the number of possible modes of failure of even a simple structure is so large as to preclude practical enumeration, let alone analysis of reliability. It seemed difficult to justify replacing a design "rationale" that is irrational but works, with another rationale, more complicated but also irrational.

The early 1960s were, accordingly, spent in the search for a way to circumvent these difficulties. For example, Turkstra (1970) presented structural design as a problem of decision making under uncertainty and risk. The significance is that in a wide variety of circumstances rational decision making is possible although information is lacking; see, e.g., Luce and Raiffa (1957). In engineering, feasibility is rarely an issue. It can be postulated that there exists a solution to almost any problem. The task is to find the "best" one. In everyday structural engineering this usually means a design that minimizes the present value of initial cost. With this postulate rational design is possible, even when uncertainty and risk are present. There exists a best design of any structure; it is relative to the existing state of knowledge, the level of technology, the cost per hour of engineering time and so on. The problem of rational structural design is to devise a procedure that produces an optimal design, i.e., a design that minimizes the expected present value of total cost.

It is immaterial whether this procedure itself is rational. It may indeed be rational to consult a dowser before drilling a well, if there is appropriate statistical evidence of reduced expected cost. Similarly, a design method is not necessarily irrational just because it contains approximations, random factors, or even irrational elements. Lind et al. (1964) defined the problem of rational design of a code as finding a set of best values of the load and resistance factors. They suggested an iterative procedure, considering the code as a "black box" control device for member sizes given the structural proportions, and hence controlling structural safety and cost. The procedure they suggested was merely a systematic version of the real trial-and-error process of development of structural design codes, but such an approach seems to have been too radical for the time.

Cornell (1967) suggested the use of a second-moment format (see Chapter 4). At that time it was widely understood that a simple Gaussian model of the random variables would be inaccurate for highly reliable systems such as structures. The format therefore appeared rather radical and it could have been expected to be ignored, like the very similar format by Su (1959). However, Lind (1973) showed that Cornell's safety index requirement could be used to derive a set of safety factors on loads and resistances. This approach relates reliability analysis to practically accepted methods of design (Ravindra et al., 1974). It has been modified and employed in many structural standards (CSA, 1974; NKB, 1977; OHBDC, 1983; etc.). Many practical developments followed shortly after, e.g., other safety indices (Rosenblueth and Esteva, 1972), and code optimization (Ravindra and Lind, 1973, 1983; see Chapter 6).

In the ensuing years, some serious difficulties with the second moment format were discovered in the development of practical examples. First, it was not obvious how to define a reliability index in cases of multiple random variables, e.g., when more than two loads were involved. More disturbingly, Ditlevsen (1973) and Lind (1973) independently discovered problems of invariance: Cornell's index was not constant when certain simple problems were reformulated in a mechanically equivalent way; yet no other safety index would remain constant under other mechanically admissible transformations (see Chapter 4). Several years were spent in the search for a way out of this dilemma without resolution. In the early 1970s, therefore, second-moment reliability-based structural design was becoming widely accepted, although at the same time it seemed impossible to develop a logically firm basis for the rationale. 1974 saw the publication of the first standard in limit states format based on a probabilistic rationale (CSA, 1974).

Second phase (1974-1984). The logical impasse of the invariance problem was overcome in the early 1970s (Hasofer and Lind, 1974), and the limitations of safety index methods became clarified (Veneziano, 1974). Several codes were developed and implemented in short succession, and the procedures were documented in guideline reports (CEB, 1976; CIRIA, 1977; CSA, 1981), reducing many aspects of probability-based design to a routine. It was to be expected, then, that the emphasis would change to extensions and development of specific details. This has actually happened. For example, random process models of strengths and loads have attracted considerable attention in the past decades. Chapters 9 and 10 present some examples of these developments.

However, an equally significant trend has grown out of a belated recognition that structural reliability theory is inadequate to account for the observed performance of real structures. The theory accounts for fairly small random fluctuations from mean loads and strengths. Most failures should occur under combinations of fairly low, but not exceptional variations of normal strength. Instead, structures usually seem to fail under loadings that they should have been able to withstand, in which case a human error is indicated, or under exceptionally high loads which they could not be expected to withstand. Thus a structural failure is a priori assigned to human error or an "act of God" — in the eyes of the law there is no room in between for the random occurrence.

Until recently, there were few systematic studies of structural failures. Some, e.g., Smith (1976), seemed compatible with reliability theory. But Matousek (1977) concluded that structural failure nearly always was caused by gross human error, and therefore connected with strengths very different from the mean (of similar structures without human error). Such failures would not be closely correlated with the loads, and hence should occur most frequently at loads around the design mean values. On the other hand, some failures are clearly caused in structures under extremely high loads. For sufficiently high loads, the structures affected would fail almost regardless of strength, so structures of average strength would be most common among the failures. Brown (1979) has shown that the failure rates predicted by the theory are too small — by a factor of

10 or more. He pointed out, for example, that the failure rate for large suspension bridges has been approximately 1 out of 40 in this century.

Thus the majority of structural failures are ascribed to human error, a factor that is not taken into account in the theory. It would seem, therefore, that the theory is of little value as a description of real-world processes. However true this may be, the purpose of the theory is not the description of structural performance, but *control* of the process intended to produce reliable structures efficiently. First, one must ask whether the apparatus of structural proportioning based on reliability theory is an effective means of control. It seems to be. Second, one must ask whether the setting of this "control knob" is correct — are the code parameters near the optimum? That is, are the marginal returns equal for the investment in the safety margin for error-free structures and the investment in inspection and quality control? This may or may not be the case. The matter is discussed further in Chapter 6.

Present structural practice gives failure rates due to human error that are perhaps 10 times as frequent as those due to inadequate safety margin. This may well be near the economic optimum. After all, it is relatively inexpensive to buy extra safety by using more material. Nevertheless, structural reliability theory is incomplete without consideration of human error. This has given rise to a minor, second change of paradigm. The theory developed to date is seen to be only part of a more extensive theory of structural quality control, accounting also for the effects of human error. The study of human error in structural production is strikingly different from the established engineering disciplines of mechanics and structural reliability, largely because it may require knowledge of methods of social science. A characteristic difficulty is the need first to develop a set of workable concepts. Such efforts are currently under way under the auspices of several research establishments around the world (Melchers and Harrington, 1983; Lind, 1983).

In periods of high economic activity, there is rapid production of structures, often of new design and often designed by engineers with less experience. It is natural that structural reliability then becomes a matter of concern. In periods of low economic activity, on the other hand, existing structures present an ever-increasing maintenance problem. Indeed, it appears that most structures do not fail — they just become a maintenance liability. Good structural design produces a reliable and durable structure, and there is some difficulty in separating these two components of quality.

"Structural quality assurance" could arise as a new discipline that unifies the considerations of maintenance and repair in design theory.

1.2 THE PROBLEM OF STRUCTURAL SAFETY

The life cycle of a structure is not unlike the life cycles of other machines or living creatures: manufacture (or birth), a service period, and retirement (or replacement, or death). During its life-time the structure is subjected to *loads* (or *actions),* i.e., forces or forced displacements. The loads may cause a change of condition or *state* of the structure, going from the undamaged state to a state of deterioration, or wear, damage in varying degrees, failure, or col-lapse. Maintenance or repair may become necessary, causing a *loss.* One may distinguish among four loss categories: economical loss, the loss of cultural value, injury, and death. The relative importance of these components varies greatly with the type of structure and its exposure, as is evident from Table 1.1. This table also shows that the risk of death from structural failure is very small. It is comparable to the risk of death from lightning and   snake   bite   in

TABLE 1.1 Risk of Death				
Exposure	Risk of Death per Hour per 10^8 Persons Exposed	Hours of Exposure per Person Exposed per Year	Risk of Death per 10^4 Exposed Person per Year	Ratio of Wounded to Number of Deaths
Mountain climbing (international)	2700	100	27	
Trawl fishing (deep sea, 1958-1972)	59	2900	17	
Flying (crew)	120	1000	12	< <1
Coal mining	21	1600	3.3	
Automobile travel	56	400	2.2	20
Construction	7.7	2200	1.7	450
Flying (passengers)	120	100	1.2	< <1
Home accidents	2.1	5500	1.1	
Factory work	2	2000	0.4	
Building fires	0.15	5500	0.08	5
Structural failure	**0.002**	**5500**	**0.001**	**6**

Source: CIRIA (1977).

Canada, for example (Allen, 1981). Economic losses from structural malfunction are small in comparison with the cost of construction. This is evident by the low insurance rates in effect for such coverages.

The condition of a structure can vary more or less continuously from the undamaged state to a state of collapse. As a simplification, it is assumed in the design stage that malfunction can occur in only a finite number of *modes*. Each mode of malfunction (or of "failure") gives rise to a design inequality. Structural design in the narrow sense is to determine a set of structural member capacities such that these design inequalities are satisfied. It is important, of course, that the set of modes considered cover all the failure possibilities that can be imagined for the structure. As a fundamental simplification it is assumed that all states of the structure with respect to each mode i can be divided into two sets: F_i, the states *failed* in mode i, and S_i, the states unfailed or *safe* in mode i. This assumption is common in reliability theory, but sometimes it may be necessary or convenient to account for a continuous loss spectrum, as in Chapter 6.

The condition of a structure depends on the actions, the material strength and workmanship, and other random quantities. It further depends on time, t, and the design parameters, **p**, which are nominal, deterministic variables selected in the design process. The actions are generally stochastic processes in time and physical space, $Z(t,x,y,z)$. In the space of the actions, the structural strength appears as limiting surfaces (see Fig. 1.1). Since strength is random, these surfaces are also random in location. Failure is the passage of the process Z through a limiting surface of the safe set. Chapters 9 and 10 give several examples of such random process models. A simpler formulation represents all uncertain quantities in terms of a set of n *basic* random variables, $Z_1, Z_2, \ldots, Z_n$. Some of these represent actions, while others represent the strength of the structure in various modes. This formulation is often made directly, or it is the outcome in the analysis of a stochastic process model.

The basic random variables are collected in a random vector $\mathbf{Z} = (Z_1, Z_2, \ldots, Z_n)$. $\mathbf{Z}$ is a random point in an n-dimensional vector space. For each failure mode i one can define a scalar point function $g_i = g_i(\mathbf{p}, \mathbf{z}, t)$ of the design parameter vector **p**, of the vector **z** of uncertain variables, and of time t, to be positive if, and only if, the structure characterized by **p** is safe in mode i at time t. $g_i(\ )$ is called a *failure function* for mode i. It is often unnecessary to make explicit reference to **p** and t, and one can simply write $g_i = g_i(\mathbf{z})$. The corresponding *safety margin* M_i for mode i is a random variable defined as $M_i = g_i(\mathbf{Z})$. The boundary between the

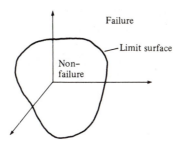

Figure.1.1 Structural strength as a limiting surface in the space of actions.

safe states S_i and the failed states F_i in z-space is the set of *limit states* in mode i. One can choose $g_i(\ )$ to be continuous, with $g_i(\mathbf{z}) = 0$ on the limit state boundary. It is often convenient to include the limit states in the failure set.

Limit states design designates design calculations using factored loads and material properties representative of failure conditions. Limit states design methods are gaining favor over working stress design methods that use unfactored loads and reduced allowable stresses, representative of an arbitrary safe state of higher probability. Limit states design is treated in Chapters 3 and 6.

A particular structure subjected to a particular history of actions can be represented by a sample point $\mathbf{z}$, called a realization of $\mathbf{Z}$. The *reliability* P_R of a structure is the probability that the sample point falls in the safe region. Conversely, the probability of failure, P_F, is the probability that the realization falls in the failure region. It follows that $P_R + P_F = 1$. The probability of failure in mode i, P_i, is defined in a similar way.

Lifetime distribution and reliability. Let t denote time elapsed since the structure was put in service, i.e., the age of the structure. For any set of structures a plot can be made of the proportion, y, of the structures that are in a failed state at age t,

$$y = F(t) \tag{1.1}$$

$F(t)$ is called the *lifetime distribution function* for the set, and its complement $G(t) = 1 - F(t)$ is called the *survival function*. The derivative $f(t) = dF(t)/dt$ is called the *failure rate function*. The *mortality*, or *hazard function*, at time t is defined as the probability of failure per unit time conditional upon survival to time t,

$$\rho(t) = \frac{f(t)}{G(t)} \tag{1.2}$$

Figure 1.2 illustrates the lifetime distribution and related functions. The curves for mechanical systems and for living organisms are similar in shape (the "bathtub curve"). There is first a brief period of high mortality, then a period of low and nearly constant mortality, and finally a period of monotonically increasing mortality.

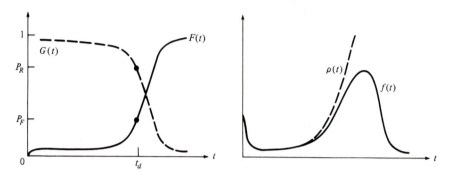

Figure 1.2   Lifetime distribution function $F(t)$, survival function $G(t)$, failure rate function $f(t)$, and hazard function $\rho(t)$.

When the set of structures is a population, these functions may be interpreted as probabilities. Let t_d be a time of particular interest, e.g., a projected life or *design life*. Then $P_F = F(t_d)$ is the *failure probability* and its complement, $P_R = 1 - P_F = G(t_d)$, is the *reliability* of the population. Structural reliability theory aims to predict or compare these probabilities from the attributes of a structure and its environment. Rarely can the reliability of a structure be determined by observation on the population. Often the structure is unique and the population is then abstract. When the population is available for testing, as in mass-produced structures or components, there is rarely the time or the money available to test a reasonable number over a lifetime. Instead, the probabilities are calculated from the probabilities of failure of the parts of the system. These, in turn, are calculated from system parameters, component loads, and component strengths. Some systems are relatively simple to analyze, as when system failure occurs if, and only if, a component fails. Such *weakest-link systems,* exemplified by a chain or a statically determinate structure, are analyzed in Chapters 5 and 9.

For a weakest-link structure, one may conveniently define a failure function $g = \min\{g_1, g_2, \ldots, g_m\}$. In most systems the relationship with component failure functions is more complex or absent (Chapter 5). There is a vast literature on the analysis of reliability of systems where the survival function of any component is independent of the state of any other component (Barlow and

Proschan, 1975; Kaufmann et al., 1977). In most structures the force on surviving components, and therefore their life expectancies, changes when a component fails. Structural reliability is therefore generally outside the scope of this extensive specialized reliability literature.

Example 1.1

To illustrate some of the concepts introduced so far, consider a simply supported reinforced roof slab. The slab is one-way reinforced and has a fixed span L and an effective depth h. It is under-reinforced with steel area A_s per unit width. Material strength f_c' and roof load (snow) are random (Fig. 1.3).

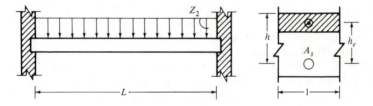

Figure 1.3 Reinforced concrete slab.

The roof can fail as a result of overload due to rain, snow, etc. The condition would deteriorate, first by cracking and later by leakage, unsightly sag, spalling, or even collapse. To simplify this spectrum of malperformance, define failure as the development of a plastic hinge near midspan. This gives only one mode of failure. To get the simplest model, consider the steel strength as one random variable, Z_1, ignore the influence of concrete compressive strength on section strength, and consider only the uniform component of the load (dead load plus live load), Z_2. The basic random variables of this model are Z_1 and Z_2. The midspan moment is $M_m = 0.125 Z_2 L^2$, while the moment capacity is $M_p = h_e A_s Z_1$. Use is therefore made of the failure function

$$g(z_1, z_2) \;=\; M_p - M_m \;=\; h_e A_s z_1 - \frac{1}{8} z_2 L^2 \quad \begin{cases} >0 & \text{safe} \\ =0 & \text{yield limit state (failure)} \\ <0 & \text{no equilibrium (failure)} \end{cases}$$

To illustrate the design, make the simplifying assumption that h_e is a constant fraction of h, and that A_s is a constant fraction of the slab cross-section, which is proportional approximately to h as well. Thus $8h_e A_s / L^2$ is taken as a constant design parameter, to be determined, giving the scaled failure function

$$g(z_1, z_2) \;=\; p\, z_1 - z_2$$

The safety margin is

$$M = g(Z_1, Z_2) = p Z_1 - Z_2$$

Suppose that statistical data are available, giving estimators for the distributions of the basic variables Z_1 and Z_2. Estimators for the statistics of M as a function of p can then be calculated, and the value of p that gives a particular reliability can be found. As a specific example, let the strength Z_1 be normally distributed with mean value μ_1 and variance σ_1^2, while the load Z_2 is normally distributed with mean value μ_2 and variance σ_2^2. A design with probability of failure less than, say, 10^{-4} is sought.

$$P_F = P(M \leqslant 0) < 10^{-4}$$

Since M is normally distributed, this requires $\Phi((0 - \mu_M)/\sigma_M) < 10^{-4}$ where $\Phi(\ )$ is the standard normal distribution function. A table of this function then gives $\mu_M/\sigma_M > 3.72$. Now, $\mu_M = p\mu_1 - \mu_2$, while $\sigma_M^2 = p^2\sigma_1^2 + \sigma_2^2$. Inserting these into $\mu_M > 3.72\,\sigma_M$ gives a lower limit on the design parameter p. Since the thickness and thus the cost of the slab increases with p, the least-cost design is obtained by setting p equal to its lower limit.

This is, of course, a rather crude model of a slab, and the calculated reliability is undoubtedly subject to considerable model error. For example, the strength varies with location and increases with hydration, i.e., with time. The strength is thus a random process $Z = Z(t,x,y,z)$, but it would make sense to neglect spatial variations and to consider the time variation as deterministic: $Z = f(t)Y$. The loading is another random process in space and time. The spatial variation is not too important. But the probability of exceeding any particular value of the snow load increases with the design life t_d. Thus the load distribution function is dependent on t_d. In general, it is not normally distributed, but it can be modeled by a normal distribution when the parameters are selected judiciously. Accurate modeling requires empirical data, of course. With finite data there is parameter uncertainty to be taken into account as well.

1.3 STRUCTURAL DESIGN

Reliability analysis is of interest mainly as a tool in the design process. For each failure mode a desirable or acceptable value of the reliability must be available to the designer. This reliability target P_R^* may be chosen by judgment or by rational analysis. Allen (1981) estimated the total number of structures (5 million in Canada) in service and the number of failures per year (100 in Canada). This gives a failure rate of 2×10^{-5} per year from all causes. Human error is estimated to account for approximately 90% of these failures, leaving a design reliability of "error-free" structures of 2×10^{-6} per year or 10^{-4} for a 50-year service life. With the aim to vary the reliability in favor of the number of persons at risk n, and considering other relevant factors, Allen (1981) suggested the formula

$$P_F^* = \frac{TA}{W\sqrt{n}} P_0 \qquad (1.3)$$

for the target failure probability P_F^* in terms of the lifetime of the structure T in years, an activity factor A ($=0.3$ for post-disaster activities, 1.0 for buildings, etc.), a warning factor W, and a basic annual failure probability $P_0 = 10^{-5}$. This formula has the advantage of simplicity. Moreover, it is based on empirical data. However, there are some limitations that preclude its use in many circumstances.   First, the variation with n is arbitrary; it reflects a compromise between taking no account or full proportional account of the number of persons exposed to risk. Also, for some structures the economic loss component is important, and the expected loss should figure in the reliability target.

In the rational analysis the target reliability P_R^* is considered as a control parameter subject to optimization.   The parameter assigns a particular investment to the material placed in the structure. The more material − invested in the right places − the less is the expected loss.   The optimum placement maximizes the expected rate of return on the investment.   Such optimization, possible when economic loss components dominate over life, injury, and culture components, leads to the principle of equal marginal returns in the case of continuous parameter space and differentiable costs.   The principle of minimum expected total cost governs the design of all parts of a structural system, whether a single bolt, a member, or a single structure in a fleet of structures, i.e., a multi-structural system (e.g., a railroad). Technically, the optimization approach can become very complicated, as in the case of a fleet of structures in continuous operation considering maintenance and replacement.

When the expected loss of life or limb is important, the optimal reliability level becomes more controversial.   Frequently, this leads to the problem of the economic equivalent of human life, which is philosophically intractable. Risk-benefit analysis is one approach that aims to circumvent this difficulty; the reliability of a system is translated into cost per life saved (Hapgood, 1979). The target reliability may then be chosen such that the cost per life saved, by means of the safety margin for a structure or substructure, is comparable to other systems in society.   This approach, while useful, neglects some physical factors.   For example, some people expect, with reason, some systems to be safer than others. Few would want their home to be as (un)safe as a motorcycle, or

vice versa. Such considerations make it necessary to use a variety
of approaches in the practical selection of safety levels; experience
with the results forms an improved basis for socially accepted
safety levels that can be defined professionally. These considerations are dealt with in detail in Chapter 6.

1.4 SUMMARY

Safety against failure is the overriding objective in structural design
calculations. Safety is a probabilistic concept, introduced gradually
as a rationale for structural design during the past half century.
Structural failures are rare events and contribute very little to the
risk of death. As probabilistic concepts have been introduced, it has
come to be understood that the majority of structural failures are
associated with gross human error, which is excluded from the
usual probabilistic theory of structural reliability. This theory is still
valid, but it concerns the strength of structures that are correctly
planned and built.
  Structural malperformance is classified into modes of failure.
With respect to such modes, it is assumed that any structure is in
either a failed state or a safe state. The state depends on uncertain
quantities such as loads, material strength, workmanship, etc. Often
these random quantities can be modeled by a finite set of basic
random variables, i.e., a random vector. Sometimes random process
modeling is necessary. The state is quantified in terms of a failure
function, which is a function of the basic random variables. The
structure passes from safe state to failed state in each failure mode
through the limit states. In limit states design the nominal values of
the random variables involved are representative of failure conditions. The salient aspects of the reliability of a structure, like that
of any system, can be discussed in terms of the lifetime distribution, the survival function, the failure rate, and the hazard function.

REFERENCES

ALLEN, D. E., "Criteria for Design Safety Factors and Quality
Assurance Expenditure," in *Structural Safety and Reliability*, ed. T.
Moan and M. Shinozuka, Proceedings ICOSSAR'81, Trondheim,
Norway, June 1981, pp. 667-678.

BARLOW, R. E. and F. PROSCHAN, *Statistical Theory of Reliability
and Life Testing*, Holt, Rinehart and Winston, New York, 1975.

BASLER, E., "Analysis of Structural Safety," paper presented to the ASCE Annual Convention, Boston, Mass., June 1960.

BROWN, C.B., "A Fuzzy Safety Measure," *Journal of the Engineering Mechanics Division,* ASCE, Vol. 105, 1979, pp. 855-872.

CEB (Comité Europeen du Béton), Joint Committee on Structural Safety CEB-CECM-FIP-IABSE-IASS-RILEM, "First Order Reliability Concepts for Design Codes," *CEB Bulletin No. 112,* July 1976.

CIRIA (Construction Industry Research and Information Association), "Rationalisation of Safety and Serviceability Factors in Structural Codes," *CIRIA Report No. 63,* London, 1977.

CORNELL, C. A., "Some Thoughts on "Maximum Probable Loads" and "Structural Safety Insurance"," *Memorandum,* Department of Civil Engineering, Massachusetts Institute of Technology, to Members of ASCE Structural Safety Committee, March 1967.

CSA (Canadian Standards Association), "Standards for the Design of Cold-Formed Steel Members in Buildings," *CSA S-136,* 1974, 1981.

DITLEVSEN, O., "Structural Reliability and the Invariance Problem," Research Report No. 22, Solid Mechanics Division, University of Waterloo, Waterloo, Canada, 1973.

FERRY-BORGES, J. and M. CASTANHETA, *Structural Safety,* Laboratorio Nacional de Engenharia Civil, Lisbon, 1971.

FORSSELL, C., "Economy and Construction," *Sunt Förnuft,* 4 (in Swedish), 1924, pp. 74-77. Translated in excerpts in *Structural Reliability and Codified Design,* ed. N. C. Lind, SM Study No. 3, Solid Mechanics Division, University of Waterloo, Waterloo, Canada, 1970.

FREUDENTHAL, A. M., "The Safety of Structures," *Trans. ASCE,* Vol. 112, 1947.

HAPGOOD, F., "Risk Benefit Analysis: Putting a Price on Life," *Atlantic,* Vol. 243, 1979, pp. 33-38.

HASOFER, A. M. and N. C. LIND, "Exact and Invariant Second Moment Code Format," *Journal of the Engineering Mechanics Division,* ASCE, Vol. 100, 1974, pp. 111-121.

JOHNSON, A. I., *Strength, Safety and Economical Dimensions of Structures,* Statens Kommitté för Byggnadsforskning, Meddelanden No. 22, Stockholm, 1953.

KAUFMANN, A., D. GROUCHKO and R. CRUON, *Mathematical Models for the Study of the Reliability of Systems,* Academic Press, New York, 1977.

LIND, N. C., "The Design of Structural Design Norms," *Journal of Structural Mechanics,* Vol. 1, 1973, pp. 357-370.

LIND, N. C., "Structural Quality and Human Error," in *Reliability Theory and its Application in Structural and Soil Mechanics,* ed. P. Thoft-Christensen, NATO ASI Series E, Martinus Nijhoff, The Hague, 1983, pp. 225-236.

LIND, N. C., C. J. TURKSTRA and D. T. WRIGHT, "Safety, Economy and Rationality in Structural Design," in *Proceedings,* IABSE 7th Congress, Rio de Janeiro, Preliminary Publication, 1964, pp. 185-192.

LUCE, R. D. and H. RAIFFA, *Games and Decisions,* John Wiley, New York, 1957.

MATOUSEK, M., "Outcome of a Survey on 800 Construction Failures," in *Proceedings,* IABSE Colloquium on Inspection and Quality Control Institute of Structural Engineering, Swiss Federal Institute of Technology, Zurich, 1977.

MAYER, M., *Die Sicherheit der Bauwerke,* Springer Verlag, Berlin, 1926.

MELCHERS, R. E. and M. V. HARRINGTON, "Structural Reliability as Affected by Human Error," in *Proceedings,* Fourth International Conference on Application of Statistics and Probability in Soil and Structural Engineering, ICASP4, University of Firenze, Italy, June 1983, pp. 683-694.

NKB (The Nordic Committee on Building Regulations), "Recommendations for Loading and Safety Regulations for Structural Design," *NKB-Report,* No. 36, Copenhagen, November 1978.

OHBDC (Ontario Highway Bridge Design Code), Ontario Ministry of Transportation and Communication, Downsview, Ontario, 1983.

PLUM, N. M., "Is the Design of Our Houses Rational When Initial Cost, Maintenance and Repair Are Taken into Account?," *Ingeniøren,* 50, 1950, p. 454.

PUGSLEY, A., *The Safety of Structures,* Edward Arnold, London, 1966.

RAVINDRA, M. K. and N. C. LIND, "Theory of Structural Code Optimization," *Journal of the Structural Division,* ASCE, Vol. 99, 1973, pp. 541-553.

RAVINDRA, M. K. and N. C. LIND, "Trends in Safety Factor Optimization," *Beams and Beam Columns,* ed. R. Narayanan, Applied Science Publishers, Barking, Essex, England, 1983, pp. 207-236.

RAVINDRA, M. K., N. C. LIND and W. W. SIU, "Illustrations of Reliability-Based Designs," *Journal of the Structural Division,* ASCE, Vol. 100, 1974, pp. 1789-1811.

ROSENBLUETH, E. and L. ESTEVA, "Reliability Basis for Some Mexican Codes," *ACI Publication SP-31,* 1972, pp. 1-41.

SMITH, D. W., "Bridge Failures," *Proceedings of the Institution of Civil Engineers,* Vol. 60, 1976, pp. 367-382.

SU, H. L., "Statistical Approach to Structural Design," *Proceedings of the Institution of Civil Engineers,* Vol. 13, 1959, pp. 353-362.

TURKSTRA, C. J., *Theory of Structural Design Decisions,* SM Study No. 2, Solid Mechanics Division, University of Waterloo, Waterloo, Canada, 1970.

VENEZIANO, D., "Contributions to Second-Moment Reliability Theory," Research Report R74-33, Department of Civil Engineering, Massachusetts Institute of Technology, Cambridge, Mass., 1974.

WEIBULL, W., "A Statistical Theory of the Strength of Materials," *Proceedings, Royal Swedish Institute of Engineering Research,* No. 151, Stockholm, Sweden, 1939.

2

RELIABILITY METHODS

2.1 RESEARCH MODELS AND TECHNICAL MODELS

Having accepted the dichotomy of structural behavior into "failure" and "nonfailure," one can proceed to consider the methods that can be used to determine the probability of each state. A reliability method, in the narrowest sense, is a method to evaluate the reliability P_R of a system. In practice, systems are so complex that it is impossible to examine all failure modes. Instead, a finite subset of failure modes is considered. Also, as a matter of necessity, the system properties are idealized. A reliability method, in a broader sense, is a method to calculate the reliability $P_R{'}$ for the idealized system with respect to a specific set of failure scenarios. $P_R{'}$ is used vicariously as a measure of P_R. Finally, a spectrum of useful design methods based on reliability concepts have become available in recent years. Many of these methods do not explicitly calculate reliabilities, yet aim to produce structures with a prescribed reliability, or aim to compare reliabilities. The methods in Chapters 3 and 4 are examples of such methods. Accordingly, a *reliability method* is defined, in a broad sense, as a method to decide if a structure is acceptably reliable.

An important objective of reliability engineering is to develop accurate and efficient reliability methods. The requirements of accuracy and efficiency are difficult to reconcile, unfortunately, and a distinction may be drawn between two kinds of models, which Duddeck (1977) calls the *research model* and the *technical model*. The former serves to give a better understanding of structural reality; the emphasis is on accuracy. The perfect research model minimizes the difference $P_R - P_R{}'$. The technical model, in contrast, is a tool in design, an aid to decision making; the emphasis is on efficiency in the decision choice. Duddeck (1977) lists some of the characteristics of a technical model, which can be summarized as follows:

1. Reality is not portrayed but substituted.
2. Validity is restricted to certain regions of application.
3. Some variables and theories are ignored if the design is insensitive to them.
4. Loads are idealized and limit states are simplified to a few representative ones.
5. The mechanical model is simplified considerably: neglect of imperfections, initial stress, secondary stress concentrations, etc.

For a given context, the engineer must select the most appropriate compromise between these two extremes of modeling, considering the conflicting demands of fidelity and simplicity of representation. Just as one should not confuse the primary stress in a truss with the real stress, one should remember the distinction between P_R and its operational value $P_R{}'$. The research model is developed and maintained solely as a reference standard for technical models.

2.2 DEFINITION AND IDEALIZATION OF THE STRUCTURAL SYSTEM

Consider some of the difficulties in evaluating the "true" reliability of a system. First, it is necessary to specify the system with precision. For example, an airplane or a power plant could be highly reliable mechanically but badly designed from the viewpoint of human engineering, so that operator error is likely. The reliability of such a system is appreciably different if the operator is considered part of the system. What to the passenger may appear as a relatively unreliable means of transport may to the pilot appear as a reliable aircraft, which however, is difficult to fly.

Second, although one may attribute reliability to a system, it is a fallacy to think that this reliability is a property of the system. For example, consider the reliability of a prefabricated roof truss designed for a 50 kN/m snow load. This reliability means the probability that a random realization of the truss does not fail within its design life. To the designer at the time of designing, this probability takes one value, P_{R0}, which among other things, reflects the uncertainty in loading history, workmanship, and material. Once the truss has been produced, the reliability takes a new value, P_{R1}, reflecting a particular realization of the material; when the truss is put into service — in a location where the design snow load is only 45 kN/m, for example — the reliability takes yet another value, P_{R2}; and so on. The reliability is not a property of the system, but an attribute of our state of knowledge of the system. It is meaningless, in particular, to talk about the "true" reliability of the system.

Instead, it is preferable to call P_R the *overall* reliability. The complement $P_F = 1 - P_R$ is called the *total probability of failure* (Ferry-Borges, 1977). The overall reliability can be estimated statistically from records of failures for sufficiently large sets of sufficiently homogeneous systems. This is commonly done using experimental records for mass-produced parts.

On the other hand, one can perhaps calculate the failure probability from the probability of failure in a finite set of failure modes. This value, called the *theoretical failure probability, P_F'*, is a function of the basic random variables and the system parameters. The difference $P_a = P_F - P_F'$ is the *adjunct probability of failure* (Ferry-Borges, 1977). Experience to date has shown that the adjunct probability of failure is some 4 to 10 times larger than the theoretical value P_F' calculated by ignoring the failure modes involving human error (Brown, 1979; Matousek, 1977).

An important element of the reliability analysis of a structure is the determination of the reliability in a single failure mode. Many reliability methods employ the idealization that such element analyses can be combined in a simple fashion to provide a measure of the theoretical reliability P_R' of the structure. For example, load and resistance factor methods generally assume that if all members individually are sufficiently reliable, the structure as a whole will be sufficiently reliable. In applications it must be verified that such a representation is sufficiently accurate; if not, more complex models of structural system reliability (Chapter 5) must be invoked, accounting for structural behavior past the linear range, correlation of failure modes, failure progression, etc.

2.3 ERROR CLASSIFICATION

To examine the reasons for the large difference between the theoretical and the total probability of failure, consider a structure that failed in mode f before the expiration of its design life under loads estimated to be represented by the point s_f in load space (Fig. 2.1a). According to the design, the structure should have been able to sustain the proportional load $s_f /E = S_f^*$ in the same failure mode. If the load-carrying capacity is dependent on the load trajectory, it is here assumed that S_f^* is calculated for the proportional trajectory $t_f^* = t_f /E$, where t_f is the estimated failure loading trajectory. *E* is called the *error factor;* it is a random variable, reflecting the uncertainty in material strength parameters and other basic variables that are not represented in load space. *E* is positive and approaching zero if the failure mode was not foreseen in the design. Usually, but not necessarily, it is less than unity when the actual load-carrying capacity was less than anticipated in the design.

Assuming as a mental experiment that many replicas of the structure were produced and subjected to the same load history, it is clear that they would not fail exactly at point s_f but at some

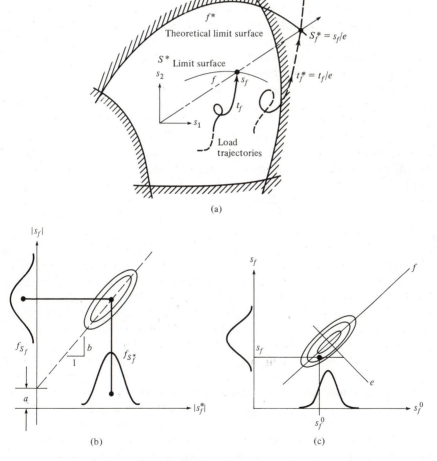

Figure 2.1 Illustration of error classification.

other points on trajectory t_f. Thus s_f may be thought of as the realization of a random variable S_f which is the true strength of the design. The replication is not thought of merely as a reproduction of the same design to an identical set of drawings, but rather to include complete redesigns to the same performance specifications by different designers, and production by different contractors. S_f^* reflects variations in design when aiming to satisfy the same target load capacity, in addition to strength variability. Figure 2.1b shows a hypothetical graph of S_f and S_f^*, actually of the magnitudes $|S_f|$ and $|S_f^*|$. The regression line $E[S_f \mid s_f^*] = a + bs_f^* = S_f^0$ indicates that there is a *systematic error* in design and production; a is an *additive bias* and b is a *bias factor.* Systematic errors are associated with the methods used in design and production. In principle these errors can be detected by experiment and eliminated. In Fig. 2.1c all systematic error has been removed, and the magnitudes of s_f and the unbiased image s_f^0 of s_f^* from Fig. 2.1a have been shown. The component of the deviation from the mean in direction f indicates variation due to material factors and other factors beyond effective control, while the component in direction e indicates design and production error $s_f - s_f^0$. This error component, also called the *human error,* is further subdivided into *gross error* and *random human error* according to criteria that are discussed in Section 2.5.

The occurrence of gross error is controlled by review, inspection, training, etc., while the influence of random fluctuations is controlled by design parameters such as safety factors. A major problem in structural engineering is to select the appropriate values of these safety factors. These parameters do not contribute effectively to limit the frequency or severity of failures attributed to gross human error. Indeed, a structure containing a serious error that would collapse before being placed into service may, if the safety margin is large enough, survive to be put into service only to fail shortly afterward, when the consequences may be much more serious. Accordingly, it is possible and expedient to separate the problems of gross error from the problem of random fluctuations. Unless otherwise specified, as in Section 2.4, gross error is assumed not to be a factor in the theoretical reliability P_R'.

2.4 IDEALIZATION OF LOADS AND INFLUENCES

Civil engineering structures are exposed to natural forces. Every structure, however strong, could experience forces larger than its capacity. A *cutoff value* is therefore established in design for every force intensity as a matter of necessity or practical choice, or both. For example, no ordinary civil engineering structure is expected to survive a direct hit by a megaton bomb or a tornado. If a force exceeds the cutoff limit, structural failure is the behavior to be expected, so it is not counted as a failure of the structure (to perform as expected of it). It is difficult to establish these limits accurately, but fortunately the question is acute only in the case of failure under extraordinary loads.

It is worth pointing out that this discounting of failure under extraordinary loads cannot be defended so readily for other structures. Unlike a civil engineering structure, a nuclear reactor containment vessel can cause numerous deaths if it fails, and the engineer cannot a priori assume that the loss of life is limited in space and time. Failure modes under extraordinary loads, however large and improbable, must be examined for such "high consequence" systems.

Many failures of structures are not structural failures. For example, a structure can be destroyed by fire, by biological degradation (rot or fungus), by lack of maintenance, or by adverse action of its environment, such as undermining of foundations by scouring. Ignoring such influences is an aspect of the idealization of the load that has important consequences for the value of the theoretical reliability.

Cornell and Larrabee (1977) have reviewed the simplifications in the representation of load processes that are commonly made in technical models. They observe that these simplifications often involve a separation of a structure-independent *load environment process* and a *load-structure interaction filter* that is only grossly structure-independent. The load environment process, which could be further decomposed into a climatic and a local environment process, is usually reduced to a simple scalar random process of time or simply an extreme load random variable for a given lifetime. The reader may find it interesting as an exercise to identify this decomposition for the case of earthquake or snow.

Example 2.1

The wind loading in building codes provides an example of load idealization (Davenport, 1982). A versatile formulation of wind pressure W, which reflects the format used in many recent building codes, is the factored product form

$$W = Q\,C_e\,C_p\,C_g\,M$$

in which Q is the velocity pressure, C_e the exposure height factor, C_p the aerodynamic shape factor, C_g the dynamic gust factor, and M is a model factor. Q and M are random processes, whereas C_e, C_p, and C_g are random factors that represent filters to produce the random process W.

Q is the macroclimatic term, representing velocity pressure in open country at a height of 10 m. C_e reflects the local environmental factors: open or urban area, height, and orientation of the building surface element in relation to the wind, while C_p and C_g in combination reflect the dynamic response of the structure. The decomposition of the various influences isolated in factors is, of course, an approximation; moreover, any real data used in describing the wind (anemometer measurements, wind tunnel data) are subject to error and numerical simplification. The model factor process M expresses the ratio between the true process W and the process $Q\,C_e\,C_p\,C_g$.

For most practical purposes, the random processes are represented by random variables. Thus Q is commonly represented by the reference velocity pressure q, the average pressure over several minutes to an hour, for a specified return period. For example, NBCC (1980) uses a return period of 10 years for deflections and 100 years for important building structures. In design, the equation above is used to yield a design value W_s in terms of design values of Q_d, C_{ed}, $(C_p, C_g)_d$, and a load factor M_d.

Extreme-value analysis of observed annual maxima is used to define the reference velocity pressure for the return periods of interest. This should properly be done separately for each direction of the wind, but is for simplicity done without regard to wind direction, and the usually strong directional effect is, instead, accounted for by adjustment of the pressure coefficient C_p. The calculation of specified values of the wind load parameters is discussed further in Section 10.1.

2.5 HUMAN ERROR

Human error is an important, perhaps dominant, cause of failure in structures. Effective control of reliability requires control of the influence of human error. Human error mechanisms differ from the physical causes of random fluctuations in loads and resistances, and they are not yet well understood.

Human error is controlled by simply taking care that mistakes are not made. With increasing care the number and magnitude of errors are reduced. Tasks of formal logic, including strictly mathematical tasks, can, with care, be performed entirely without error. All other tasks, including numerical analysis, will have small remanent errors which it is inexpedient or uneconomical to control. These are called random human errors and their influence on the reliability of the structure is, like other random errors, controlled by safety margin allowances.

Logic errors are relatively common in structures and are often serious in consequence. For example, lateral bracing may be missing in formwork, or a cage of reinforcement steel in a beam may be placed upside down. The effect of such errors can be modeled mathematically by a binary error model (see Example 2.2). The causes of logic error are psychological and sociological factors that are obscure and perhaps complex. The most effective approach to control of logic error in structures is checking. Checking of the individual's own task is promoted by emphasis in learning (reward and penalty), social pressure, and good working conditions (clear organization of responsibility, skillful supervision, ample time to perform the work, etc.). Independent checking by another person is believed to be very effective, although there may be a certain "halo effect": If a piece of work is generally very competently done, the checker may tend to lose his or her critical attitude — may become less alert and more likely to miss an isolated error.

The nature and complexity of the processes of design and construction of a structure prevent effective study and control by the methods of the structural engineer. Exceptions may be high-energy plants, because of the risks involved, mass-produced structures, and some simpler unit operations of ordinary structural production if carried out invariably by a standard procedure. In such cases, it may be possible to use a cause-consequence graph analysis (Barlow et al., 1975).

Example 2.2

Consider a structure with one limit state and two basic random variables R and S and safe domain $R > S$ (Lind, 1983). Without loss of generality the variables are scaled so that S has unit mean. Then the mean value of R equals the central safety factor θ (Fig. 2.2a).

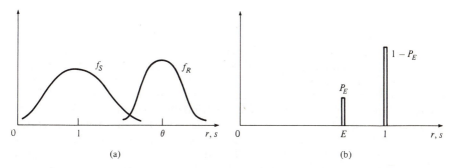

Figure 2.2 Illustration of error model: (a) error-free; (b) error factor.

Assume that R and S are normally distributed and independent. Then the probability of failure for the error-free structure P_{F0} is

$$P_{F0} = \Phi\left[-\frac{\mu_{R-S}}{\sigma_{R-S}}\right] = \Phi\left[-\frac{\theta-1}{(\theta^2 V_R^2 + V_S^2)^{1/2}}\right] = \Phi(-\beta_0)$$

in which V denotes the coefficient of variation. Assume next that an error occurs with probability P_E, lowering R by a factor E (Fig. 2.2b). Then the probability of failure of the error-free structure drops to $(1-P_E)P_{F0}$, while the probability of failure with error becomes $P_E\Phi(-\beta_E)$, in which

$$\beta_E = \frac{\theta E - 1}{(\theta^2 E^2 V_0^2 + V_S^2)^{1/2}}$$

The total probability of failure changes to

$$P^* = (1-P_E)\Phi(-\beta_0) + P_E\Phi(-\beta_E)$$

The proportion of structures with error in the failed population is

$$\alpha = \frac{P_E\Phi(-\beta_E)}{P^*}$$

A set of values that can be representative of common civil engineering structures is: $V_R = 0.15$ and $V_S = 0.30$ with a central safety factor of $\theta = 3$, giving $\beta_0 = 4.02$. If one considers a binary error that reduces the strength by 2 standard deviations if it occurs, then $E = 1 - 2 \times 0.15 = 0.7$. Figure 2.3 shows the variation of the failure probability P^* with error frequency P_E for $\theta = 3$ and other values of the central safety factor. The figure shows — not surprisingly — that highly reliable designs are more sensitive to the presence of binary human error.

On the basis of scanty information from the aircraft industry, the frequency of human error can be estimated, very roughly, as $P_E = 10\%$. With a central safety factor of 3, the equations above indicate that approximately 1 out of 1000 structures should fail in a lifetime, and that some 85% of these failures could be ascribed to human error. Both of these observations are in general agreement with facts.

Example 2.3

A filter model of human error processes is suggested by the observation that errors are committed with high frequency but are mostly eliminated shortly after they are made, through a continuous checking process (feedback) (Lind, 1983). Somewhat similar to the process of steering a vehicle, it could be considered as a process of continuous monitoring and correction of observed error. The resultant error in the process is the net error surviving the continuous elimination process. Moreover, error is often detected because something "does not look quite right," rather than by formal checking or independent repetition of the task in which the error was originally committed. This suggests that the likelihood of error elimination increases significantly with the magnitude of the error. Small errors should have a better chance of survival than large ones.

Consider a system with scalar capacity R if correctly designed and built but with actual capacity $R + E$, where E is an additive expression of human error, a random variable. Consider the filtering of the distribution of E in the inspection process. Let $p_E(x)$ be the density of error; i.e., $p_E(x)dx$ is the probability of an error between x and $x + dx$ magnitude:

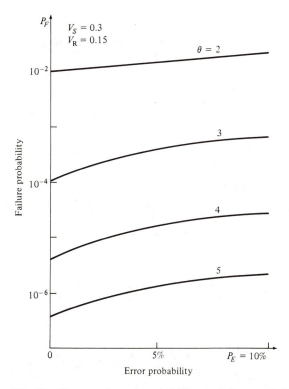

Figure 2.3   Significance of error probability on failure probability.

$$p_E(x)\,dx \;=\; P(x \leqslant E < x + dx)$$

Let t be a measure of the amount of checking, e.g., the time invested in inspection. Then $p_E(x)$ decreases with t, except in the case where there is zero error, $x = 0$. The rate of decrease, dp_E/dt, is assumed proportional to the amount of error present, i.e., p_E. Assuming, furthermore, that it is a function of the magnitude $h(x)$ of the error gives

$$\frac{dp_E(x)}{dt} \;=\; -c_0\,h(x)\,p_E(x)$$

where c_0 is a constant, yielding

$$p_E(x) \;=\; p_{EO}(x)\exp[-h(x)\,c_0\,t]$$

Consider first the special case that the initial error density is uniform, $p_{EO}(x)=p_0$, and that $h(x)$ is quadratic, $h(x)=c_1 x^2$. Then

$$p_E(x)=p_0\exp\!\left(-\frac{x^2}{2}\,(2c_0 c_1 t)\right)$$

This equation shows that the error is normally distributed for $t>0$. Choosing a normally distributed initial error gives the same result, of course, but avoids the complication of infinite error probability for $t=0$. Thus assume that

$$p_{EO}(x,0) = \frac{1}{s(2\pi)^{1/2}} \exp\left(-\frac{x^2}{2s^2}\right)$$

That is, at $t=0$ error is certain to occur and be normally distributed with zero mean and variance s^2. Then

$$p_E(x,t) = \frac{1}{s(2\pi)^{1/2}} \exp\left(-\frac{x^2}{2s^2}\right) \exp\left(-\frac{x^2}{2s^2}\frac{t}{t_0}\right)$$

$$= \frac{1}{s(2\pi)^{1/2}} \exp\left|-\frac{x^2}{2s^2}\left(1+\frac{t}{t_0}\right)\right|$$

in which t_0 is a constant. With the notation

$$p_E(t) = \left(1+\frac{t}{t_0}\right)^{-1/2}$$

$$\sigma_F = s\,p_E(t)$$

one obtains

$$p_E(x,t) \in p_E(t)\,N(0,\sigma_F^2)$$

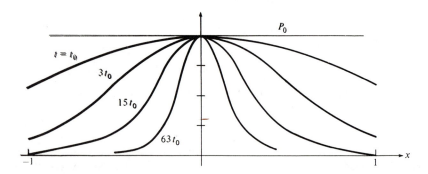

Figure 2.4 Probability density of error magnitude as a function of amount of inspection.

The model implies exponential decay with t of the density of any error of fixed magnitude x (Fig. 2.4). The distribution shape remains normal, with a concentrated probability mass of zero error. The total error probability $p_E(t)$ decays slower than exponential with t. At the unit $t=t_0$ the total error probability has decreased from 1 to $1/\sqrt{2}$.

2.6 MODEL UNCERTAINTY

A structure is produced through a sequence of decisions made by designers, contractors, inspectors, material suppliers, and by writers of codes and standards. These decisions are based on conceptual and mathematical models that link data obtained by scientific observations of loads and strengths with the outcome of the design decisions. The form that the final structure takes depends on the entire conceptual construct behind these decisions, called the *model.* The reliability of the structure is therefore also influenced by the model. The influence can only be determined empirically by replication or, in part, by comparison with more exact models. It is not often possible or expedient to compare models deductively; experiment is required and if this is not possible, judgment is invoked instead.

A perfect model would reflect the influence of many random variables and random processes, synthesizing an enormous amount of data by means of a complex mathematical apparatus into a payoff function over all possible decisions in the production of the structure. This process would result in an unmanageable decision problem, of course. The perfect model would not be functional. Many of the variables and processes are ignored in the practical model because of lack of information or for reasons of economy. The influence of those random variables that remain is simplified into tractable mathematics, and their distribution and correlation structure are simplified as well.

The perfect model can be represented conceptually by a deterministic surface in the space of all random variables that have an influence on the reliability, together with a probability density function for these variables. In the practical representation, the perfect model is projected into the formulation subspace of basic random variables, with the corresponding marginal probability density function and a random failure surface. This surface, which is assumed topologically equivalent to a hypersphere, is then simplified mathematically, replacing it with an approximate fixed surface plus a random vector process mapping the failure surface into the fixed surface. The parameters of this mapping constitute the *basic modeling random variables.* The space of basic random variables is now augmented with the basic model random variables, and model uncertainty is thus formally included in the standard finite random variable vector formalism of structural reliability theory. Ditlevsen (1982) has provided some examples.

2.7 LEVELS OF RELIABILITY METHODS

The great variety of idealizations in reliability models of structures, and the numerous ways in which it is possible to combine these idealizations to suit a particular design problem, make it desirable

to have a classification. Structural reliability methods are divided
into *levels,* characterized by the extent of information about the
structural problem that is used and provided.

Reliability methods that employ only one "characteristic"
value of each uncertain parameter are called *level I methods.* Load
and resistance factor formats, including the allowable stress formats,
are examples of level I methods.

Reliability methods that employ two values of each uncertain
parameter (commonly mean and variance), supplemented with a
measure of the correlation between the parameters (usually covariance), are called *level II methods.* Reliability index methods are
examples of level II methods.

Reliability methods that employ probability of failure as a
measure, and which therefore require a knowledge of the joint distribution of all uncertain parameters, are called *level III methods.*

Finally, a reliability method that compares a structural prospect with a reference prospect according to the principles of
engineering economic analysis under uncertainty, considering costs
and benefits, of construction, maintenance, repair, consequences of
failure, and interest on capital, etc., is called a *level IV method.*
Such methods are appropriate for structures that are of major
economic importance if the prospects of loss of life, limb, and
intangibles (e.g., cultural values) are minor. Highway bridges,
transmission towers, and nuclear power plant structures are suitable
objects of level IV design. The design methods are still in the process of development.

This classification of reliability methods is not exhaustive. For
example, a method could employ more information than a level II
method and yet not employ the complete distribution information
of a level III method; it might employ some of the concepts of
economic prospect comparison level IV, and so on. The classification has, however, proved to be very useful in practical discourse
on reliability methods.

The *rationale* of a reliability method is a justification in terms
of a higher level. Thus a level I method may be justified on level
II, in that it provides a reliability index that in some sense is close
to a target value (see Chapter 6). The parameters of the method are
determined by calibration to approximate the higher level. Level I
methods in new structural design codes and standards are now routinely calibrated by Level II or Level III methods (see Chapter 6).
Level IV methods can be justified by recourse to broader economic
principles, or to socioeconomic principles, if other than strictly
economic values are taken into consideration.

2.8 SUMMARY

A reliability method serves to decide whether or not a design is judged to be adequately reliable. Reliability methods can be divided broadly into research models, which aim to calculate a precise value of the reliability from probability data, and technical models, which aim to be efficient in practical design decision making. Probability data are assigned, not measured, and there is no "true" reliability. One distinguishes between the theoretical reliability, calculated from a reliability model, and the overall reliability that (in principle) is the observable failure frequency limit of a design. The difference — the adjunct probability of failure — is associated with gross error.

Loads are idealized in reliability modeling. Loads are often represented as a random process emanating from a source and modified by passage through "filters" to the structural element under analysis. Structural resistance to loads is also idealized; the theoretical resistance calculated by reliability theory and mechanics is modified by model error and by gross human error. According to the degree of idealization, structural reliability methods can be classified into levels. Level I employs specific load and resistance factors; each basic random variable can be represented by one characteristic value. In level II methods, the random variables are represented by their means and covariance, and design aims to achieve a specified value of a reliability index. In level III design, the probability of failure in a technical model is aimed at a prescribed value, while in level IV design there is an attempt to achieve a reliability that is in harmony with that of other technological apparatus in society.

REFERENCES

BARLOW, R. E., J. B. FUSSEL and N. D. SINGPURWALLA, eds., *Theoretical and Applied Aspects of System Reliability and Safety Assessment,* Society for Industrial and Applied Mathematics, Philadelphia, 1975.

BROWN, C. B., "A Fuzzy Safety Measure," *Journal of the Engineering Mechanics Division,* ASCE, Vol. 105, 1979, pp. 855-872.

CORNELL, C. A. and R. D. LARRABEE, "Representation of Loads for Code Purposes," in *Proceedings,* ICOSSAR'77, ed. H. Kupfer et al., Werner Verlag, Düsseldorf, 1977, pp. 135-148.

DAVENPORT, A. G., "On the Assessment of the Reliability of Wind Loading on Low Buildings," in *Proceedings, 5th Colloquium on Industrial Aerodynamics, Aachen, West Germany, June 14-16, 1982.*

DITLEVSEN, O., "Model Uncertainty in Structural Reliability," *Structural Safety,* Vol. 1, 1982, pp. 73-86.

DUDDECK, H., "The Role of Research Models and Technical Models in Engineering Science," in *Proceedings, ICOSSAR'77,* ed. H. Kupfer et al., Werner Verlag, Düsseldorf, 1977, pp. 115-118.

FERRY-BORGES, J., "Implementation of Probabilistic Safety Concepts in International Codes," in *Proceedings, ICOSSAR'77,* ed. H. Kupfer et al., Werner Verlag, Düsseldorf, 1977, pp. 121-133.

LIND, N. C., "Management of Gross Errors," in *Proceedings, Fourth International Conference on Application of Statistics and Probability in Soil and Structural Engineering, ICASP4, University of Firenze, Italy, June 1983, pp. 669-682.*

LIND, N. C., "Models of Human Error in Structural Reliability," *Structural Safety,* Vol. 1, 1983, pp. 167-175.

MATOUSEK, M., "Outcome of a Survey on 800 Construction Failures," in *Proceedings, IABSE Colloquium on Inspection and Quality Control Institute of Structural Engineering, Swiss Federal Institute of Technology, Zurich, 1977.*

NBCC (National Building Code of Canada), National Research Council of Canada, Ottawa, Ontario, 1975, 1977, 1980.

3

MULTIPLE SAFETY FACTOR FORMATS

(level I)

3.1 ORIGIN AND PURPOSE

Modern structural engineering, using the scientific method of experiment and prediction, began in Italy during the Renaissance with tests on the strength of trusses and beams. The test results could be interpreted rationally in terms of stress as an intensive force quantity, leading naturally to the discovery of material strength as an invariant property. Stone and brick masonry were then dominant civil engineering structural materials; dead load dominated as a result. Thus the major load component was strictly limited in magnitude by the law of gravity and by the invariance of the specific gravity of the material. Under these circumstances it is logical to choose stress as the formulation variable and to design on the basis of an *allowable stress.*

Due to the high variability of the strength of some materials, such as mortar or cast iron, it was natural, according to judgment, to use higher safety factors for materials with greater dispersion. In times of high economic expansion and rapid technological change, however, there was less engineering experience available for investment in the average structure, and as a result the basis for judgment was not always sound and the rates of failure were high. In the late nineteenth century, codes and design standards were introduced and were immediately successful in reducing the rates of failure of steel structures, including pressure vessels. In effect, these codes imposed the judgment of the most experienced engineers on all designers in matters of allowable stress as well as in many other matters of proper procedure of design.

The introduction of structural design codes, in response to a public demand, had the effect of transferring a part of the responsibility for structural integrity from the individual engineer, making the profession collectively responsible for the safety of those structures that conform to the code and are built according to prudent engineering procedure. This is one manifestation of a general trend toward higher accountability in society. The trend is also apparent within the profession; there is a growing demand that code requirements be discussed in professional forum and justified. Thus it is now common that new codes and standards are first issued for public comment and then published together with a commentary justifying the requirements. Allowable stress code formats are, however, difficult to justify in terms of reliability because structures with relatively high live load, which has high dispersion, are systematically biased toward higher probability of failure. Only by adopting *code-specified values* of the live load that are conservative in varying degree and therefore difficult to reconcile with reality could more uniform reliability be achieved with an allowable stress format. For more complex load effects the allowable stress formats become impractical as more information about the load is available. Consider wind loading, for example; for a single locality a modern code must reflect differences in exposure, orientation, local terrain, and aerodynamic and structural dynamic effects, and would need to present a tabulation of all important combination values of the wind pressure. In times of rapid growth of technical information, such tabulations need frequent extensive revision as well. Moreover, for structures that behave nonlinearly, whether for reasons of geometry or material, it is not possible to determine a uniform allowable stress for a constant specified load; the safety margin depends on the load level in relation to the characteristics of the structure. Finally, there are structural design considerations (e.g., overturning of buildings or retaining walls) that do not involve stresses and cannot be modeled realistically in terms of allowable effects.

All these shortcomings can be reduced or eliminated in design formats that employ multiple safety factors. Such formats were introduced in Denmark in the 1940s in reinforced concrete and geotechnical standards under the name of *partial safety factor formats;* in the reinforced concrete design standard of the American Concrete Institute in the 1950s under the name of *ultimate strength design;* in the National Building Code of Canada since 1977 as the *limit states design* (LSD) option (NBCC, 1980); and is being introduced worldwide as the preferred format of structural design codes (NKB, 1978). In the United States the term *load and resistance factor format* is currently used (Ravindra and Galambos, 1978; Ellingwood et al., 1980). It has, evidently, been difficult to find a terminology that is universally suitable. From the viewpoint of reliability-based optimality, it appears, however, that the most powerful characteristic of the new method is the flexibility that the presence of many adjustable factors gives. For this reason all these methods are jointly called *multiple-factor formats.*

3.2 MULTIPLE-FACTOR FORMATS

In multiple-factor formats the various possibilities of structural malperformance are classified, and each is associated with a type of limit state. Two types are common: *ultimate limit states* and *serviceability limit states.* Ultimate limit states correspond to the limit of load-carrying capacity of a member or the structure as a whole. Examples are: formation of a plastic yield mechanism, brittle fracture, fatigue fracture, instability, buckling, and overturning. Serviceability limit states imply deformations in excess of tolerance without exceeding the load-carrying capacity. Examples are cracks, permanent deflections, or vibrations. Other limit states classes are possible, e.g., accidental limit states (CEB, 1976) or progressive limit states (NKB, 1978), relating to total failure of an accidentally damaged structure (due to explosion, fire or vehicle impact, etc.).

Verification of a structure with respect to a limit state uses a mathematical model of the limit state in the form of a failure or limit state function $g(\ )$ of the relevant load parameters Q_i, strength parameters R_j, and nominal geometric parameters L_k defining the structure, chosen such that

$$g(Q_i, R_j, L_k) = 0 \qquad (3.1)$$

characterizes the limit state. Conventionally, $g < 0$ signifies exceeding the limit state, i.e., failure.

The *design values* of the basic random variables are denoted by q_{id}, r_{jd}, and l_{kd}, respectively. The design values may be written as

$$q_{id} = \gamma_i q_{ic} \qquad (3.2)$$

$$r_{jd} = \varphi_j r_{jc} \qquad (3.3)$$

$$l_{kd} = \zeta_k l_{kc} \qquad (3.4)$$

in which q_{ic}, r_{jc}, and l_{kc} are *characteristic values*, γ_i are *load factors*, φ_j are *resistance factors*, and ζ_k are *geometrical factors*. Verification of the structure with respect to the limit state is done by inserting the design values and the dimension parameters into the limit state function to ascertain that

$$g(q_{id}, r_{jd}, l_{kd}) \geqslant 0 \qquad (3.5)$$

The characteristic values are often the mean value for dead load, the 98% fractile in the distribution of the annual maxima for movable loads, the 2 or 5% fractile for strength parameters, and the mean value for geometrical parameters. The load factors are usually greater than unity except when the load has a stabilizing effect. The resistance factors are usually less than unity. The geometrical factors are generally equal to unity.

The selection of the characteristic values must be done judiciously to ensure that a near-constant reliability can be obtained. The selection can in many cases be done on the basis of optimization, also taking into account relevant factors external to the model. For example, the fractiles of concrete strength are determined by standard cylinder tests; the cylinder strength f_c' is a *vicarious* strength parameter because there is a systematic difference between the strength of the concrete in place in the structure at the time of failure and the value of f_c'.

Load and resistance factors can be decomposed further according to basic sources of uncertainty. The wind load in NBCC (1980), for example, takes the form $QC_e C_a C_g$, where the last three factors are load factors partial to exposure and orientation, aerodynamic effects, and structural dynamic effects, respectively. The strength factor φ_j in (3.3) is in NKB (1978) written as $1/(\gamma_{m1}\gamma_{m2}\gamma_{m3}\gamma_{n1}\gamma_{n2})$, in which γ_{m1} accounts for the dispersion in material strength, γ_{m2} accounts for the model uncertainty, γ_{m3} accounts for the uncertainty due to use of a vicarious strength parameter, γ_{n1} accounts for the type and consequences of failure, and γ_{n2} accounts for the degree of inspection and control other than the statistical quality control of the material.

Multiple-factor codes may differ somewhat in practice from (3.2) to (3.4). Thus NKB (1978) writes the design values of the geometric parameters as the additive form

$$l_{kd} = l_{kc} + \Delta_k \qquad (3.6)$$

in which Δ_k is the prescribed tolerance. This form is just more convenient numerically; it does not represent a deviation in principle from limit states design.

eg. area × (yield limit)

char. value of material

The NBCC (1980) limit states design format specifies the formula

$$\varphi R \geqslant \alpha_D D + \gamma \psi (\alpha_L L + \alpha_Q Q + \alpha_T T) \tag{3.7}$$

in which the capital letters symbolize the characteristic values of strength, R, dead load, D; live load, L; snow, wind, and earthquake load, Q; and thermal effects, T; φ is the strength factor, the α's are load factors, ψ is a *load combination factor,* and γ is an *importance factor.* Equation (3.7) is a special case of the limit states design format, (3.5). First, load and resistance effects have been separated on the right- and left-hand sides of the inequality. This is not restrictive, since the mechanical analysis of the limit state naturally is made by comparison of demand (load or stress) with capacity (member or section resistance, or strength). If there is more than one basic strength factor (as in component structures), the form of (3.7) is not adequate. Second, the load effect appears to be obtained as the sum of individual components in (3.7), which is highly restricted in comparison with (3.5). However, the right-hand side of (3.7) is symbolic and is meant to express the effect of the loads acting simultaneously, rather than the sum of the individual load effects. Moreover, the symbols in (3.7) are meant to take on a variety of values, to generate various cases of loading: i.e., dead load only, dead plus live load, dead plus live plus snow load, etc. The code specifies $\alpha_D = 1.25$ (or 0.85 if this is more critical), $\alpha_L = \alpha_Q = \alpha_T = 1.5$, and $\psi = 1$, 0.75 or 0.67 for one, two, or three load effects other than dead load acting simultaneously. Truly, the right-hand side of (3.7) is therefore not an algebraic expression. It is more general than (3.5) because the meaning is that the maximum effect of the load is to be calculated for several load combinations: D, $D+L$, $D+Q$, . . . , $D+L+Q+T$. At the same time the format is more restrictive than (3.5) since the load combination factor ψ is the same independent of the origin of the loads.

The format of (3.5) together with the wording of the code is a legal-technical compromise, expressing the many loading cases to be examined succinctly but with precision. A limit state format for verification of structural adequacy takes the general form

$$g_R(\varphi_j r_{jc}, l_{kc} + \Delta_k) \geqslant \max_m g_S(\gamma_{im} q_{ic}, l_{kc} + \Delta_k), \quad m = 1,2,...,n \tag{3.8}$$

with the resistance function isolated on the left-hand side and the loading function as the maximum effect of n load combinations isolated on the right-hand side. The matrix of load factors, if determined rationally according to an objective can be very complex. For example, OHBDC (1983) prescribes 24 basic load variables in

$n = 17$ load combinations. Since engineers until quite recently have proceeded by memory in many calculations, there has been a need for simplicity in the system of load factors. This is reflected in (3.7) as well as in OHBDC (1983), which uses only 14 different nonzero constants in the load factor matrix, and thus is committed to memory relatively easily. Of course, the simplifications mean that the load factors matrix can only approximate a rationally optimized matrix, leading to designs that are less than optimal. However, the load factors currently in service are largely based on judgment, and more complex versions would not be more convincing if determined in the same fashion. It is to be expected that optimal load factors matrices can be determined in the near future by means of modern load process models (Chapter 10). Such matrices can be used advantageously for better safety and economy in computer-aided design; indeed, they are easily implemented on many programmable hand-held calculators.

3.3 CHARACTERISTIC VALUES

The characteristic values of load parameters are chosen to be high but measurable fractiles. Commonly, the characteristic value is the 98th percentile of the annual maximum load. For strengths, the characteristic values are usually the 2nd, 5th, or 10th percentile. Often the producer's "guaranteed" or "specified" minimum values are used. For structural steel, unpublished test results suggest that the guaranteed yield point corresponds approximately to the 5th percentile. The characteristic value of the strength serves as the primary quality control parameter in the relationship between producer and owner. By the choice of a low fractile as the characteristic value, producers who produce with a small variation in strength properties are rewarded. Aspects of control of characteristic values are treated in Chapter 6.

To demonstrate the estimation of a characteristic value consider the common case of a random strength parameter with a two-parameter distribution. By appropriate mapping (see Example 3.1) the parameter is transformed into a normally distributed variable R, with unknown mean μ and unknown variance σ^2. The characteristic strength value r_c is defined as the α-fractile of R, which may be written as

$$r_c = \mu + u_\alpha \sigma \qquad (3.9)$$

in which

$$u_\alpha = \Phi^{-1}(\alpha) \qquad (3.10)$$

For $\alpha = 1, 2, 5$, and 10%, u_α equals -2.33, -2.05, -1.64, and -1.28, respectively. For estimation of r_c a random sample of transformed test values $r_1, r_2, \ldots, r_n$ is given. The mean value and the variance are estimated by the *sample mean* $\bar{r}$ and the *sample variance* s^2, respectively

$$\bar{r} = \frac{1}{n} \sum_{i=1}^{n} r_i \tag{3.11}$$

$$s^2 = \frac{1}{n-1} \sum_{i=1}^{n} (r_i - \bar{r})^2 \tag{3.12}$$

$\bar{r}$ is the outcome of a random variable $\bar{R}$, which is normally distributed with mean value μ and variance σ^2/n. s^2 is the outcome of a random variable S^2, which is independent of $\bar{R}$. $(n-1)S^2/\sigma^2$ has a chi-square distribution with $n-1$ degrees of freedom. The characteristic value can be estimated by the value r_α,

$$r_\alpha = \bar{r} + u_\alpha s \tag{3.13}$$

This estimate is the outcome of a random variable R_α,

$$R_\alpha = \bar{R} + u_\alpha S \tag{3.14}$$

R_α is greater or smaller than the true characteristic value r_c with about equal probability. However, one often wishes to use an estimate $r_{s\alpha}$, such that r_c is greater than $r_{s\alpha}$ with probability K. This probability is called the *confidence level* and is generally specified in the code, e.g., as 75% or 90% (NKB, 1978). Clearly, $r_{s\alpha}$ is close to r_α if K equals 50%, and $r_{s\alpha}$ is less than r_α if K is greater. $r_{s\alpha}$ is written as

$$r_{s\alpha} = \bar{r} + k_\alpha(n)s \tag{3.15}$$

$r_{s\alpha}$ is the outcome of a random variable $R_{s\alpha}$ and the probability K, that $R_{s\alpha}$ is less than r_c is

$$P(R_{s\alpha} \leqslant r_c) = P(\bar{R} + k_\alpha(n) S \leqslant \mu + u_\alpha \sigma) \tag{3.16}$$

$$= P\left[\frac{\dfrac{\bar{R} - \mu}{\sigma/\sqrt{n}} - u_\alpha \sqrt{n}}{\sqrt{S^2/\sigma^2}} \leqslant -k_\alpha(n)\sqrt{n} \right]$$

To compute $k_\alpha(n)$ use is made of the non-central t-distribution. When U is standard normally distributed and Z is independent of U with a $\chi^2(n)$-distribution, then the random variable

$$T = \frac{U + \lambda}{\sqrt{Z/n}} \tag{3.17}$$

has a non-central t-distribution with n degrees of freedom and non-centrality parameter λ. The distribution is denoted by $t(n,\lambda)$ and is tabulated in, e.g., Johnson and Welch (1940). $k_\alpha(n)$ is thus determined such that $-k_\alpha(n)\sqrt{n}$ is the Kth fractile in the distribution $t(n-1,-u_\alpha\sqrt{n})$. It is noted that $k_\alpha(n)$ is independent of μ and σ. Table 3.1 presents some values of $k_\alpha(n)$.

TABLE 3.1 $-k_\alpha(n)$ for the 10th Percentile and 75% Confidence Level for a Normal Distribution

n	σ unknown	σ known
3	2.50	1.67
4	2.13	1.62
5	1.96	1.58
6	1.86	1.56
10	1.67	1.49
20	1.53	1.43
30	1.47	1.40
∞	1.28	1.28

In some cases the standard deviation σ is known since the sample comes from a large production with a constant standard deviation. The value $k_\alpha(n)$ is then simply

$$k_\alpha(n) = u_\alpha - \Phi^{-1}(K)/\sqrt{n} \tag{3.18}$$

Values for this case are also shown in Table 3.1.

When tables for the non-central t-distribution are not available, then close approximations to $k_\alpha(n)$ can be computed from a result due to Hald (1952). Hald has shown that $R_{s\alpha}$ is approximately normally distributed with mean value and standard deviation

$$E[R_{s\alpha}] \approx \mu + k_\alpha(n)\sigma \tag{3.19}$$

$$D[R_{s\alpha}] \approx \sigma \left| \frac{1}{n} + \frac{k_\alpha(n)^2}{2(n-1)} \right|^{1/2} \tag{3.20}$$

The approximation to $k_\alpha(n)$ is thus the solution to

$$\Phi \left| \frac{u_\alpha - k_\alpha(n)}{\left| \dfrac{1}{n} + \dfrac{k_\alpha(n)^2}{2(n-1)} \right|^{1/2}} \right| = K \tag{3.21}$$

leading to the approximation

$$k_\alpha(n) \approx \frac{u_\alpha - \Phi^{-1}(K)\left| \dfrac{1}{n}\left(1 - \dfrac{(\Phi^{-1}(K))^2}{2(n-1)}\right) + \dfrac{u_\alpha^2}{2(n-1)} \right|^{1/2}}{1 - \dfrac{(\Phi^{-1}(K))^2}{2(n-1)}} \tag{3.22}$$

In Chapter 6 estimation of distribution parameters and characteristic values is addressed in more detail.

Example 3.1

Independent random sampling of the strength of a material gave the following values (MPa):

$$223 \quad 250 \quad 298 \quad 310 \quad 317$$
$$349 \quad 351 \quad 360 \quad 382 \quad 385$$

With $n = 10$, (3.11) and (3.12) give $\bar{r} = 322.5$ MPa and $s = 54.1$ MPa, respectively. Assuming that the strength distribution is normal, one finds the 75% confidence level of the 10% strength fractile by Table 3.1 as $r_{s,0.1} = 322.5 - 1.67 \times 54.1 = 232.2$ MPa.

The sample size is insufficient to judge if the distribution is indeed normal. If it is known a priori that the distribution is lognormal, one can transform the observations by taking the logarithm, giving

$$2.348 \quad 2.398 \quad 2.474 \quad 2.491 \quad 2.501$$
$$2.543 \quad 2.545 \quad 2.556 \quad 2.582 \quad 2.585$$

and $\bar{r} = 2.5023$, $s = 0.0783$, and $r_{s,0.1} = 2.5023 - 1.67 \times 0.0783 = 2.3715$, giving the 75% confidence limit on the characteristic value as $10^{2.3715} = 235.3$ MPa. One observes that there is a small influence of the assumption of distribution shape in this example. Also, it is noticed that increasing the sample size could be expected to increase the 75% confidence limit. If the mean and standard deviation remain the same, the confidence limit could be expected to approach $\bar{r} + u_{0.1}s = 322.5 - 1.28 \times 54.1 = 253.2$ MPa.

3.4 CALIBRATION

The characteristic values, confidence levels, safety factors, load combination factors, etc., — in short, all numbers specified in the design code (other than physical and mathematical constants) — are called the *code parameters*. Selection of a set of numerical values for the code parameters is called *calibration* of the code format. Formal calibration seeking to minimize a specified objective function is called *code optimization* (Chapter 6). Calibration of a new multiple-factor code may consist simply of the selection of a set of parameters such that the new code gives approximately the same dimensions as an existing code, or the parameters may be chosen by judgment to give the desired reliability spectrum over future designs, judged on the experience with an old code. It is desirable that a level I format be calibrated by a reliability method on a higher level. Many multiple-factor codes in service were calibrated at least in part by level II reliability methods, described in the next chapter.

3.5 SUMMARY

Until recently, most structural design calculations have employed an allowable stress format. In the earliest designs the allowable stress was selected by the engineer, taking into consideration the dispersion and uncertainty in loads, strength, and analysis. To broaden the judgment base embodied in each design, the method of analysis and the safety factor were standardized and laid down in design codes during the nineteenth century. The allowable stress format employs, ostensibly, only one safety factor and it lacks the flexibility to adjust the safety margin according to differences in load dispersion, likelihood of load combinations, consequences of failure, and uncertainty in analysis.

Multiple-factor formats were introduced to give the desired flexibility in safety margin. Each basic random variable is represented in the design inequalities by a characteristic value, multiplied by a load or a resistance factor as appropriate. The design inequalities take the general form of (3.8), separating demand on one side from capacity on the other. The design inequalities further contain factors or terms that relate to particular circumstances of the limit state, such as tolerances or importance factors.

The characteristic values of the loads are used in a simplified representation of the random process of the load by means of multiple load combinations, each with a particular set of load factors. The characteristic values of the strength parameters serve as quality control parameters and have contractual significance for supplier and owner; they are usually chosen as low fractiles of the random strength parameters. In the calibration of the design format the remaining code parameters are selected to achieve a particular objective. Calibration may be informal (by "judgment"), seeking to achieve an appropriate estimated degree of safety in each failure mode. Or it may be more formal, in the process of code optimization, optimizing a specified objective function that reflects the cost of material and the expected cost of failure.

REFERENCES

CEB (Comité Europeen du Béton), Joint Committee on Structural Safety CEB-CECM-FIP-IABSE-IASS-RILEM, "Common Unified Rules for Different Types of Construction and Material," *CEB Bulletin No. 116E,* 1976.

ELLINGWOOD, B. et al., "Development of a Probability Based Load

Criterion for American National Standard A58," *National Bureau of Standards Publication 577,* Washington, D.C., 1980.

HALD, A., *Statistical Theory with Engineering Applications,* John Wiley, New York, 1952.

JOHNSON, N. L. and B. L. WELCH, "Applications of the Non-Central *t*-Distribution," *Biometrika,* Vol. 31, 1940, pp. 362-389.

NBCC (National Building Code of Canada), National Research Council of Canada, Ottawa, Ontario, 1975, 1977, 1980.

NKB (The Nordic Committee on Building Regulations), "Recommendations for Loading and Safety Regulations for Structural Design," *NKB-Report,* No. 36, Copenhagen, November 1978.

OHBDC (Ontario Highway Bridge Design Code), Ontario Ministry of Transportation and Communication, Downsview, Ontario, 1983.

RAVINDRA, M. K. and T. V. GALAMBOS, "Load and Resistance Factor Design for Steel," *Journal of the Structural Division,* ASCE, Vol. 104, 1978, pp. 1337-1353.

4

SECOND MOMENT
RELIABILITY INDEX

(level II)

4.1 INTRODUCTION

Second-moment reliability methods were introduced to avoid the tail sensitivity problem described in Chapter 1. Presently, these methods are widely used for reliability calculations as well as for code calibration purposes. The idea behind second-moment reliability theory is that all uncertainties concerning the structural reliability are expressed solely in terms of expected values (first moments) and covariances (second moments) of the entering parameters (Cornell, 1969; Ditlevsen, 1981). These parameters are called the *basic variables* and they are denoted here by Z_i. The basic variables include loading parameters; strength parameters; and geometrical, statistical, and model uncertainty variables. For the application of second-moment reliability methods, the number of basic variables must be finite. Further, it must be possible for each set of values of the basic variables to state whether or not the structure has failed (see also Chapter 2). This leads to a unique division of z-space into two sets, called the *safe set S* and the *failure set F*, respectively. The two sets are separated by the *failure surface* (or *limit state surface*). In this chapter various reliability measures called second-moment reliability indices are defined, based on the failure surface and the second-moment representation of the basic variables. By *second-moment representation* is meant the set of expected values

$E[Z_i]$, and the set of covariances $Cov[Z_i, Z_j]$ including the variances $Var[Z_i]$, or any set from which these values can be determined uniquely. *i.e. distribution forms are not specified.*

4.2 THE CORNELL RELIABILITY INDEX

The failure surface separating the safe set and the failure set is denoted by L_Z. A function $g(z_i)$ is called a *failure function* (or *limit state function*) if

$$g(z_i) > 0, \quad z_i \in S$$
$$g(z_i) = 0, \quad z_i \in L_Z \qquad (4.1)$$
$$g(z_i) < 0, \quad z_i \in F$$

According to this definition a failure function specifies the failure surface L_Z and satisfies a sign convention outside L_Z. The failure surface, on the other hand, does not define a unique failure function and care must be taken not to introduce a certain arbitrariness through the failure function. A simple choice for the failure function is

$$g(z_i) = \begin{cases} 1, & z_i \in S \\ 0, & z_i \in L_Z \\ -1, & z_i \in F \end{cases} \qquad (4.2)$$

For computational reasons a differentiable g-function is generally chosen whenever possible. The g-function usually results from the use of a mechanical analysis method for the structure.

The random variable obtained by replacing the parameters z_i in the failure function with the corresponding random variables Z_i is called a *safety margin* and is denoted by M:

$$M = g(Z_i) \qquad (4.3)$$

This safety margin, by definition, reflects the arbitrariness introduced by the choice of a failure function.

Cornell (1969) defined a *reliability index* (or *safety index*) β_C as

$$\beta_C = \frac{E[M]}{D[M]} = \frac{1}{c.v.(M)} \text{std. dev.} \qquad (4.4)$$

This definition is illustrated geometrically in Fig. 4.1. In this one-dimensional case the failure surface is simply the point $m = 0$. The idea behind the reliability index definition is that the distance from the location measure $E[M]$ to the limit state surface provides a good measure of reliability. The distance is measured in units of the uncertainty scale parameter $D[M]$.

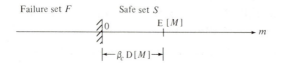

Figure 4.1 Geometrical illustration of the Cornell reliability index β_C.

In the original formulation by Cornell the failure function was, written as the difference between a resistance r and the corresponding load effect s:

$$g(r,s) = r - s \qquad (4.5)$$

The corresponding safety margin is

$$M = R - S \qquad (4.6)$$

and if R and S are uncorrelated, the reliability index (4.4) becomes

$$\beta_C = \frac{E[R] - E[S]}{\sqrt{\text{Var}[R] + \text{Var}[S]}} = \frac{\mu_R - \mu_S}{\sqrt{\sigma_R^2 + \sigma_S^2}} \qquad (4.7)$$

where the symbols μ_R, μ_S, σ_R, and σ_S are self-explanatory. For $\mu_S > 0$ the reliability index is bounded by $1/V_R = \mu_R/\sigma_R$, where V_R is the coefficient of variation of R.

If the failure surface is a hyperplane, it is possible to define a linear failure function

$$g(z_i) = a_0 + \sum_{i=1}^{n} a_i z_i = a_0 + \mathbf{a}^T \mathbf{z} \qquad (4.8)$$

Here a vector notation has been introduced; $\mathbf{a}^T$ is a row vector with elements a_i, and $\mathbf{z}$ is a column vector with elements z_i. The safety margin corresponding to the failure function in (4.8) is

$$M = a_0 + \sum_{i=1}^{n} a_i Z_i = a_0 + \mathbf{a}^T \mathbf{Z} \qquad (4.9)$$

The reliability index (4.4) takes the value

$$\beta_C = \frac{a_0 + \mathbf{a}^T E[\mathbf{Z}]}{\sqrt{\mathbf{a}^T \mathbf{C}_Z \mathbf{a}}} \qquad (4.10)$$

Cornell?

where $E[\mathbf{Z}]$ is the vector of expected values and $\mathbf{C}_Z$ is the matrix of covariances of $\mathbf{Z}$. It can be shown that β_C in (4.10) is invariant under any linear transformation of the basic variables.

If the failure surface is not a hyperplane, it is not possible to formulate a linear safety margin in terms of the basic variables. This means that the expected value and variance of M cannot be calculated solely from the second-moment representation of the basic variables. Definition (4.4) therefore applies only to hyperplane failure surfaces. In the next sections the reliability index definition is extended to cover structures with nonlinear failure surfaces, but first an example will illustrate the calculation of β_C in (4.10).

Example 4.1

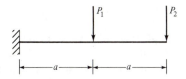

Figure 4.2   Cantilever beam loaded by two concentrated forces.

Figure 4.2 shows a cantilever beam loaded by two concentrated forces P_1 and P_2. The moment capacity of the beam at the support is B. The second-moment representation of the set of basic variables $\mathbf{Z}=(B,P_1,P_2)$ is

$$E[\mathbf{Z}] = \begin{vmatrix} 250 \text{ kNm} \\ 10 \text{ kN} \\ 10 \text{ kN} \end{vmatrix}$$

$$\mathbf{C}_Z = \begin{vmatrix} 900 \text{ (kNm)}^2 & 0 & 0 \\ 0 & 9 \text{ kN}^2 & 6 \text{ kN}^2 \\ 0 & 6 \text{ kN}^2 & 9 \text{ kN}^2 \end{vmatrix}$$

The beam fails if the moment capacity is exceeded, so a linear safety margin M is

$$M = B - aP_1 - 2aP_2$$

With a equal to 4 m the reliability index becomes

$$\beta_C = \frac{250 - 4 \times 10 - 8 \times 10}{\sqrt{900 + 4^2 \times 9 + 8^2 \times 9 + 2 \times 4 \times 8 \times 6}} = 2.90$$

4.3 FIRST-ORDER SECOND-MOMENT RELIABILITY INDEX

The failure function in (4.5) is not unique for the failure surface; many alternatives exist. For physical reasons the basic variables R and S are often restricted to positive values. A simple alternative to (4.5) is then

$$g(r,s) = \log(r/s) \qquad (4.11)$$

in which form the limits on R and S are emphasized. The corresponding safety margin is

$$M = \log(R/S) \qquad (4.12)$$

and the reliability index (4.4) becomes

$$\beta_{RE} = \frac{E[\log(R/S)]}{D[\log(R/S)]} \qquad (4.13)$$

This formulation was proposed by Rosenblueth and Esteva (1972).

The safety margin $\log(R/S)$ is a nonlinear function of R and S, and consequently its mean value and standard deviation cannot be calculated solely from the second-moment representation of (R,S). One way to circumvent this problem is to linearize the safety margin. This makes a choice of linearization procedure necessary. The simplest choice is to use the linear terms in a Taylor series expansion around a point. Linearizing around the point of expected values (μ_R,μ_S) results in a safety margin M_{FO} given by

$$M_{FO} = \log\mu_R - \log\mu_S + \frac{R-\mu_R}{\mu_R} - \frac{S-\mu_S}{\mu_S} \qquad (4.14)$$

The use of this linear safety margin expression yields the reliability index β_{FO} by (4.10) as

$$\beta_{FO} = \frac{\log\mu_R - \log\mu_S}{\sqrt{V_R^2 + V_S^2}} \qquad (4.15)$$

where V_R and V_S are the coefficients of variation of R and S, respectively, and where R and S are assumed to be uncorrelated. The value of β_{FO} is different from the value of the reliability index given in (4.7). This lack of uniqueness is due to the ambiguity in the choice of failure function. The surfaces given by the equation

$$m = g(r,s) \qquad (4.16)$$

are different for the two choices of failure function. The approximating tangent hyperplanes at the point $(\mu_R,\mu_S,m(\mu_R,\mu_S))$ do not intersect the (r,s)-plane in the same failure surface and correspond to different values of the reliability index. The reliability index β_{FO} thus depends on the specific choice of failure function, not just on the failure surface. The ambiguity was first explained by Ditlevsen (1973).

For a general nonlinear failure function, linearization of the safety margin around the point z gives a linear safety margin M_{FO} as

$$M_{FO} = g(\mathbf{z}) + \sum_{i=1}^{n} \frac{\partial g}{\partial z_i}(\mathbf{z})(Z_i - z_i) \qquad (4.17)$$

with the corresponding reliability index β_{FO}:

$$\beta_{FO} = \frac{g(\mathbf{z}) + \displaystyle\sum_{i=1}^{n} \frac{\partial g}{\partial z_i}(\mathbf{z})(\mathrm{E}[Z_i] - z_i)}{\left[\displaystyle\sum_{i=1}^{n}\sum_{j=1}^{n} \frac{\partial g}{\partial z_i}(\mathbf{z}) \frac{\partial g}{\partial z_j}(\mathbf{z}) \mathrm{Cov}[Z_i, Z_j]\right]^{1/2}} \qquad (4.18)$$

This reliability index is called a *first-order second-moment reliability index*. If the linearization point is the mean-value point, then the reliability index should further be called a *mean-value first-order second-moment reliability index*.

To avoid arbitrariness in β_{FO} both the failure function and the linearization point need to be specified, which means a severe drawback for the applicability of the reliability index. In particular, the necessity of a standardized failure function implies that only one out of many mechanically equivalent descriptions can be allowed for a particular problem. This is illustrated further in Example 4.2. There is no arbitrariness due to the choice of failure function if only information about the failure surface is used, i.e., if the linearization point is selected as a point on the failure surface. This fact also follows directly from (4.18). Among the points on the failure surface a natural choice is the point which in some sense has the smallest distance from the mean value point, since the reliability index is meant somehow to measure the distance to the failure surface. The next section shows how the linearization point can be defined, leading to the reliability index definition proposed by Hasofer and Lind (1974).

Example 4.2

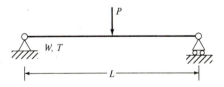

Figure 4.3 Simply supported beam loaded at the midpoint.

Figure 4.3 shows a simply supported steel beam loaded at the midpoint by a concentrated force P. The length of the beam is L, and the bending moment capacity at any point along the beam is WT, where W is the plastic section modulus and T is the yield stress. The second-moment representation of the basic variable vector $\mathbf{Z} = (P, L, W, T)$ is

$$E[\mathbf{Z}] = \begin{vmatrix} 10 \text{ kN} \\ 8 \text{ m} \\ 100 \times 10^{-6} \text{ m}^3 \\ 600 \times 10^3 \text{ kN/m}^2 \end{vmatrix}$$

$$\mathbf{C_Z} = \begin{vmatrix} 4 \text{ kN}^2 & 0 & 0 & 0 \\ 0 & 10 \times 10^{-3} \text{ m}^2 & 0 & 0 \\ 0 & 0 & 400 \times 10^{-12} \text{ m}^6 & 0 \\ 0 & 0 & 0 & 10 \times 10^9 \text{ (kN/m}^2)^2 \end{vmatrix}$$

If safety checking is done on the basis of load effects, the safety margin is

$$M_1 = g_1(P, L, W, T) = WT - \frac{PL}{4}$$

The safety margin $M_{FO,1}$ obtained by a linearization around the mean-value point is

$$M_{FO,1} = E[W]E[T] - \frac{E[P]E[L]}{4} + E[T](W - E[W]) + E[W](T - E[T])$$

$$- \frac{E[L](P - E[P])}{4} - \frac{E[P](L - E[L])}{4}$$

Based on this linearized safety margin, the first-order reliability index has the value $\beta_{FO,1} = 2.48$.

If, on the other hand, safety checking is done on the basis of stresses, the safety margin is

$$M_2 = g_2(P, L, W, T) = T - \frac{PL}{4W}$$

The safety margin $M_{FO,2}$ obtained by a linearization around the mean-value point is now

$$M_{FO,2} = T - \frac{E[P]E[L]}{4E[W]} - \frac{E[L]}{4E[W]}(P - E[P])$$

$$- \frac{E[P]}{4E[W]}(L - E[L]) + \frac{E[P]E[L]}{4E[W]^2}(W - E[W])$$

and the corresponding reliability index is $\beta_{FO,2} = 3.48$.

4.4 THE HASOFER AND LIND RELIABILITY INDEX

Figure 4.1 shows how the reliability index can be interpreted as a measure of the distance to the failure surface. In the one-dimensional case the standard deviation of the safety margin was conveniently used as the scale. To obtain a similar scale in the case of more basic variables, Hasofer and Lind (1974) proposed a nonhomogeneous linear mapping of the set of basic variables into a set of normalized and uncorrelated variables X_i, i.e., into a set with the vector of expected values

$$E[X] = 0 \qquad (4.19)$$

and with the covariance matrix

$$C_X = \text{Cov}[X,X^T] = I \qquad (4.20)$$

where I is the unit matrix. The transformation can be written as

$$X = A(Z - E[Z]) \qquad (4.21)$$

whereby (4.19) is always fulfilled. It follows from (4.20) for the transformation matrix A that

$$\text{Cov}[X,X^T] = A C_Z A^T = I \qquad (4.22)$$

A can thus be determined by standard techniques from linear algebra, e.g., based on eigenvalues and corresponding eigenvectors for C_Z. Since C_Z is nonnegative definite and symmetric, A can also be determined in a unique way as a lower-triangular matrix. A simple algorithm for this can be found in Rubinstein (1981). Any orthogonal transformation of the X-set results in a new set of normalized and uncorrelated variables. The distribution of the X-set is thus rotationally symmetric with respect to the second-moment distribution.

By the same mapping,

$$x = A(z - E[Z]) \qquad (4.23)$$

the mean-value point in z-space is mapped into the origin of x-space, and the failure surface L_Z in z-space is mapped onto the corresponding failure surface L_X in x-space as shown in Fig. 4.4. Due to the rotational symmetry of the second-moment representation of the X-set, it therefore follows that the geometrical distance from the origin in x-space to any point on L_X is simply the number of standard deviations from the mean-value point in z-space to the corresponding point on L_Z. The distance to the failure surface can then be measured by the *reliability index function* (Veneziano, 1974):

$$\beta(x) = (x^T x)^{1/2}, \quad x \in L_X \qquad (4.24)$$

or equivalently,

$$\beta(z) = [(z - E[Z])^T C_Z^{-1} (z - E[Z])]^{1/2}, \quad z \in L_Z \qquad (4.25)$$

The smallest distance from the origin to a point on the failure surface was proposed by Hasofer and Lind (1974) as the definition of a reliability index

$$\beta_{HL} = \min_{z \in L_Z} \{(z - E[Z])^T C_Z^{-1} (z - E[Z])\}^{1/2} \qquad (4.26)$$

The solution for z is here denoted z^*, and this point is traditionally named the *design point*.

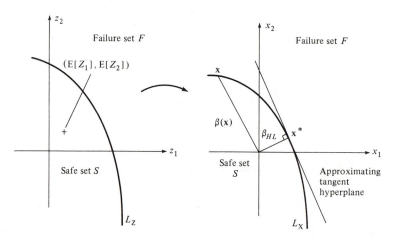

Figure 4.4 Geometrical illustration of the reliability index β_{HL}.

It is a straightforward exercise to show that the values of the reliability indices β_C in (4.10) and β_{HL} in (4.26) coincide when the failure surface is a hyperplane. The definition in (4.26) thus provides a generalization of the definition by Cornell to structures with nonlinear failure surfaces.

The Hasofer and Lind reliability index can also be interpreted as a first-order second-moment reliability index. It follows from Fig. 4.4 that the value of β_{HL} is the same for the true failure surface as for the approximating tangent hyperplane at the design point. The ambiguity in the value of the first-order reliability index is thus resolved when the design point is taken as the linearization point. The resultant reliability index will be a sensible measure for the distance to the failure surface.

4.4.1 Determination of the Design Point

It follows from (4.26) that the reliability index β_{HL} is the solution to a nonlinear optimization problem with one constraint:

Lagrange problem

$$\beta_{HL} = \min_{g(\mathbf{z})=0} \{(\mathbf{z} - E[\mathbf{Z}])^T \mathbf{C_Z}^{-1} (\mathbf{z} - E[\mathbf{Z}])\}^{1/2} \qquad (4.27)$$

Numerous general iterative algorithms are available to solve such an optimization problem. The iteration method presented in this section is very simple and has proved to work well for practical problems in connection with the algorithm of Rackwitz and Fiessler (1978) described in Section 5.2. There is, however, no guarantee that the algorithm converges in all situations.

For simplicity the iteration method is first explained in the normalized x-space. The method is illustrated in Fig. 4.5. The point $\mathbf{x}^*$ corresponds to the design point $\mathbf{z}^*$ and, as illustrated in the figure, the unit normal vector $\boldsymbol{\alpha}^*$ to the failure surface at $\mathbf{x}^*$ is proportional to the vector from the origin to $\mathbf{x}^*$, with the reliability index β_{HL} as the constant of proportionality:

$$\mathbf{x}^* = \beta_{HL} \boldsymbol{\alpha}^* \qquad (4.28)$$

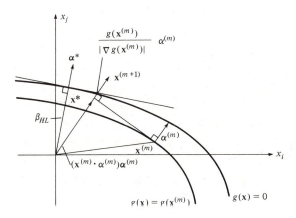

Figure 4.5 Illustration of solution method for the design point.

The point $\mathbf{x}^*$ is determined as the limit of a sequence $\mathbf{x}^{(0)}, \mathbf{x}^{(1)}, \ldots, \mathbf{x}^{(m)}, \ldots$. The unit normal vector to the trajectory $g(\mathbf{x}) = g(\mathbf{x}^{(m)})$ at the point $\mathbf{x}^{(m)}$ is denoted $\boldsymbol{\alpha}^{(m)}$. The vector $\boldsymbol{\alpha}^{(m)}$ is parallel to the gradient vector of the trajectory at $\mathbf{x}^{(m)}$ and is directed toward the failure set:

$$\boldsymbol{\alpha}^{(m)} = - \frac{\nabla g(\mathbf{x}^{(m)})}{|\nabla g(\mathbf{x}^{(m)})|} \qquad (4.29)$$

Here $\nabla g(\mathbf{x}^{(m)})$ is the gradient vector which is assumed to exist:

$$\nabla g(\mathbf{x}) = \left(\frac{\partial g}{\partial x_1}(\mathbf{x}), \ldots, \frac{\partial g}{\partial x_n}(\mathbf{x}) \right) \qquad (4.30)$$

The initial point in the sequence can be taken as the origin, and the iteration method is thereafter based on linearization in each step. At step m the point is $\mathbf{x}^{(m)}$ and the surface $x_{n+1} = g(\mathbf{x})$ is replaced by its tangent hyperplane at $\mathbf{x}^{(m)}$. The intersection between this hyperplane and the plane $x_{n+1} = 0$ has the equation

$$g(\mathbf{x}^{(m)}) + \sum_{i=1}^{n} \frac{\partial g}{\partial x_i}(\mathbf{x}^{(m)})(x_i - x_i^{(m)}) = 0 \qquad (4.31)$$

The point $\mathbf{x}^{(m+1)}$ is now taken as the point on the intersection which is closest to the origin. This point is obtained as the sum of two terms. First $\mathbf{x}^{(m)}$ is projected on the normal $\boldsymbol{\alpha}^{(m)}$ to the trajectory and then an additional term is introduced to account for the fact that $g(\mathbf{x}^{(m)})$ may be different from zero:

$$\mathbf{x}^{(m+1)} = (\mathbf{x}^{(m)^T}\boldsymbol{\alpha}^{(m)})\boldsymbol{\alpha}^{(m)} + \frac{g(\mathbf{x}^{(m)})}{|\nabla g(\mathbf{x}^{(m)})|} \boldsymbol{\alpha}^{(m)} \qquad (4.32)$$

It follows from (4.32) that if the sequence converges toward a point $\mathbf{x}^*$, then

$$\mathbf{x}^* = \beta\boldsymbol{\alpha}^*, \quad g(\mathbf{x}^*) = 0 \qquad (4.33)$$

and a stationary value β for the reliability index function (4.24) has been found. The failure surface may contain several points corresponding to stationary values of the reliability index function. It is therefore necessary to use several starting points to find all stationary values $\beta_1, \beta_2, \ldots, \beta_r$. The Hasofer and Lind reliability index is

$$\beta_{HL} = \min\{\beta_1, \beta_2, \ldots, \beta_r\} \qquad (4.34)$$

From the definition of $\beta(\mathbf{x})$ as the length of the vector $\mathbf{x}$ it follows that

$$\frac{\partial \beta(\mathbf{x})}{\partial x_i} = \frac{\partial}{\partial x_i}\left(\sum_{j=1}^{n} x_j^2\right)^{1/2} = \alpha_i \qquad (4.35)$$

The numerical value of α_i^* is thus a measure of the sensitivity of the reliability index to inaccuracies in the value of x_i at the point of minimum distance $\mathbf{x}^*$. By applying (4.23) it is possible to measure the sensitivity of β_{HL} to inaccuracies in the second-moment representation of the basic variables. This is demonstrated in detail in Section 5.6. For independent variables it is shown that β_{HL} is increased by a factor of approximately $1/(1 - \alpha_i^2)$ if σ_i is set equal to zero. For basic variables with small numerical values α_i^*, no great error is therefore introduced by considering these variables as fixed and equal to their mean values. The values α_i^* are often referred to as the *sensitivity factors*.

It may be more convenient to formulate the iteration algorithm in the original z-space. Then the iterative rule (4.32) reads

$$\mathbf{z}^{(m+1)} = E[\mathbf{Z}] + \mathbf{C_Z}\,\nabla g(\mathbf{z}^{(m)})\,\frac{(\mathbf{z}^{(m)} - E[\mathbf{Z}])^T\,\nabla g(\mathbf{z}^{(m)}) - g(\mathbf{z}^{m})}{\nabla g(\mathbf{z}^{(m)})^T\,\mathbf{C_Z}\,\nabla g(\mathbf{z}^{(m)})} \quad (4.36)$$

Example 4.3 shows the calculation of the reliability index in a case of practical relevance.

Example 4.3

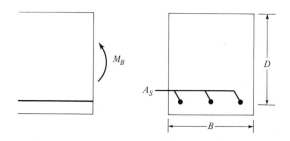

Figure 4.6   Reinforced concrete cross section subjected to pure bending.

The cross section of a reinforced concrete beam is shown in Fig. 4.6. The sectional bending moment is M_B. The ultimate bending moment is

$$M_U = \left|1 - K\frac{A_S T_S}{BDT_C}\right|A_S DT_S$$

where A_S is the area of reinforcement, T_S the yield stress of the reinforcement, T_C the maximum compressive strength of the concrete, B the width of the beam, D the effective depth of the reinforcement, and K is a factor related to the stress-strain relation of concrete. For an ideal plastic stress-strain curve, K equals 0.5, and for a linear elastic stress-strain curve, K equals 2/3.

The set of basic variables is $\mathbf{Z} = (M_B, D, T_S, A_S, K, B, T_C)$. The basic variables are mutually uncorrelated. The mean values, standard deviations and coefficients of variation are given in Table 4.1. A safety margin M is the difference between M_U and M_B:

$$M = Z_2 Z_3 Z_4 - \frac{Z_5 Z_3^2 Z_4^2}{Z_6 Z_7} - Z_1$$

The set of normalized and uncorrelated variables $\mathbf{X}$ is simply obtained by the relations

$$X_i = \frac{Z_i - E[Z_i]}{D[Z_i]}, \quad i = 1, 2, \dots, 7$$

The failure function corresponding to the safety margin M in terms of the parameters x_i is then, in units of MN and m,

<table>
<tr><td colspan="5" align="center">**TABLE 4.1   Basic Variables**</td></tr>
</table>

Variable	Symbol	Mean Value	Standard Deviation	Coefficient of Variation
M_B	Z_1	0.01 MNm	0.003 MNm	0.30
D	Z_2	0.30 m	0.015 m	0.05
T_S	Z_3	360 MPa	36 MPa	0.10
A_S	Z_4	226×10^{-6} m^2	11.3×10^{-6} m^2	0.05
K	Z_5	0.5	0.05	0.10
B	Z_6	0.12 m	0.006 m	0.05
T_C	Z_7	40 MPa	6 MPa	0.15

$$g(\mathbf{x}) = 0.3(1+0.05x_2)\,360(1+0.1x_3)\,226 \times 10^{-6}(1+0.05x_4)$$
$$- \frac{0.5(1+0.1x_5)\,360^2(1+0.1x_3)^2\,226^2 \times 10^{-12}(1+0.05x_4)^2}{0.12(1+0.05x_6)\,40(1+0.15x_7)} - 0.01(1+0.3x_1)$$

The gradient vector is easily determined and the iteration with the origin as starting point goes as shown in Table 4.2.

<table>
<tr><td colspan="8" align="center">**TABLE 4.2   Iteration in x-Space**</td></tr>
</table>

	$\mathbf{x}^{(0)}$	$\mathbf{x}^{(1)}$	$\mathbf{x}^{(2)}$	$\mathbf{x}^{(3)}$	$\mathbf{x}^{(4)}$	$\mathbf{x}^{(5)}$	$\boldsymbol{\alpha}^{(5)}$
x_1	0	2.401	2.628	2.614	2.615	2.615	0.766
x_2	0	-0.977	-0.832	-0.832	-0.831	-0.831	-0.244
x_3	0	-1.843	-1.848	-1.866	-1.866	-1.866	-0.547
x_4	0	-0.921	-0.790	-0.792	-0.790	-0.790	-0.232
x_5	0	0.055	0.037	0.037	0.037	0.037	0.011
x_6	0	-0.028	-0.019	-0.019	-0.019	-0.019	-0.006
x_7	0	-0.083	-0.057	-0.056	-0.056	-0.056	-0.016
β	0	3.31	3.41	3.41	3.41	3.41	

The iteration is stopped at $\mathbf{x}^{(5)}$, and the reliability index is

$$\beta_{HL} = |\mathbf{x}^{(5)}| = 3.41$$

It follows from the numerical values in $\boldsymbol{\alpha}^{(5)}$ that the uncertainties in the factor K, the beam width B, and the maximum compressive strength of concrete T_C have little influence on the reliability index.

4.5 THE GENERALIZED RELIABILITY INDEX

The concept of a reliability index is introduced to make a comparison between the reliabilities of different structures possible. It is therefore necessary that the ordering of structures according to their reliability indices is both consistent and at the same time not too crude. The Hasofer and Lind reliability index generally provides a satisfactory ordering, but examples can be constructed in which the ordering is not reasonable. One such example is shown in Fig. 4.7, which shows the failure surfaces for two structures a and b in the normalized x-space. The Hasofer and Lind reliability index is larger for structure a than for structure b. It seems, however, unreasonable to state that structure a is more safe than structure b. The shortcoming of β_{HL} in this case is due to its independence of the curvature of the failure surface at the design point. Unless the radius of curvature is large compared to β_{HL}, the approximation of the failure surface by a hyperplane may be too crude. Another example where β_{HL} does not lead to a reasonable ordering is given in Example 4.4. In this example the reliability index function has many local minima which are not reflected in the value of β_{HL}.

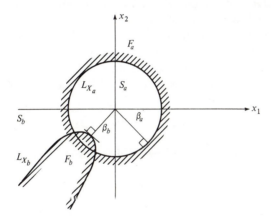

Figure 4.7   Illustration of a shortcoming of the reliability index β_{HL}.

The conclusion is thus that in certain cases there may be a need for a reliability index that is more selective than β_{HL}. One such reliability index was proposed by Ditlevsen (1979). The idea is to introduce a weight function $\psi_n(\mathbf{x})$ in x-space. A reliability measure γ is then obtained by integrating the weight function over the safe set S,

$$\gamma = \int_S \psi_n(\mathbf{x}) \, dS \qquad (4.37)$$

The reliability measure γ will be consistent and sufficiently selective if the weight function is rotationally symmetric and positive (Ditlevsen, 1979). An obvious choice is then the n-dimensional standardized normal probability density function:

$$\psi_n(x_1, x_2, \ldots, x_n) = \frac{1}{(2\pi)^{n/2}} \exp\left(-\frac{1}{2}(x_1^2 + x_2^2 + \ldots + x_n^2)\right) \quad (4.38)$$

$$= \varphi(x_1)\varphi(x_2) \cdots \varphi(x_n); \quad \mathbf{x} \in R^n$$

where $\varphi(x)$ is the probability density function for a normally distributed random variable with zero mean and unit variance.

$$\varphi(x) = \frac{1}{\sqrt{2\pi}} \exp\left(-\frac{1}{2}x^2\right) \quad (4.39)$$

The reliability measure γ can also be computed from a formulation in z-space. By inversion of (4.21) the Z_j-variables are expressed as

$$\mathbf{Z} = \mathbf{A}^{-1}\mathbf{X} + \mathbf{E}[\mathbf{Z}] \quad (4.40)$$

The variables X_i are taken as independent, normalized, and normally distributed, as given by (4.38). It is straightforward to show that the sum of two independent normal variables is normal. As a consequence of this and of (4.40), the variables Z_j are all normally distributed. The joint probability density function of the variables Z_j, $j = 1, 2, \ldots, n$, follows from that of the variables X_i, $i = 1, 2, \ldots, n$, by a direct variable transformation including the volume element, i.e.,

$$f_{\mathbf{Z}}(\mathbf{z})\, dz_1 \cdots dz_n = \varphi(x_1) \cdots \varphi(x_n)\, dx_1 \cdots dx_n \quad (4.41)$$

$$= \frac{1}{(2\pi)^{n/2}} \exp\left(-\frac{1}{2}\mathbf{x}^T\mathbf{x}\right) dx_1 \cdots dx_n$$

Introducing (4.21) and (4.22) gives

$$f_{\mathbf{Z}}(\mathbf{z}) = \frac{1}{(2\pi)^{n/2}\, |\, \mathbf{C}_{\mathbf{Z}}\, |^{1/2}} \exp\left(-\frac{1}{2}(\mathbf{z} - \mathbf{E}[\mathbf{Z}])^T \mathbf{C}_{\mathbf{Z}}^{-1}(\mathbf{z} - \mathbf{E}[\mathbf{Z}])\right) \quad (4.42)$$

where $|\, \mathbf{C}_{\mathbf{Z}}\, |$ is the determinant of $\mathbf{C}_{\mathbf{Z}}$ and where it follows from (4.22) that

$$|\, \mathbf{A}\, | = |\, \mathbf{C}_{\mathbf{Z}}\, |^{-1/2} \quad (4.43)$$

Corresponding to (4.37), the reliability measure γ can be defined as the integral of $f_{\mathbf{Z}}(\mathbf{z})$ over the safe set S_z in z-space:

$$\gamma = \int_{S_z} \frac{1}{(2\pi)^{n/2}\, |\, \mathbf{C}_{\mathbf{Z}}\, |^{1/2}} \exp\left(-\frac{1}{2}(\mathbf{z} - \mathbf{E}[\mathbf{Z}])^T \mathbf{C}_{\mathbf{Z}}^{-1}(\mathbf{z} - \mathbf{E}[\mathbf{Z}])\right) dS \quad (4.44)$$

For a hyperplane failure surface the value of γ is different from the value of β_C. To obtain coinciding values of the two reliability measures in this important case, the *generalized reliability index* β_G is defined as a monotonically increasing function of γ,

$$\beta_G = \Phi^{-1}(\gamma) = \Phi^{-1}\left|\int_S \varphi(x_1)\,\varphi(x_2)\,\cdots\,\varphi(x_n)\,dS\right| \qquad (4.45)$$

Here Φ^{-1} is the inverse cumulative distribution function of a normally distributed random variable with zero mean and unit variance. It is a simple matter to show that the reliability measures β_C and β_G coincide for any hyperplane failure surface. Alternatively, β_G can be expressed as

$$\beta_G = -\Phi^{-1}(1-\gamma) = -\Phi^{-1}\left|\int_F \varphi(x_1)\,\varphi(x_2)\,\cdots\,\varphi(x_n)\,dF\right| \qquad (4.46)$$

where F denotes the failure set corresponding to the safe set S.

It must be strongly emphasized that the choice of the normal density function is purely pragmatic. The choice does not imply any assumption that the distribution of the basic variables is normal. The fundamental assumption is still that only second-moment information is available for the basic variables and the normal density function is introduced solely to extend the reliability index definition to structures with nonlinear failure surfaces. The generalized reliability index provides a more consistent and selective measure of reliability than the Hasofer and Lind reliability index for nonlinear failure surfaces. In most cases of practical relevance the numerical values of β_G and β_{HL} are, however, almost coinciding. Only in cases where the radius of curvature of the failure surface at the design point in x-space is not large compared to β_{HL} or when the reliability index function has several local minima of almost the same magnitude can the values of the two reliability indices deviate significantly.

The evaluation of the generalized reliability index is much more involved than the evaluation of β_{HL} because the n-dimensional normal density must be integrated over either the safe set or over the failure set. This integration is simple in only very few cases, and numerical schemes such as the one of Milton (1972) are limited to problems with few basic variables due to an exponential increase in computation time with dimension. In practical cases either approximations or bounds are calculated, and this can be done quite efficiently. An approximation of the failure surface by the tangent hyperplane at the design point x^* leads to β_{HL} as a first-order approximation to β_G. If this first-order approximation is not satisfactory, better approximations can be obtained, e.g., by approximating the failure surface better locally by a quadratic surface at the design point or globally by a set of hyperplanes, i.e., a polyhedral surface. The following two sections therefore give some results for the generalized reliability index for convex polyhedral safe sets and for safe sets bounded by quadratic surfaces.

4.5.1 Convex Polyhedral Safe Set

Let the safe set in z-space be bounded by k hyperplanes. For each hyperplane a linear safety margin is defined as in (4.9) by

$$M_i = a_{0i} + \mathbf{a}_i^T \mathbf{Z}, \qquad i = 1, 2, \ldots, k \tag{4.47}$$

Since the basic variables Z_j are joint normally distributed, the safety margins M_i are also joint normally distributed with

$$E[M_i] = a_{0i} + \mathbf{a}_i^T E[\mathbf{Z}] \tag{4.48}$$

$$\mathrm{Var}[M_i] = \mathbf{a}_i^T \mathbf{C}_\mathbf{Z} \mathbf{a}_i \tag{4.49}$$

$$\mathrm{Cov}[M_i, M_j] = \mathbf{a}_i^T \mathbf{C}_\mathbf{Z} \mathbf{a}_j \tag{4.50}$$

A point $\mathbf{z}$ is in the safe set if all $M_i(\mathbf{z}) > 0$. The reliability measure γ is thus

$$\gamma = P \left| \bigcap_{i=1}^{k} (M_i(\mathbf{Z}) > 0) \right| \tag{4.51}$$

A set of normalized variables Y_i is defined by

$$Y_i = \frac{E[M_i] - M_i}{D[M_i]}, \qquad i = 1, 2, \ldots, k \tag{4.52}$$

The Y_i-variables are normally distributed with zero mean values, unit variances, and covariances equal to

$$\rho[Y_i, Y_j] = \mathrm{Cov}[Y_i, Y_j] = \frac{\mathrm{Cov}[M_i, M_j]}{D[M_i] D[M_j]} = \rho[M_i, M_j] \tag{4.53}$$

Indices b_i corresponding to individual face reliability indices are introduced as

$$b_i = \frac{E[M_i]}{D[M_i]} \tag{4.54}$$

Based on these indices and the correlation matrix $\mathbf{R} = \{\rho[Y_i, Y_j]\}$, the reliability measure γ is

$$\gamma = P \left| \bigcap_{i=1}^{k} (Y_i \leqslant b_i) \right| = \Phi_k(\mathbf{b}; \mathbf{R}) \tag{4.55}$$

$$= \int_{-\infty}^{b_1} \int_{-\infty}^{b_2} \cdots \int_{-\infty}^{b_k} \varphi_k(y_1, y_2, \ldots, y_k; \mathbf{R}) \, dy_1 dy_2 \cdots dy_k$$

$\Phi_k(\ ; \mathbf{R})$ denotes the cumulative distribution function for a set of k normalized normal variables with correlation matrix $\mathbf{R}$. $\varphi_k(\ ; \mathbf{R})$ is the corresponding probability density function. It follows from (4.55) that γ depends only on the reliability indices corresponding to each face of the failure surface and on the corresponding safety margin correlation coefficients.

The normal distribution function $\Phi_k(\mathbf{b};\mathbf{R})$ cannot be evaluated in a simple way for a general choice of $\mathbf{R}$. Several bounding techniques have been proposed as well as various methods giving approximations. A bounding technique which is computationally very simple and requires only evaluation of the one-dimensional normal distribution function is given in Section 5.4 in connection with calculation of failure probabilities for series systems. This bounding technique is good in particular if all b_i are large. Hohenbichler (1982) has presented a computationally more complicated method which gives good approximations for all values of b_i. Ditlevsen (1984) has suggested an approximation by the first three terms in a Taylor expansion from a correlation matrix corresponding to equicorrelation. Deák (1980) has suggested a simulation procedure which is generally efficient. This section presents an asymptotic expansion for $\Phi_k(\mathbf{b};\mathbf{R})$ and shows how bounds on $\Phi_k(\mathbf{b};\mathbf{R})$ can be obtained based on an exact result for a particular form of $\mathbf{R}$.

Asymptotic expansion of the normal distribution function. For the univariate normal distribution function an asymptotic expansion was given by Laplace as

$$\Phi(-b) \sim \frac{\varphi(b)}{b}\left(1 - \frac{1}{b^2} + \frac{1 \times 3}{b^4} - \frac{1 \times 3 \times 5}{b^6} + \cdots\right), \quad b \to \infty \quad (4.56)$$

The expansion is an enveloping one in that $\Phi(-b)$ is smaller than every summand with an odd number of terms and larger than every summand with an even number of terms. The truncation error after any number of terms is numerically less than the first term neglected.

The expansion has been generalized by Ruben (1964) to the multivariate normal distribution function, i.e., to a multivariate normal integral over an infinitely extended rectangle. The multivariate normal integral is denoted by $\Phi_k(-\mathbf{b};\mathbf{R})$, where $-\mathbf{b}$ is the vertex of the rectangle. The first term in the asymptotic expansion is

$$\Phi_k(-\mathbf{b};\mathbf{R}) \sim \frac{\exp\left(-\frac{1}{2}\mathbf{b}^T\mathbf{R}^{-1}\mathbf{b}\right)}{(2\pi)^{1/2}\sqrt{\det\mathbf{R}}\prod_{i=1}^{k}a_i}, \quad |\mathbf{b}| \to \infty \quad (4.57)$$

The coefficients a_i are elements in a vector $\mathbf{a}$ defined as

$$\mathbf{a} = \mathbf{R}^{-1}\mathbf{b} \quad (4.58)$$

The expansion is valid provided all coefficients a_i defined by (4.58) are positive. The right-hand side of (4.57) is an upper bound on the multinormal integral. Additional terms in the asymptotic expansion are given by Ruben (1964), but these terms are somewhat more complicated to compute. The first term in the expansion was also given by Savage (1962).

Bounds on the normal distribution function. The normal distribution function $\Phi_k(\mathbf{b};\mathbf{R})$ is easily evaluated for a correlation matrix of the form

$$\mathbf{R} = \mathbf{R}_\lambda = \begin{vmatrix} 1 & \lambda_1\lambda_2 & \cdots & \lambda_1\lambda_k \\ \lambda_1\lambda_2 & 1 & \cdots & \lambda_2\lambda_k \\ . & . & \cdots & . \\ \lambda_1\lambda_k & \lambda_2\lambda_k & \cdots & 1 \end{vmatrix} \qquad (4.59)$$

where $|\lambda_i| \leqslant 1$. $k+1$ independent and standardized variables V_i are introduced in terms of the k variables Y_j as

$$Y_j = \lambda_j V_{k+1} + \sqrt{1-\lambda_j^2}\, V_j, \quad j=1,2,\ldots,k \qquad (4.60)$$

Conditioning upon $V_{k+1}=v$, the distribution function is then computed as

$$\begin{aligned} \Phi_k(\mathbf{b};\mathbf{R}_\lambda) &= \int_{-\infty}^{\infty} \varphi(v)\, \Phi_k(\mathbf{b};\mathbf{R}_\lambda \mid V_{k+1}=v)\, dv \\ &= \int_{-\infty}^{\infty} \varphi(v)\, \Phi_k \left| \{\frac{b_i-\lambda_i v}{\sqrt{1-\lambda_i^2}}\};\mathbf{I} \right| dv \\ &= \int_{-\infty}^{\infty} \varphi(v) \prod_{i=1}^{k} \Phi \left| \frac{b_i-\lambda_i v}{\sqrt{1-\lambda_i^2}} \right| dv \end{aligned} \qquad (4.61)$$

This result can be used to give bounds on the distribution function for a general correlation matrix $\mathbf{R}$. First it is observed that the cumulative distribution function, when considered a function of the correlation coefficients, increases with each argument. This follows from the identity

$$\begin{aligned} \Phi_k(\mathbf{b};\mathbf{R}) &= \Phi_k(\mathbf{b};\mathbf{R}_{\rho_{ij}=0}) + \int_0^{\rho_{ij}} \frac{\partial\Phi_k(\mathbf{b};\mathbf{R}_{\rho_{ij}=z})}{\partial z}\, dz \\ &= \Phi_k(\mathbf{b};\mathbf{R}_{\rho_{ij}=0}) + \int_0^{\rho_{ij}} \frac{\partial^2\Phi_k(\mathbf{b};\mathbf{R}_{\rho_{ij}=z})}{\partial b_i \partial b_j}\, dz \end{aligned} \qquad (4.62)$$

$\mathbf{R}_{\rho_{ij}=z}$ is the correlation matrix $\mathbf{R}$ with the elements ρ_{ij} and ρ_{ji} replaced by z, and use has been made of the identity

$$\frac{\partial^2\Phi_k}{\partial\rho_{ij}} = \frac{\partial^2\Phi_k}{\partial b_i \partial b_j}, \quad i \neq j \qquad (4.63)$$

The statement is true since the integrand is positive. Constants $\lambda_1, \lambda_2, \ldots, \lambda_k$ can now be chosen such that $\lambda_i \lambda_j = \rho_{ij}$ for three different pairs (i,j) and such that either $\lambda_i \lambda_j \geq \rho_{ij}$ or $\lambda_i \lambda_j \leq \rho_{ij}$ for the remaining ρ_{ij}. Using the first set of λ_i-factors in (4.61) gives an upper bound on the distribution function and using the second set gives a lower bound. This bounding technique has been suggested by Rackwitz (1978).

As an example, let all ρ_{ij} be positive. $\lambda_1, \lambda_2,$ and λ_3 are determined such that $\lambda_1 \lambda_2 = \rho_{12}$, $\lambda_1 \lambda_3 = \rho_{13}$, and $\lambda_2 \lambda_3 = \rho_{23}$. The solution is

$$\lambda_1 = \left(\frac{\rho_{12}\rho_{13}}{\rho_{23}}\right)^{1/2}, \quad \lambda_2 = \left(\frac{\rho_{12}\rho_{23}}{\rho_{13}}\right)^{1/2}, \quad \lambda_3 = \left(\frac{\rho_{13}\rho_{23}}{\rho_{12}}\right)^{1/2} \quad (4.64)$$

The remaining λ-values are then selected as either

$$\lambda_i = \max_{j < i} \left\{ \frac{\rho_{ij}}{\lambda_j} \right\}, \quad i = 4, 5, \ldots, k \quad (4.65)$$

or as

$$\lambda_i = \min_{j < i} \left\{ \frac{\rho_{ij}}{\lambda_j} \right\}, \quad i = 4, 5, \ldots, k \quad (4.66)$$

corresponding to an upper and lower bound, respectively, on the distribution function. Naturally, it must be checked that all λ_i obtained in this way are numerically less than or equal to 1. The closeness of the bounds depends on **b** as well as on how close **R** is to a matrix of the form (4.59). A good choice appears to select λ-values as given by the example above with indices 1, 2, and 3 corresponding to the three smallest elements in **b**.

Example 4.4

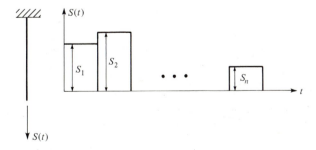

Figure 4.8   Bar loaded by time varying axial force.

Figure 4.8 shows a bar loaded by a time-varying axial load $S(t)$, $t = 1, 2, \ldots, n$. The resistance of the bar is a deterministic constant r and the loading varies as shown in the figure. The variables $S_1, S_2, \ldots, S_n$ have the second-moment representation

$$E[S_i] = \mu_S$$

$$\text{Var}[S_i] = \sigma_S^2$$

$$\text{Cov}[S_i, S_j] = \rho\sigma_S^2, \quad i \neq j$$

The failure function can be selected as

$$g(s_1, s_2, \ldots, s_n) = r - \max\{s_1, s_2, \ldots, s_n\}$$

or equivalently

$$g(s_1, s_2, \ldots, s_n) = \min\{r - s_1, r - s_2, \ldots, r - s_n\}$$

The safe set in s-space is thus an n-dimensional corner. In the normalized space the safe set is still an n-dimensional corner, but the sides are no longer perpendicular except for $\rho = 0$. Because of symmetry the distance from the origin to each side of the corner in the normalized space is the same. β_{HL} is equal to this common distance, and

$$\beta_{HL} = \frac{r - \mu_S}{\sigma_S}$$

independent of ρ and n. The reliability, however, is expected to decrease by decreasing ρ and by increasing n, which is not reflected in β_{HL}. In this case β_{HL} is thus a poor measure of reliability. The generalized reliability index can be evaluated exactly in this simple case. To do this the n variables S_i are written as in (4.60) (with $\lambda_i = \lambda = \sqrt{\rho}$ for $\rho > 0$)

$$S_i = \mu_S + \sigma_S(\sqrt{\rho}X_{n+1} + \sqrt{1-\rho}X_i)$$

in terms of $n+1$ normalized and uncorrelated variables. By first conditioning upon $X_{n+1} = u$ it follows that the generalized reliability index becomes

$$\beta_G = \Phi^{-1}\left|\int_{-\infty}^{\infty} \varphi(u)\Phi\left(\frac{\beta_{HL} - \sqrt{\rho}u}{\sqrt{1-\rho}}\right)^n du\right|$$

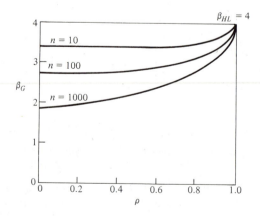

Figure 4.9 Variation of the generalized reliability index β_G with ρ and n.

Figure 4.9 shows how β_G varies as a function of ρ and n. The generalized reliability index can also be calculated for $\rho < 0$ by a recursive technique based on a result by Johnson and Kotz (1972).

4.5.2 Quadratic Safe Set

The failure function can be brought into the standardized form

$$g(\mathbf{x}) = \sum_{i=1}^{m-1} a_i(x_i - \delta_i)^2 + \sum_{i=m}^{n} b_i x_i - c \tag{4.67}$$

by a suitable rotation of the coordinate system. The safety margin M is expressed as

$$M = A + B - c \tag{4.68}$$

where the random variables A and B are defined as

$$A = \sum_{i=1}^{m-1} a_i(X_i - \delta_i)^2 \tag{4.69}$$

$$B = \sum_{i=m}^{n} b_i X_i \tag{4.70}$$

respectively. Since $\mathbf{X}$ is a vector of uncorrelated standardized normal variables, it follows that A and B are independent. A is a linear combination of noncentral chi-square distributed variables, while B is normally distributed. The reliability measure γ is

$$\gamma = \int_{-\infty}^{\infty} f_B(b)[1 - F_A(c - b)]\, db = \int_{-\infty}^{\infty} f_A(a)[1 - F_B(c - a)]\, da \tag{4.71}$$

When the distribution of A is known, γ is thus simple to compute. Unfortunately, only very few exact and convenient results are available for the distribution of A (Fiessler et al., 1979). An efficient complex numerical integration method has been proposed by Rice (1980).

Due to the limited knowledge on the distribution of A general results for quadratic safe sets are nor pursued any further. Instead, the special case of a hyperparabolic failure surface is analyzed in more detail. The hyperparabolic failure surface is obtained as a second-order approximation to a general failure surface.

First an asymptotic result is given for a general failure surface. It is assumed that the reliability index function (4.24) has its global minimum at $\mathbf{x} = \mathbf{x}^*$ only. Breitung (1984) has derived the asymptotic result for the probability content in the failure set:

$$1 - \gamma \sim \Phi(-\beta) \prod_{i=1}^{n-1} (1 - \beta \kappa_i)^{-1/2}, \qquad \beta \to \infty \tag{4.72}$$

$\beta = \beta_{HL}$ is the length of $\mathbf{x}^*$ and κ_i are the main curvatures of the failure surface at $\mathbf{x}^*$. The result is asymptotic in a sense defined by $\beta \to \infty$, with $\beta \kappa_i$ fixed. The corresponding asymptotic result for the generalized reliability index β_G in (4.45) is

$$\beta_G \sim \beta_{HL}, \qquad \beta_{HL} \to \infty \qquad (4.73)$$

Asymptotically, the Hasofer and Lind reliability index obtained by linearization of the failure surface at the design point therefore coincides with the generalized reliability index when the reliability index function has a unique minimum.

The asymptotic formula (4.72) yields good approximations to γ for large values of β. For small to moderate values, better approximations can be derived and one such approximation is given here. The second-order approximation to the failure surface at the design point is given through

$$0 = g(\mathbf{x}) \approx \nabla g(\mathbf{x}^*)^T (\mathbf{x} - \mathbf{x}^*) + \frac{1}{2}(\mathbf{x} - \mathbf{x}^*)^T \mathbf{D}(\mathbf{x}^*)(\mathbf{x} - \mathbf{x}^*) \quad (4.74)$$

$\nabla g(\mathbf{x}^*)$ is the gradient vector at the design point and the symmetric matrix $\mathbf{D}$ is the matrix of second-order partial derivatives at this point:

$$D_{ij}(\mathbf{x}^*) = \frac{\partial^2 g(\mathbf{x}^*)}{\partial x_i \partial x_j} \qquad (4.75)$$

As described in Section 4.4.1, the gradient vector and the vector from the origin are proportional at the design point:

$$\mathbf{x}^* = \beta \boldsymbol{\alpha}^*, \qquad \nabla g(\mathbf{x}^*) = - |\nabla g(\mathbf{x}^*)| \boldsymbol{\alpha}^* \qquad (4.76)$$

where $\boldsymbol{\alpha}^*$ is the unit directional vector to $\mathbf{x}^*$. Applying an orthogonal transformation,

$$\mathbf{x} = \mathbf{H}\mathbf{y} \qquad (4.77)$$

with the nth row in $\mathbf{H}$ equal to $\boldsymbol{\alpha}^*$, (4.74) is rewritten as

$$y_n = \beta - \frac{1}{2|\nabla g(\mathbf{x}^*)|} \tilde{\mathbf{y}}^T \mathbf{H}\mathbf{D}\mathbf{H}^T \tilde{\mathbf{y}} \qquad (4.78)$$

The $\tilde{\mathbf{y}}$-vector in this equation is

$$\tilde{\mathbf{y}} = (y_1, y_2, \ldots, y_n - \beta) \qquad (4.79)$$

The approximating surface is thus a general second-order surface. The solution with respect to y_n keeping up to second-order terms in y_i, $i = 1, 2, \ldots, n-1$, yields a hyperparabolic surface

$$y_n = \beta + \mathbf{y}^T \mathbf{A}\mathbf{y} \qquad (4.80)$$

with $\mathbf{y} = (y_1, y_2, \ldots, y_{n-1})$ and

$$A_{ij} = -\frac{(\mathbf{HDH}^T)_{ij}}{2\,|\,\nabla g(\mathbf{x}^*)\,|}, \qquad i, j = 1, 2, \ldots, n-1 \qquad (4.81)$$

A further orthogonal transformation,

$$\mathbf{y} = \mathbf{H_1 v} \qquad (4.82)$$

brings (4.80) into a diagonalized form:

$$y_n = \beta + \sum_{i=1}^{n-1} \lambda_i v_i^2 \qquad (4.83)$$

The λ-factors are related to the main curvatures κ_i as

$$\lambda_i = -\frac{1}{2}\kappa_i \qquad (4.84)$$

Since the transformations (4.77) and (4.82) are orthogonal, the probability content in the failure set corresponding to the second-order approximation of the failure surface is

$$1 - \gamma = \int_{-\infty}^{\infty} \cdots \int_{-\infty}^{\infty} \varphi(y_1) \cdots \varphi(y_{n-1}) \qquad (4.85)$$

$$\int_{\beta + \mathbf{y}^T \mathbf{Ay}}^{\infty} \varphi(y_n)\, dy_n\, dy_1 \cdots dy_{n-1}$$

or alternatively, if the transformation (4.82) has also been carried out:

$$1 - \gamma = \int_{-\infty}^{\infty} \cdots \int_{-\infty}^{\infty} \varphi(v_1) \cdots \varphi(v_{n-1}) \qquad (4.86)$$

$$\int_{\beta + \sum \lambda_i v_i^2}^{\infty} \varphi(y_n)\, dy_n\, dv_1 \cdots dv_{n-1}$$

Tvedt (1983) has derived a three-term approximation to (4.85) by a power series expansion in terms of $\mathbf{y}^T \mathbf{Ay}$, ignoring terms of order higher than two. The resultant approximation for $1 - \gamma$ is

$$1 - \gamma \approx A_1 + A_2 + A_3 \qquad (4.87)$$

with the three terms A_1, A_2, and A_3:

$$A_1 = \Phi(-\beta)[\det(\mathbf{I} + 2\beta\mathbf{A})]^{-1/2} \qquad (4.88)$$

$$A_2 = [\beta\Phi(-\beta) - \varphi(\beta)]\{[\det(\mathbf{I} + 2\beta\mathbf{A})]^{-1/2} \qquad (4.89)$$

$$- [\det(\mathbf{I} + 2(\beta + 1)\mathbf{A})]^{-1/2}\}$$

$$A_3 = (\beta + 1)[\beta\Phi(-\beta) - \varphi(\beta)]\{[\det(\mathbf{I} + 2\beta\mathbf{A})]^{-1/2} \qquad (4.90)$$

$$- \mathrm{Re}[(\det(\mathbf{I} + 2(\beta + i)\mathbf{A}))^{-1/2}]\}$$

or, in terms of the main curvatures κ_j:

$$A_1 = \Phi(-\beta) \prod_{j=1}^{n-1}(1-\beta\kappa_j)^{-1/2} \tag{4.91}$$

$$A_2 = [\beta\Phi(-\beta)-\varphi(\beta)]\left\{\left|\prod_{j=1}^{n-1}(1-\beta\kappa_j)^{-1/2}\right.\right. \tag{4.92}$$

$$\left.\left. - \prod_{j=1}^{n-1}(1-(\beta+1)\kappa_j)^{-1/2}\right|\right\}$$

$$A_3 = (\beta+1)[\beta\Phi(-\beta)-\varphi(\beta)]\left\{\left|\prod_{j=1}^{n-1}(1-\beta\kappa_j)^{-1/2}\right.\right. \tag{4.93}$$

$$\left.\left. - \text{Re}\left[\prod_{j=1}^{n-1}(1-(\beta+i)\kappa_j)^{-1/2}\right]\right|\right\}$$

The first term, A_1, is the asymptotic result (4.72). Re[ ] denotes the real part and i is the imaginary unit. The formula assumes the matrices $\mathbf{I}+2\beta\mathbf{A}$ and $\mathbf{I}+2(\beta+1)\mathbf{A}$ to be positive definite. The formula has been found to give very good approximations in almost all cases. The asymptotic behavior of the three terms can be compared in the asymptotic sense used in (4.72). It may be shown that the ratio of the second term to the first term is

$$\frac{A_2}{A_1} \sim \frac{1}{2\beta^2}\sum_{j=1}^{n-1}\frac{K_j}{1-K_j}, \quad \beta\to\infty \tag{4.94}$$

where $K_j=\beta\kappa_j$. Similarly, for $\beta\to\infty$, the ratio of the third to the first term is

$$\frac{A_3}{A_1} \sim -\frac{3}{8\beta^2}\sum_{j=1}^{n-1}\left(\frac{K_j}{1-K_j}\right)^2 - \frac{1}{2\beta^2}\sum_{j=1}^{n-1}\sum_{m=j+1}^{n-1}\frac{K_jK_m}{(1-K_j)(1-K_m)} \tag{4.95}$$

Examples of the application of formula (4.87) are presented in Chapter 5.

4.6 SUMMARY

In this chapter various reliability indices are defined for structures with parameters described by their second-moment representation and with the set of possible values of the parameters divided by a limit state surface into a failure set and a safe set. For hyperplane limit state surfaces a simple reliability index is defined as the ratio of the mean and the standard deviation of a linear safety margin. This definition is generalized for systems with a nonhyperplane limit state surface, which is approximated by a hyperplane at a suitably selected point on the failure surface. In some cases a further generalization is needed and this is introduced in the form of the generalized reliability index. A formal probability structure is imposed on the set of possible observations of the parameters and a formal failure probability is calculated. The computational efforts are increased at each generalization, and in particular the generalized reliability index is therefore seldomly used. The computation of the generalized reliability index is demonstrated for safe sets bounded by a quadratic surface and for convex polyhedral safe sets.

REFERENCES

BREITUNG, K., "Asymptotic Approximations for Multinormal Integrals," *Journal of the Engineering Mechanics Division,* ASCE, Vol. 110, 1984, pp. 357-366.

CORNELL, C. A., "A Probability-Based Structural Code," *Journal of the American Concrete Institute,* Vol. 66, No. 12, 1969, pp. 974-985.

DEÁK, I., "Three Digit Accurate Multiple Normal Probabilities," *Numerische Mathematik,* Vol. 35, Springer Verlag, Berlin, 1980, pp. 369-380.

DITLEVSEN, O., "Structural Reliability and the Invariance Problem," Research Report No. 22, Solid Mechanics Division, University of Waterloo, Waterloo, Canada, 1973.

DITLEVSEN, O., "Generalized Second-Moment Reliability Index," *Journal of Structural Mechanics,* Vol. 7, 1979, pp. 435-451.

DITLEVSEN, O., *Uncertainty Modeling with Applications to Multidimensional Civil Engineering Systems,* McGraw-Hill, New York, 1981.

DITLEVSEN, O., "Taylor Expansion of Series System Reliability," *Journal of the Engineering Mechanics Division,* ASCE, Vol. 110, 1984, pp. 293-307.

FIESSLER, B., H.-J. NEUMANN and R. RACKWITZ, "Quadratic Limit States in Structural Reliability," *Journal of the Engineering Mechanics Division,* ASCE, Vol. 105, 1979, pp. 661-676.

HASOFER, A. M. and N. C. LIND, "Exact and Invariant Second-Moment Code Format," *Journal of the Engineering Mechanics Division,* ASCE, Vol. 100, 1974, pp. 111-121.

HOHENBICHLER, M., "An Approximation to the Multivariate Normal Distribution," in *Proceedings, EUROMECH 155 Reliability*

Theory of Structural Engineering Systems, DIALOG 6-82, Danish Engineering Academy, Lyngby, Denmark, 1982, pp. 79-100.

JOHNSON, N. L. and S. KOTZ, *Distributions in Statistics: Continuous Multivariate Distributions,* John Wiley, New York, 1972.

MILTON, R. C., "Computer Evaluation of the Multivariate Normal Integral," *Technometrics,* Vol. 14, 1972, pp. 881-889.

RACKWITZ, R., "Close Bounds for the Reliability of Structural Systems," *Berichte zur Zuverlässigkeitstheorie der Bauwerke,* Heft 29, LKI, Technische Universität München, 1978, pp. 67-78.

RACKWITZ, R. and B. FIESSLER, "Structural Reliability under Combined Random Load Sequences," *Computers & Structures,* Vol. 9, 1978, pp. 489-494.

RICE, S. O., "Distribution of Quadratic Forms in Normal Random Variables — Evaluation by Numerical Integration," SIAM, *J. Sci. Stat. Comp.,* Vol. 1, 1980, pp. 438-448.

ROSENBLUETH, E. and L. ESTEVA, "Reliability Basis for Some Mexican Codes," *ACI Publication SP-31,* 1972, pp. 1-41.

RUBEN, H., "An Asymptotic Expansion for the Multivariate Normal Distribution and Mill's Ratio," *Journal of Research NBS,* Vol. 68B (Mathematics and Mathematical Physics), No. 1, 1964, pp. 3-11.

RUBINSTEIN, Y., *Simulation and the Monte Carlo Method,* John Wiley, New York, 1981.

SAVAGE, I. R., "Mill's Ratio for Multivariate Normal Distributions," *Journal of Research NBS,* Vol. 66B (Mathematics and Mathematical Physics), No. 3, 1962, pp. 93-96.

TVEDT, L., "Two Second-Order Approximations to the Failure Probability," Veritas Report RDIV/20-004-83, Det norske Veritas, Oslo, Norway, 1983.

VENEZIANO, D., "Contributions to Second-Moment Reliability Theory," Research Report R74-33, Department of Civil Engineering, Massachusetts Institute of Technology, Cambridge, Mass., April 1974.

5

LEVEL III
RELIABILITY METHODS
AND SYSTEM RELIABILITY

5.1 CALCULATION OF FAILURE PROBABILITIES

Level III reliability methods can be applied when both the distribution of the basic variables Z_i and the limit state surface dividing the z-space into a safe and a failure set are known. From the limit state surface a failure function is defined as in (4.1). The reliability measure is the probability of survival P_R, which is calculated as

$$P_R = \int_S f_{\mathbf{Z}}(\mathbf{z}) d\mathbf{z} \tag{5.1}$$

Here $f_{\mathbf{Z}}(\mathbf{z})$ is the joint probability density function of the basic variables, and S is the safe set. In terms of a safety margin M defined as in (4.3) the reliability is

$$P_R = P(M > 0) \tag{5.2}$$

As an alternative to P_R, a reliability index β_R is often defined as

$$\beta_R = \Phi^{-1}(P_R) = -\Phi^{-1}(P_F) \tag{5.3}$$

where ↖ Normal cdf

$$P_F = 1 - P_R = \int_F f_{\mathbf{Z}}(\mathbf{z}) d\mathbf{z} \tag{5.4}$$

is the failure probability and F is the failure set.

In general the integral (5.1) cannot be computed analytically. Alternative evaluation techniques such as numerical integration or simulation are generally very time consuming. The main interest in structural engineering is, however, on structures of high reliability, for which it is often sufficient to calculate the failure probability within a factor of, say, 2 to 5. This section presents a method to evaluate the integral (5.1) efficiently with such accuracy.

One case in which the integral (5.1) can be evaluated analytically deserves special attention. This occurs when the basic variables are jointly normally distributed and the failure surface is a hyperplane. Then the failure probability is simply

$$P_F = \Phi(-\beta_C) \qquad (5.5)$$

where β_C is the second-moment reliability index for the safety problem. If the basic variables are standardized, independent normally distributed but the failure surface is not a hyperplane, the failure probability is as in (5.5) with β_C replaced by the generalized reliability index β_G. In this case very good approximations to P_F can be obtained as explained in Section 4.5. In particular, $\Phi(-\beta_{HL})$, where β_{HL} is the Hasofer and Lind second-moment reliability index for the case, can often provide a good approximation to P_F. This will be the case at least if the point on the limit state surface closest to the origin is the only stationary point of the reliability index function (4.24) and if the principal curvatures of the limit state surface at this point are not too large in magnitude. The reason for this is that the probability density function of the basic variables decreases very quickly, namely as $\exp(-r^2/2)$, with the distance r from the origin. The area of integration giving the major contribution to the failure probability is therefore located close to the point on the failure surface closest to the origin, and the failure surface is well approximated by the tangent hyperplane around this point (see Fig. 5.1).

In general the basic variables are not normally distributed. The fact that probability contents in various sets are well approximated in a standardized normal space leads to the idea of finding a one-to-one transformation

$$\mathbf{T}: \mathbf{Z} = (Z_1, Z_2, \ldots, Z_n) \rightarrow \mathbf{U} = (U_1, U_2, \ldots, U_n) \qquad (5.6)$$

where the random variables $U_1, U_2, \ldots, U_n$ are uncorrelated and standardized normally distributed. The limit state surface in z-space is mapped on the corresponding limit state surface in u-space. To evaluate the probability content in the failure set in u-space, search is first for the minimum distance β from the origin to a point on the failure surface. The failure probability is then approximated to

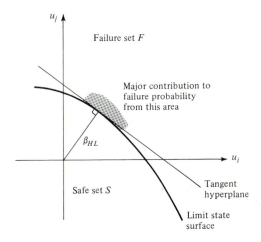

Figure 5.1 Illustration of approximate failure probability calculation.

a first-order by $\Phi(-\beta)$, corresponding to a linearization of the failure surface. The linearization point is the design point $\mathbf{u}^*$. A reliability method based on this procedure is called a *first-order reliability method (FORM)*, and β is the *first-order reliability index*. Better approximation can be obtained by improved approximation of the failure surface, e.g., by a quadratic surface or a set of hyperplanes. The probability content in the approximating failure set can be evaluated by the methods described in Section 4.5. A reliability method that uses a quadratic approximation to the failure surface at the design point is called a *second-order reliability method (SORM)*.

The simplest definition of the transformation $\mathbf{T}$ appears when the basic variables are mutually independent with distribution functions $F_{Z_1}, F_{Z_2}, \ldots, F_{Z_n}$. Each variable can then be transformed separately with the transformation defined by the identities

$$\Phi(u_i) = F_{Z_i}(z_i), \quad i = 1, \ldots, n \tag{5.7}$$

The transformation is thus

$$\mathbf{T}: u_i = \Phi^{-1}(F_{Z_i}(z_i)), \quad i = 1, \ldots, n \tag{5.8}$$

with the inverse transformation

$$\mathbf{T}^{-1}: z_i = F_{Z_i}^{-1}(\Phi(u_i)), \quad i = 1, \ldots, n \tag{5.9}$$

The failure function g_u in u-space is given in terms of the failure function g in z-space as

$$g(\mathbf{z}) = g(\mathbf{T}^{-1}(\mathbf{u})) = g_u(\mathbf{u}) \tag{5.10}$$

The design point $\mathbf{u}^*$ is the solution to a minimization problem with one constraint:

$$\min_{g_u(\mathbf{u})=0} |\mathbf{u}| \qquad (5.11)$$

Alternatively, $\mathbf{u}^*$ is determined as $\mathbf{u}^* = \mathbf{T}(\mathbf{z}^*)$, where $\mathbf{z}^*$ is the solution to

$$\min_{g(\mathbf{z})=0} |\mathbf{T}(\mathbf{z})| \qquad (5.12)$$

Any algorithm solving this optimization problem can be used. If the algorithm from Section 4.4.1 is used, the value of the g_u-function and the gradient vector must be computed. The value of the g_u-function is given in (5.10) and the partial derivatives are

$$\frac{\partial g_u}{\partial u_i} = \frac{\partial g}{\partial z_i}\frac{\partial z_i}{\partial u_i} = \frac{\partial g}{\partial z_i}(\mathbf{z})\frac{\varphi(u_i)}{f_{Z_i}(z_i)} \qquad (5.13)$$

$$= \frac{\partial g}{\partial z_i}(\mathbf{z})\frac{\varphi(\Phi^{-1}(F_{Z_i}(z_i)))}{f_{Z_i}(z_i)}$$

where u_i and z_i are related through (5.7). Using the search algorithm of Section 4.4.1, the failure function therefore need not be expressed explicitly in terms of the u-variables, but all calculations are based on the g-function in terms of the original variables z_i. After each step in the algorithm, a new approximation to $\mathbf{u}^*$ is computed and the corresponding point $\mathbf{z} = \mathbf{T}^{-1}(\mathbf{u})$ is determined before the next step in the algorithm. The inverse transformation $\mathbf{T}^{-1}$ is often only given numerically, and this step can thus cause practical problems.

Let the point on the failure surface closest to the origin have the coordinates u_i^*. The tangent hyperplane to the failure surface at this point has the equation

$$\sum_{i=1}^{n} \frac{\partial g_u}{\partial u_i}(\mathbf{u}^*)(u_i - u_i^*) = 0 \qquad (5.14)$$

and the first-order approximation to the failure probability P_F is

$$P_F \approx \Phi(-\beta) = \Phi(-|\mathbf{u}^*|) \qquad (5.15)$$

The transformation (5.8) is only one out of an infinite set of possibilities, and the approximation in (5.15) depends on the choice of transformation. Practical experience indicates that the transformation in (5.8) is a good choice. Example 5.1 will show an example where this transformation is superior to another possibility.

Example 5.1   *for 29)*

Consider a safety problem with two <mark>independent basic variables</mark> Z_1 and Z_2, <mark>both uniformly distributed</mark> on the interval [0,1]:

$$F_{Z_i}(z_i) = z_i, \quad z_i \in [0,1], \quad i = 1,2$$

The failure function is (see Fig. 5.2a)

$$g(z_1,z_2) = 1.8 - z_1 - z_2$$

$$P_F = \frac{1}{2}\left(.2 \times .2\right) = \underline{0.02}$$

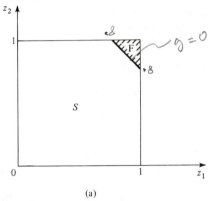

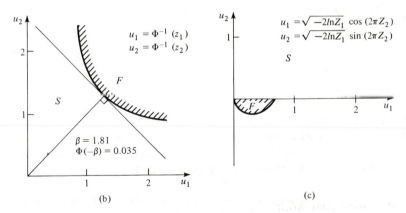

Figure 5.2   Transformation into standard normal space.

The failure probability is easily computed by integration:

$$P_F = 0.02$$

The failure surface after the transformation (5.8) is shown in Fig. 5.2b. The approximation by (5.15) is

$$P_F \approx \Phi(-1.81) = 0.035$$

which is within a factor of two of the exact result even though the original distributions are far from normal.

Another possible choice for **T** is in this case

$$\mathbf{T}: \quad \begin{aligned} U_1 &= \sqrt{-2\log Z_1}\,\cos\,(2\pi Z_2) \\ U_2 &= \sqrt{-2\log Z_1}\,\sin\,(2\pi Z_2) \end{aligned}$$

By this choice the failure set is transformed as shown in Fig. 5.2c, and the approximation (5.15) makes no sense.

2b)   When the basic variables are not mutually independent, the *Rosenblatt transformation* (Rosenblatt, 1952) has been suggested by Hohenbichler and Rackwitz (1981) as a good choice. The transformation is defined similar to (5.8) as

Necessary for dependent basic vars with full distribution

$$\mathbf{T}: \quad \begin{aligned} u_1 &= \Phi^{-1}(F_1(z_1)) \\ u_2 &= \Phi^{-1}(F_2(z_2 \mid z_1)) \\ &\;\;\vdots \\ u_i &= \Phi^{-1}(F_i(z_i \mid z_1, z_2, \ldots, z_{i-1})) \\ &\;\;\vdots \\ u_n &= \Phi^{-1}(F_n(z_n \mid z_1, z_2, \ldots, z_{n-1})) \end{aligned} \qquad (5.16)$$

where $F_i(z_i \mid z_1, \ldots, z_{i-1})$ is the distribution function of Z_i conditional upon $(Z_1 = z_1, \ldots, Z_{i-1} = z_{i-1})$:

$$F_i(z_i \mid z_1, \ldots, z_{i-1}) = \frac{\int_{-\infty}^{z_i} f_{Z_1, \ldots, Z_{i-1}, Z_i}(z_1, \ldots, z_{i-1}, t)\,dt}{f_{Z_1, \ldots, Z_{i-1}}(z_1, \ldots, z_{i-1})} \qquad (5.17)$$

The transformation therefore first transforms Z_1 into a standardized normal variable. Then all conditional variables of $Z_2 \mid Z_1 = z_1$ are transformed into a standardized normal variable, and so forth. The Rosenblatt transformation is identical to the transformation in (5.8) when the basic variables are mutually independent. It can also be shown that the transformation is linear and of the form (4.23) when the basic variables are jointly normal.

The inverse transformation can be obtained in a stepwise manner as

$$\mathbf{T}^{-1}: \quad \begin{aligned} z_1 &= F_1^{-1}(\Phi(u_1)) \\ z_2 &= F_2^{-1}(\Phi(u_2) \mid z_1) \\ &\;\;\vdots \\ z_n &= F_n^{-1}(\Phi(u_n) \mid z_1, \ldots, z_{n-1}) \end{aligned} \qquad (5.18)$$

In practical situations the transformations **T** and its inverse $\mathbf{T}^{-1}$ must often be determined numerically.

In u-space the failure function is given as in (5.10) and the partial derivatives are

$$\frac{\partial g}{\partial u_i}(\mathbf{u}) \;=\; \sum_{j=1}^{n} \frac{\partial g}{\partial z_j} J_{ji} \tag{5.19}$$

$\mathbf{J} = \{J_{ij}\}$ is the Jacobian matrix. The inverse matrix is given by

$$J_{ij}^{-1} \;=\; \frac{\partial u_i}{\partial z_j} \;=\; \begin{cases} 0, & i < j \\[2mm] \dfrac{f_i(z_i \mid z_1, \, \ldots \, , z_{i-1})}{\varphi(u_i)}, & i = j \\[4mm] \dfrac{\dfrac{\partial F_i}{\partial z_j}(z_i \mid z_1, \, \ldots \, , z_{i-1})}{\varphi(u_i)}, & i > j \end{cases} \tag{5.20}$$

u_i can be inserted from (5.16) and the Jacobian is then given in terms of $\mathbf{z}$. $\mathbf{J}$ and $\mathbf{J}^{-1}$ are lower-triangular matrices.

The search algorithm of Section 4.4.1 for the point $\mathbf{u}^*$ on the failure surface in u-space can now be applied. The search algorithm faces the same problems in determining the inverse transformation $\mathbf{T}^{-1}$ as for the case with mutually independent basic variables. An approximate linear backward substitution of $\mathbf{u}^{(m+1)}$ into $\mathbf{z}^{(m+1)}$ is

$$\mathbf{z}^{(m+1)} \;=\; \mathbf{z}^{(m)} + \mathbf{J}(\mathbf{u}^{(m+1)} - \mathbf{u}^{(m)}) \tag{5.21}$$

Use of this approximation can, however, lead to a point $\mathbf{z}^{(m+1)}$ beyond the set of possible values for $\mathbf{Z}$. The first-order approximation to the failure probability is given by (5.15), and the tangent hyperplane to the failure surface at $\mathbf{u}^*$ has the equation (5.14). It must be noted that the value of the approximate failure probability in general depends on the ordering of the basic variables, but the difference is seldomly significant. The approximation can be improved in the same way as for independent variables by a better approximation of the failure surface.

The application of the Rosenblatt transformation is illustrated by three examples. The first example is simple, while the other examples are more involved but illustrate the potential of first-order reliability methods.

Example 5.2   for 2 b)

Consider a safety problem with two dependent basic variables Z_1 and Z_2 having a two-dimensional exponential distribution function

$$F_{Z_1 Z_2}(z_1, z_2) \;=\; 1 - \exp(-z_1) - \exp(-z_2) + \exp[-(z_1 + z_2 + z_1 z_2)], \quad z_1 > 0,\; z_2 > 0$$

The corresponding probability density function is

$$f_{Z_1 Z_2}(z_1, z_2) = (z_1 + z_2 + z_1 z_2) \exp[-(z_1 + z_2 + z_1 z_2)], \quad z_1 > 0, \ z_2 > 0$$

The failure function is as shown in Fig. 5.3.

$$g(z_1, z_2) = 18 - 3z_1 - 2z_2$$

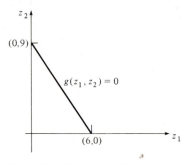

Figure 5.3 Failure surface in z-space.

Numerical integration gives the failure probability as

$$P_F = 2.94 \times 10^{-3}$$

corresponding to a reliability index of

$$\beta_R = -\Phi^{-1}(P_F) = 2.755$$

The marginal distribution functions are

$$F_{Z_1}(z_1) = 1 - \exp(-z_1), \quad z_1 > 0$$

$$F_{Z_2}(z_2) = 1 - \exp(-z_2), \quad z_2 > 0$$

and the conditional distribution function $F_2(z_2 \mid z_1)$ is

$$F_2(z_2 \mid z_1) = 1 - (1 + z_2) \exp[-(z_2 + z_1 z_2)], \quad z_1 > 0, \ z_2 > 0$$

The transformation (5.16) into a set of independent, standardized normal variables becomes

$$u_1 = \Phi^{-1}(1 - \exp(-z_1))$$

$$u_2 = \Phi^{-1}(1 - (1 + z_2)\exp[-(z_2 + z_1 z_2)])$$

The failure surface in u-space is shown in Fig. 5.4a. The reliability index function has two local minima at the points u_1^* and u_2^*. These points have the coordinates

$$u_1^* = (2.782, 0.100), \qquad z_1^* = (5.913, 0.130)$$

$$u_2^* = (-1.296, 3.253), \qquad z_2^* = (0.103, 8.846)$$

The corresponding points in z-space z_1^* and z_2^* are thus close to, but not coinciding with, the points $(6,0)$ and $(0,9)$ on the failure surface in z-space for which the probability density function has a local maximum. The

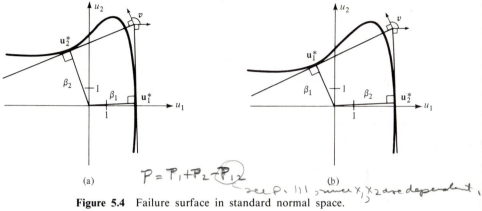

$$P = P_1 + P_2 - P_{12}$$

see p. 111, since x_1, x_2 are dependent,

(a)   (b)

Figure 5.4 Failure surface in standard normal space.

if indep't, $P_{12} = P_1 \cdot P_2$,

first-order approximation to P_F is

$$P_F \approx \Phi(-\beta_1) = \Phi(-2.784) = 2.68 \times 10^{-3}$$

A better approximation is obtained by approximating the safe set as the set bounded by the tangents at $\mathbf{u}_1^*$ and $\mathbf{u}_2^*$. The angle v between the normal vectors is given by

$$\cos v = \frac{\mathbf{u}_1^{*T} \mathbf{u}_2^*}{|\mathbf{u}_1^*|\, |\mathbf{u}_2^*|} = \frac{1}{\beta_1 \beta_2} \mathbf{u}_1^{*T} \mathbf{u}_2^* = -0.337$$

The approximate failure probability is then (see also Section 5.4) p. 110, 111

$$P_F \approx 1 - \Phi(\beta_1, \beta_2; \cos v) = 1 - \Phi(\underset{\beta_1}{2.784}, \underset{\beta_2}{3.501}; \underset{P_{12}}{-0.337}) = 2.94 \times 10^{-3}$$

corresponding to the reliability index

$$\beta_R = \Phi^{-1}(P_h) \approx 2.755$$

The difference between the two approximations is thus small and the second approximation coincides with the exact result.

To demonstrate the effect of the ordering of basic variables, consider the alternative transformation into a set of independent standardized normal variables:

$$u_1 = \Phi^{-1}(1 - \exp(-z_2))$$
$$u_2 = \Phi^{-1}(1 - (1 + z_1)\exp[-(z_1 + z_1 z_2)])$$

The failure surface in u-space is shown in Fig. 5.4b. The stationary points for the reliability index function have the coordinates

$$\mathbf{u}_1^* = (-1.123, 2.400), \qquad \mathbf{z}_1^* = (5.907, 0.140)$$
$$\mathbf{u}_2^* = (3.630, 0.142), \qquad \mathbf{z}_2^* = (0.091, 8.863)$$

The first-order approximation to P_F is

$$P_F \approx \Phi(-\beta_1) = \Phi(-2.649) = 4.04 \times 10^{-3}$$

The approximation of the safe set as the set between the tangents at $\mathbf{u}_1^*$ and $\mathbf{u}_2^*$ leads to

$$P_F \approx 1 - \Phi(2.649, 3.633; -0.388) = 4.18 \times 10^{-3}$$

corresponding to the reliability index

$$\beta_R = -\Phi^{-1}(P_F) \approx 2.638$$

Depending on the ordering of the basic variables, the approximations of β differ by 5% in this case. This is quite satisfactory, in particular for an example such as this for which β is relatively small and the basic variables are far from normally distributed.

Example 5.3

This example should be read in connection with Section 9.2 on fatigue. A specimen is subjected to time-varying loading, resulting in a constant-amplitude stress variation with stress range S at a critical point. The number of stress cycles necessary to cause fatigue failure is denoted N. The relation between S and N is

$$N = KS^{-m}$$

in which m is a deterministic material constant and K is a material parameter varying from specimen to specimen.

The material constants m and K are not known in advance but are estimated from test results on identical specimens. Figure 5.5 shows available test results (s_i, n_i), $i = 1, 2, \ldots, r$, for a typical situation. The results are shown in a transformed $(\log s, \log n)$ diagram, and these transformed variables are denoted x and y.

$$x = \log s, \quad y = \log n$$

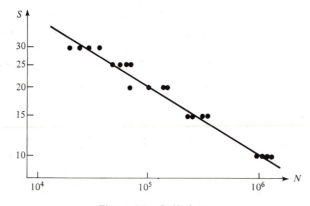

Figure 5.5 S-N data.

A linear normal regression model with constant variance is formulated with

x as the independent and y as the dependent variable. The value of Y for a given value $X = x_i$ is written as

$$Y_i \mid X = x_i] = \alpha - \beta(x_i - \bar{x}) + \sigma I_i$$

where I_i is a standardized normal variable. The mean value and variance are

$$E[Y_i \mid X = x_i] = \alpha - \beta(x_i - \bar{x})$$

$$\text{Var}[Y_i \mid X = x_i] = \sigma^2$$

$\bar{x}$ is the average x-value for the r test results:

$$\bar{x} = \frac{1}{r} \sum_{i=1}^{r} x_i$$

Due to the limited number of tests the parameters α, β, and σ cannot be determined with certainty. Rather, α, β, and σ must be considered as outcomes of random variables A, B, and Σ. The distributions of A, B, and Σ are obtained by combining prior knowledge with the results of the tests. The resulting posterior distribution is obtained by use of Bayes' theorem (Lindley, 1965), which states that

$$f_{AB\Sigma}(\alpha,\beta,\sigma \mid x_1,...,x_r,y_1,...,y_r) = C f_{Y_1,...,Y_r}(y_1,...,y_r \mid x_1,...,x_r,\alpha,\beta,\sigma) f_{AB\Sigma}(\alpha,\beta,\sigma)$$

The left-hand side is the posterior probability density function. It is proportional to the product of the likelihood function and the prior probability density function. The constant of proportionality C is a normalizing factor. The random variables $Y_i \mid X = x_i]$ are assumed mutually independent, corresponding to independent tests. The likelihood function for this example is thus

$$f_{Y_1,...,Y_r}(y_1,...,y_r \mid x_1,...,x_r,\alpha,\beta,\sigma) = \prod_{i=1}^{r} f_{Y_i}(y_i \mid x_i,\alpha,\beta,\sigma)$$

$$= \prod_{i=1}^{r} \frac{1}{\sigma} \varphi\left(\frac{y_i - \alpha + \beta(x_i - \bar{x})}{\sigma}\right)$$

$$\propto \frac{1}{\sigma^r} \exp\left| -\frac{1}{2\sigma^2} \sum_{i=1}^{r} (y_i - \alpha + \beta(x_i - \bar{x}))^2 \right|$$

$$= \frac{1}{\sigma^r} \exp\left| -\frac{1}{2\sigma^2}(D^2 + r(a - \alpha)^2 + S_{xx}(b + \beta)^2) \right|$$

where the constants a, b, S_{xx}, and D are determined as

$$a = \bar{y} = \frac{1}{r} \sum_{i=1}^{r} y_i$$

$$S_{xx} = \sum_{i=1}^{r} (x_i - \bar{x})^2 = \sum_{i=1}^{r} x_i^2 - r\bar{x}^2$$

$$S_{xy} = \sum_{i=1}^{r} (x_i - \bar{x})(y_i - \bar{y}) = \sum_{i=1}^{r} x_i y_i - r\bar{x}\,\bar{y}$$

$$S_{yy} = \sum_{i=1}^{r}(y_i - \bar{y})^2 = \sum_{i=1}^{r} y_i^2 - r\bar{y}^2$$

$$b = \frac{S_{xy}}{S_{xx}}$$

$$D^2 = S_{yy} - \frac{S_{xy}^2}{S_{xx}}$$

The prior probability density function for (A,B,Σ) is assumed to correspond to mutually independent variables, i.e.,

$$f_{AB\Sigma}(\alpha,\beta,\sigma) = f_A(\alpha)f_B(\beta)f_\Sigma(\sigma)$$

In this example it is assumed that the only prior information is that $B \in [m_1;m_2]$ and that $\Sigma > 0$. A diffuse (noninformative) prior distribution for A and $\log \Sigma$ and a uniform prior distribution for B in the interval $[m_1;m_2]$ are selected:

$$f_{AB\Sigma}(\alpha,\beta,\sigma) \; \propto \; \frac{1}{\sigma}\frac{1}{m_2-m_1}, \quad \sigma>0, \quad m_1 \leqslant \beta \leqslant m_2$$

The posterior distribution function is

$$f_{AB\Sigma}(\alpha,\beta,\sigma \mid x_1, \ldots ,x_r,y_1, \ldots ,y_r)$$

$$\propto \frac{1}{\sigma^{r+1}}\exp\left[-\frac{1}{2\sigma^2}(D^2+r(a-\alpha)^2+S_{xx}(b+\beta)^2)\right], \quad \sigma>0, \quad m_1 \leqslant \beta \leqslant m_2$$

The joint posterior distribution is thus rather complicated. Some simplification is achieved by introducing random variables T_1, T_2, and T_3 as

$$T_1 = \frac{\sqrt{r}(A-a)}{\Sigma}$$

$$T_2 = \frac{\sqrt{S_{xx}}(B+b)}{\Sigma}$$

$$T_3 = \frac{D^2}{\Sigma^2}$$

The joint posterior distribution of these variables is

$$f_{T_1T_2T_3}(t_1,t_2,t_3) \; \propto \; \exp\left(-\frac{1}{2}t_1^2\right)\exp\left(-\frac{1}{2}t_2^2\right)t_3^{[(r-2)/2]-1}\exp\left(-\frac{1}{2}t_3\right)$$

$$-\infty < t_1 < \infty, \quad 0 < t_3 < \infty, \quad \frac{\sqrt{S_{xx}\,t_3}(m_1+b)}{D} < t_2 < \frac{\sqrt{S_{xx}\,t_3}(m_2+b)}{D}$$

From this equation it follows that T_1 is independent of (T_2,T_3) and

$$T_1 \in N(0,1)$$

It further follows that

$$T_3 \in \chi^2(r-2)$$

where $\chi^2(r-2)$ denotes a chi-square distribution with $r-2$ degrees of freedom. The probability density function of T_2 conditional upon T_3 is a truncated normal distribution:

$$f_{T_2}(t_2 \mid t_3) = \frac{\varphi(t_2)}{\Phi\left|\dfrac{\sqrt{S_{xx}\, t_3}(m_2+b)}{D}\right| - \Phi\left|\dfrac{\sqrt{S_{xx}\, t_3}(m_1+b)}{D}\right|},$$

$$\frac{\sqrt{S_{xx}\, t_3}(m_1+b)}{D} < t_2 < \frac{\sqrt{S_{xx}\, t_3}(m_2+b)}{D}$$

The corresponding cumulative distribution function is

$$F_{T_2}(t_2 \mid t_3) = \frac{\Phi(t_2) - \Phi\left|\dfrac{\sqrt{S_{xx}\, t_3}(m_1+b)}{D}\right|}{\Phi\left|\dfrac{\sqrt{S_{xx}\, t_3}(m_2+b)}{D}\right| - \Phi\left|\dfrac{\sqrt{S_{xx}\, t_3}(m_1+b)}{D}\right|},$$

$$\frac{\sqrt{S_{xx}\, t_3}(m_1+b)}{D} < t_2 < \frac{\sqrt{S_{xx}\, t_3}(m_2+b)}{D}$$

If the prior distribution for B is diffuse, i.e., $m_1 = -\infty$ and $m_2 = \infty$, then T_2 and T_3 are independent and T_2 has a standardized normal distribution. The material parameters m and K can now be expressed as

$$m = B = -b + \frac{DT_2}{\sqrt{S_{xx}\, T_3}}$$

$$K = \exp(A + B\bar{x} + \Sigma I) = \exp\left| a + \frac{DT_1}{\sqrt{r T_3}} + \left(-b + \frac{DT_2}{\sqrt{S_{xx}\, T_3}}\right)\bar{x} + \frac{DI}{\sqrt{T_3}}\right|$$

where as mentioned earlier

$$I \in N(0,1)$$

The number of stress cycles to failure is expressed as

$$N = \exp\left| a + \frac{DT_1}{\sqrt{r T_3}} + \left(-b + \frac{DT_2}{\sqrt{S_{xx}\, T_3}}\right)\bar{x} + \frac{DI}{\sqrt{T_3}}\right| S^{b - (DT_2 / \sqrt{S_{xx}\, T_3})}$$

T_1 and I are independent standardized normal variables independent of (T_2, T_3). T_3 has a $\chi^2(r-2)$ distribution, while the distribution of T_2 conditional on T_3 is a truncated normal distribution as given above.

A numerical example is included to exemplify the analysis. A specimen used in a structure is subjected to a loading giving constant-amplitude stresses with stress range S. Due to various uncertainties in the stress analysis, S is not known exactly but is modeled as a lognormal random variable with expected value 20 MPa and coefficient of variation 10%. To estimate the parameters in the S-N relation, 20 test results are available (see Table 5.1). The slope parameter m is known to be in the interval

TABLE 5.1 Test Data				
i	S_i (MPa)	N_i	i	S_i (MPa)	N_i
1	10	1,207,532	11	20	107,811
2	10	1,001,329	12	20	144,877
3	10	1,164,251	13	25	53,675
4	10	1,052,142	14	25	63,114
5	15	191,809	15	25	66,566
6	15	355,136	16	25	46,152
7	15	251,138	17	30	19,547
8	15	320,859	18	30	24,162
9	20	68,162	19	30	27,696
10	20	138,848	20	30	35,947

[3.0;4.0]. The constants $\bar{x}$, a, b, D, and S_{xx} are computed.

$$r = 20, \quad \bar{x} = 2.925, \quad a = 11.832, \quad b = -3.335, \quad D = 0.9758, \quad S_{xx} = 3.008$$

The distribution function $F_N(n)$ for the number N of stress cycles to failure is expressed as

$$F_N(n) \;=\; P(N \leqslant n) \;=\; P(\log N \leqslant \log n) \;=\; 1 - P(\log N > \log n)$$

For small arguments the probability $P(\log N \leqslant \log n)$ is small and good approximations can be computed by the FORM. The failure criterion is

$$g(N,S) \;=\; \log N - \log n \;\leqslant\; 0$$

or in terms of T_1, T_2, T_3, I, and S:

$$g(T_1,T_2,T_3,I,S) = a + \frac{DT_1}{\sqrt{rS_{xx}}} + \left(-b + \frac{DT_2}{\sqrt{S_{xx}T_3}}\right)(\bar{x} - \log S) + \frac{DI}{\sqrt{T_3}} - \log n \leqslant 0$$

For large arguments the probability $P(\log N > \log n)$ is similarly small, and good approximations can be computed by the FORM. For arguments n where neither $P(\log N \leqslant \log n)$ nor $P(\log N > \log n)$ is small, the FORM can still be applied, but the results are expected to be less accurate. Figure 5.6 shows the results of the first-order reliability calculation using the Rosenblatt transformation. The results are labeled I,FORM. The FORM reliability index $\beta(n)$ corresponding to the failure event $N \leqslant n$ is shown together with the expected number of stress cycles to failure. The expected value μ_N is computed as

$$\mu_N \;=\; \int_0^\infty n f_N(n)\,dn \;=\; \int_0^\infty (1 - F_N(n))\,dn \;\approx\; \int_0^\infty \Phi(-\beta(n))\,dn$$

Figure 5.6 also contains results with a label II,FORM. These are results based on the same analysis but with an additional 20 test results to estimate the material parameters in the S-N relation. The increase in the number of test results decreases the statistical uncertainty in the estimates. A significantly lower failure probability is observed for small arguments n, corresponding to large values of β, in this case. The additional 20 test results are given in Table 5.2.

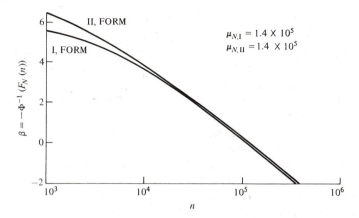

Figure 5.6 FORM results for number of cycles to failure.

i	S_i (MPa)	N_i	i	S_i (MPa)	N_i
TABLE 5.2		**Additional Test Data**			
21	10	1,314,332	31	20	174,580
22	10	927,816	32	20	113,936
23	10	859,915	33	25	55,601
24	10	981,501	34	25	63,908
25	15	426,880	35	25	46,279
26	15	190,376	36	25	43,226
27	15	330,713	37	30	49,134
28	15	313,015	38	30	25,054
29	20	158,600	39	30	30,502
30	20	89,558	40	30	41,359

The numerical constants in the failure criterion are then

$$r = 40, \quad \bar{x} = 2.925, \quad a = 11.870, \quad b = -3.300, \quad D = 1.5151, \quad S_{xx} = 6.0162$$

Example 5.4

This example should be read in connection with Chapter 7 on extreme value theory and Section 10.2 on wave climate description and wave loading. The sea surface elevation $\eta(t)$ above mean level is modeled as a sequence of mean-zero stationary Gaussian processes each described by two parameters, the significant wave height H_S and the mean zero-crossing period T_Z. The duration of one stationary period, t, is taken as 8 hours in this example, and the distribution of the largest sea elevation η_m in such an 8-hour period is computed.

The one-sided power spectral density $S_\eta(\omega)$ for $\eta(t)$ is taken as the ISSC spectrum (ISSC, 1964),

$$S_\eta(\omega) = \frac{1}{4\pi}\left(\frac{2\pi}{T_Z}\right)^4 H_S^2 \omega^{-5} \exp\left[-\frac{1}{\pi}\left(\frac{2\pi}{T_Z}\right)^4 \omega^{-4}\right], \quad \omega > 0$$

The cumulative distribution function for η_m can then be approximated as in (7.150) with the rate of qualified envelope crossings from (7.155):

$$F_{\eta_m}(\eta \mid H_S, T_Z) = \left|1 - \exp\left(-\frac{\eta^2}{2\lambda_0}\right)\right| \exp\left\{-t\frac{1}{2\pi}\frac{\sqrt{\lambda_2}}{\sqrt{\lambda_0}}\exp\left(-\frac{\eta^2}{2\lambda_0}\right)\right.$$

$$\left.\times \left|1 - \exp\left(-\sqrt{2\pi}\frac{\eta}{\sqrt{\lambda_0}}\sqrt{1 - \lambda_1^2/\lambda_0\lambda_2}\right)\right|\right\}$$

where the spectral moments λ_i are defined by

$$\lambda_i = \int_0^\infty \omega^i S_\eta(\omega)d\omega, \quad i = 0,1,2, \ldots$$

The three lowest-order spectral moments are computed upon inserting the expression for $S_\eta(\omega)$ and integrating:

$$\lambda_0 = \frac{1}{16}H_S^2$$

$$\lambda_1 = \frac{1}{16}H_S^2\frac{2\pi}{T_Z}\frac{1}{\pi^{1/4}}\Gamma\left(\frac{3}{4}\right)$$

$$\lambda_2 = \frac{1}{16}H_S^2\left(\frac{2\pi}{T_Z}\right)^2$$

In this example the distribution of (H_S, T_Z) is taken as a joint lognormal distribution with

$$E[H_S] = 3.00 \text{ m}$$
$$E[T_Z] = 7.00 \text{ s}$$
$$\text{Var}[H_S] = 3.60 \text{ m}^2$$
$$\text{Var}[T_Z] = 1.80 \text{ s}^2$$
$$\text{Cov}[H_S, T_Z] = 2.00 \text{ ms}$$

The distribution of $(\log H_S, \log T_Z)$ is joint normal with

$$E[\log H_S] = \mu_1 = 0.9304$$
$$E[\log T_Z] = \mu_2 = 1.9279$$
$$\text{Var}[\log H_S] = \sigma_1^2 = 0.5801^2$$
$$\text{Var}[\log T_Z] = \sigma_2^2 = 0.1899^2$$
$$\text{Cov}[\log H_S, \log T_Z] = \rho\sigma_1\sigma_2 = 0.8258 \times 0.5801 \times 0.1899$$

where the numerical values have been determined from the following relations valid for lognormally random variables

$$E[H_S] = \exp\left(\mu_1 + \frac{1}{2}\sigma_1^2\right)$$

$$E[T_Z] = \exp\left(\mu_2 + \frac{1}{2}\sigma_2^2\right)$$

$$\text{Var}[H_S] = E[H_S]^2[\exp(\sigma_1^2) - 1]$$
$$\text{Var}[T_Z] = E[T_Z]^2[\exp(\sigma_2^2) - 1]$$
$$\text{Cov}[H_S, T_Z] = E[H_S]E[T_Z][\exp(\rho\sigma_1\sigma_2) - 1]$$

The probability density function for (H_S, T_Z) is

$$f_{H_S T_Z}(H_S, T_Z) = \frac{1}{H_S T_Z \sigma_1 \sigma_2} \varphi \left| \frac{\log H_S - \mu_1}{\sigma_1}, \frac{\log T_Z - \mu_2}{\sigma_2}; \rho \right|$$

where $\varphi(\ ,\ ;\rho)$ is the standardized joint normal probability density function with correlation coefficient ρ. The unconditional distribution function for η_m is then

$$F_{\eta_m}(y) = \int_{T_Z=0}^{\infty} \int_{H_S=0}^{\infty} F_{\eta_m}(y \mid H_S, T_Z) f_{H_S T_Z}(H_S, T_Z) dH_S \, dT_Z$$

and it is seen that for each value of y a double integration is needed. This integration must be done numerically.

A considerable amount of computational work can be avoided by the use of the first-order reliability method. The complementary cumulative distribution function

$$G_{\eta_m}(y) = 1 - F_{\eta_m}(y) = P(\eta_m > y)$$

is determined for any value of y. Using the FORM terminology, the limit state function is

$$g(\eta_m, H_S, T_Z) = y - \eta_m$$

H_S is taken as variable 1, T_Z as variable 2, and η_m as variable 3. The conditional distribution functions needed for the transformation (5.16) are

$$F_{H_S}(H_S) = \Phi \left| \frac{\log H_S - \mu_1}{\sigma_1} \right|$$

$$F_{T_Z}(T_Z \mid H_S) = \Phi \left| \frac{\log T_Z - \left(\mu_2 + \rho \dfrac{\sigma_2}{\sigma_1}(\log H_S - \mu_1)\right)}{\sigma_2 \sqrt{1-\rho^2}} \right|$$

$$F_{\eta_m}(\eta \mid H_S, T_Z) = \left| 1 - \exp \left| -\frac{1}{2} \left(\frac{4\eta}{II_S}\right)^2 \right| \right| \exp \left| -\frac{t}{T_Z} \exp \left| -\frac{1}{2} \left(\frac{4\eta}{II_S}\right)^2 \right| \right|$$

$$\times \left| 1 - \exp \left(-\sqrt{2\pi} \frac{4\eta}{H_S} \sqrt{1 - \pi^{-1/2} \Gamma(3/4)^2} \right) \right|$$

The first-order reliability method can now be applied directly and numerical results are shown in Fig. 5.7.

The example can easily be extended to include statistical or measurement uncertainty in some of the distributional parameters in the joint distribution for (H_S, T_Z). For example, let the expected values $E[H_S]$ and $E[T_Z]$ be modeled as independent random variables with distribution functions $F_{\mu_H}(\ )$ and $F_{\mu_T}(\ )$, respectively. The distribution parameters μ_1 and μ_2 in the lognormal distribution are then also random variables. The total number of random variables is thus 5, and the computation of the unconditional distribution function for η_m involves a quadruple integration. When the first-order reliability technique is applied, this extension involves only a slight modification of the computational work. The five distribution functions needed in the transformation **T** are given as

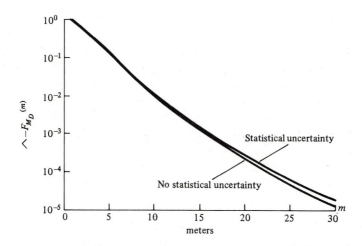

Figure 5.7   FORM results for maximum sea elevation.

$$F_{\mu_1}(\mu_1) = F_{\mu_H}\left|\exp\left(\mu_1 + \frac{1}{2}\sigma_1^2\right)\right|$$

$$F_{\mu_2}(\mu_2 \mid \mu_1) = F_{\mu_T}\left|\exp\left(\mu_2 + \frac{1}{2}\sigma_2^2\right)\right|$$

$$F_{H_S}(H_S \mid \mu_1,\mu_2) = \Phi\left|\frac{\log H_S - \mu_1}{\sigma_1}\right|$$

$$F_{T_Z}(T_Z \mid \mu_1,\mu_2,H_S) = \Phi\left|\frac{\log T_Z - \left(\mu_2 + \rho\dfrac{\sigma_2}{\sigma_1}(\log H_S - \mu_1)\right)}{\sigma_2\sqrt{1-\rho^2}}\right|$$

$$F_{\eta_m}(\eta \mid \mu_1,\mu_2,H_S,T_Z) = \left|1 - \exp\left|-\frac{1}{2}\left(\frac{4\eta}{H_S}\right)^2\right|\right| \exp\left|-\frac{t}{T_Z}\exp\left|-\frac{1}{2}\left(\frac{4\eta}{H_S}\right)^2\right|\right|$$
$$\times \left|1 - \exp\left(-\sqrt{2\pi}\frac{4\eta}{H_S}\sqrt{1-\pi^{-1/2}\Gamma(3/4)^2}\right)\right|$$

Based on these five distribution functions, the FORM applies directly. Figure 5.7 shows results where the statistical uncertainty has been included. μ_H and μ_T are taken as independent and normally distributed with expected values as given earlier and coefficients of variation of 10%. A further development of the analysis presented in this example can be found in Madsen and Bach-Gansmo (1984).

An application very similar to the one presented occurs if the extreme response of a linear structure under stationary Gaussian loading is sought and the stiffness and damping parameters are not known precisely in advance. The theory developed in Chapters 7 and 8 can then be used to compute the extreme-value distribution conditioned upon a given set of stiffness and damping properties and combined with the methods presented here be used to compute the unconditional distribution.

The basic idea in this section has been to transform the basic *Rosenblatt* variables into a space of independent standardized normal variables since effective techniques for computation of probability contents in various sets in such a space are available. Transformations to other spaces in which probability contents are easy to compute can, however, also be envisioned. One such transformation has been proposed by Chen and Lind (1983). For independent basic variables the transformation is

$$F_{Z_i}(z_i) = q_i + p_i \Phi(u_i), \quad i = 1, 2, \ldots, n \tag{5.22}$$

where q_i is either 0 or $1 - p_i$. For dependent variables the transformation is similarly

$$F_{Z_i}(z_i \mid z_1, z_2, \ldots, z_{i-1}) = q_i + p_i \Phi(u_i), \quad i = 1, 2, \ldots, n \tag{5.23}$$

The p_i-factors can be selected freely except that $p_i \geqslant 0$. For $p_i < 1$ the distribution on the right-hand side is a mixed distribution with a probability mass $1 - p_i$ at $+\infty$ and $-\infty$ for $q_i = 0$ and $1 - p_i$, respectively. For $p_i > 1$ it is a truncated normal distribution on the interval $(-\infty, \Phi^{-1}(1/p_i)]$ and $[-\Phi^{-1}(1/p_i), \infty)$ for $q_i = 0$ and $1 - p_i$, respectively. This is illustrated in Fig. 5.8. The joint cumulative distribution function in the transformed space is

$$F_{\mathbf{U}}(\mathbf{u}) = \prod_{i=1}^{n} [q_i + p_i \Phi(u_i)], \quad \text{for} \quad 0 \leqslant q_i + p_i \Phi(u_i) \leqslant 1 \tag{5.24}$$

with the corresponding joint density function

$$f_{\mathbf{U}}(\mathbf{u}) = p \prod_{i=1}^{n} \varphi(u_i) \tag{5.25}$$

where the constant p is

$$p = \prod_{i=1}^{n} p_i \tag{5.26}$$

The case for which all $p_i \leqslant 1$ is first considered. The failure surface in the transformed space is shown in Fig. 5.9 together with the approximating tangent hyperplane at the point $\mathbf{u}^*$ closest to the origin. If it is required that

$$q_i = \begin{cases} 0, & u_i^* < 0 \\ 1 - p_i, & u_i^* \geqslant 0 \end{cases} \tag{5.27}$$

then none of the discrete probabilities in the marginal distributions

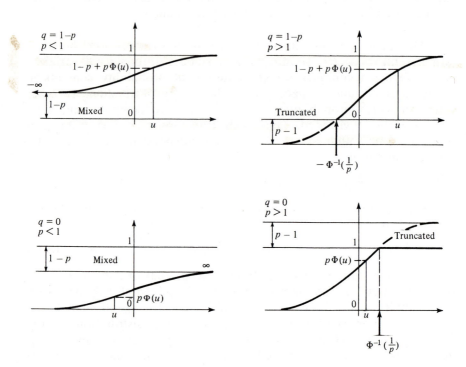

Figure 5.8 Mixed and truncated normal distribution, from Ditlevsen (1983).

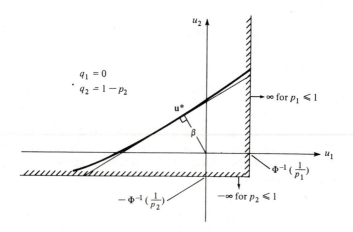

Figure 5.9 Failure surface approximation in transformed mixed normal space.

contribute to the probability content in the approximate failure set bounded by the tangent hyperplane. The first-order approximation to the failure probability is therefore

$$P_F \approx p\Phi(-\beta) \tag{5.28}$$

If, on the other hand, (5.27) is not fulfilled, the failure probability has contributions from discrete probabilities and from an integration of a continuous probability density function:

$$P_F \approx p\Phi(-\beta) + \sum_{u_i^* \geqslant 0} q_i \tag{5.29}$$

For $p_i > 1$ the computation of the first-order approximation to the failure probability is more complicated. For the case where (5.27) is fulfilled, Ditlevsen (1983) has shown that $p\Phi(-\beta)$ is an upper bound on the first-order approximation and has also calculated a lower bound.

The approximations (5.28) and (5.29) depend on the choice for p_i. In the light of the asymptotic result (4.72), a good choice appears to be one that makes all curvatures at $\mathbf{u}^*$ equal to zero. Then the first-order approximation yields an asymptotic correct result for the failure probability. In practice, such a set of p_i-values is not easy to obtain. Instead, it has been suggested by Ditlevsen (1983) to choose a set of values that makes the curvatures at $\mathbf{u}^*$ zero for a hyperplane failure surface in the original z-space. For independent basic variables the solution is

$$p_i = -\frac{u_i^* f_{Z_i}(z_i^*)^2}{\varphi(u_i^*) f_{Z_i}'(z_i^*)}, \quad i = 1, 2, \ldots, n \tag{5.30}$$

provided that the solution is nonnegative. This solution is valid for any hyperplane. p_i depends on $\mathbf{u}^*$, and vice versa. An iterative procedure must therefore be used to determine p_i and $\mathbf{u}^*$ in such a way that (5.30) is fulfilled and $\mathbf{u}^*$ is the point on the failure surface closest to the origin. Chen and Lind (1983) have presented such a procedure based on a generalization of the normal tail approximation technique presented in the following section.

Very little experience with the transformation (5.22) is available, and conclusions about its usefulness cannot be drawn at present. The procedure, however, demonstrates that many formulations of first-order reliability methods can be constructed depending on the space in which the probability calculation is carried out. A comparison between the use of the transformations (5.8) and (5.22) is shown in Example 5.5.

Example 5.5

Consider a safety problem with two basic variables Z_1 and Z_2. The basic variables are independent with the probability density function

$$f_{Z_i}(z_i) = 2(1-z_i), \quad z_i \in [0,1], \quad i=1,2$$

The limit state function is as in Example 5.1,

$$g(z_1,z_2) = 1.8 - z_1 - z_2$$

The failure probability is

$$P_F = \int_{0.8}^{1} \int_{1.8-z_1}^{1} 4(1-z_1)(1-z_2)dz_2 dz_1 = 2.67 \times 10^{-4}$$

corresponding to a reliability index

$$\beta_R = -\Phi^{-1}(P_F) = 3.46$$

The mapping into the standardized normal space is

$$u_1 = \Phi^{-1}(F_{Z_1}(z_1)) = \Phi^{-1}(1-(1-z_1)^2)$$

$$u_2 = \Phi^{-1}(F_{Z_2}(z_2)) = \Phi^{-1}(1-(1-z_2)^2)$$

The failure function in u-space is given in its parametric form as

$$(u_1,u_2) = (u_1(t),u_2(t))$$
$$= (\Phi^{-1}(1-(1-t)^2),\Phi^{-1}(1-(t-0.8)^2)), \quad 0.8 \le t \le 1.0$$

The failure surface is shown in Fig. 5.10.

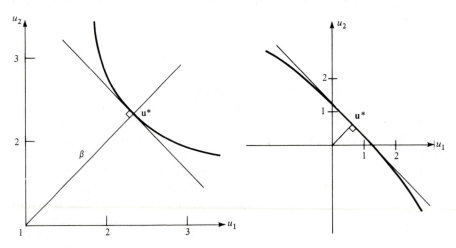

Figure 5.10 Failure surface in standard normal and mixed normal space.

Due to symmetry, the design point is on the symmetry line, $u_1 = u_2$. The design point has coordinates

$$(u_1, u_2) = (u_1(0.9), u_2(0.9)) = (2.32, 2.32)$$

corresponding to a first-order approximation

$$\beta_R \approx 3.29$$

$$P_F \approx \Phi(-\beta_R) = 5.01 \times 10^{-4}$$

The curvature κ at the design point is

$$\kappa = \frac{\begin{vmatrix} \dfrac{du_1}{dt}(0.9) & \dfrac{du_2}{dt}(0.9) \\[2mm] \dfrac{d^2u_1}{dt^2}(0.9) & \dfrac{d^2u_2}{dt^2}(0.9) \end{vmatrix}}{\left(\dfrac{du_1}{dt}(0.9)^2 + \dfrac{du_2}{dt}(0.9)^2\right)^{3/2}} = \frac{\begin{vmatrix} 7.512 & -7.512 \\ 56.2326 & 56.2326 \end{vmatrix}}{(7.512^2 + 7.512^2)^{3/2}} = 0.7046$$

The second-order approximation to the failure probability is given in (4.87). The three terms A_1, A_2, and A_3 are:

$$A_1 = \Phi(-3.29)\,(1 + 3.29 \times 0.7046)^{-1/2} = 2.75 \times 10^{-4}$$

$$A_2 = [3.29\Phi(-3.29) - \varphi(3.29)]\,\{(1 + 3.29 \times 0.7046)^{-1/2}$$
$$- (1 + 4.29 \times 0.7046)^{-1/2}\} = -6.66 \times 10^{-6}$$

$$A_3 = 4.29\,[3.29\Phi(-3.29) - \varphi(3.29)]\,\{(1 + 3.29 \times 0.7046)^{-1/2}$$
$$- \mathrm{Re}[(1 + (3.29 + i)0.7046)^{-1/2}]\} = -5.10 \times 10^{-6}$$

This leads to the approximation

$$P_F \approx 2.75 \times 10^{-4} - 6.66 \times 10^{-6} - 5.10 \times 10^{-6} = 2.63 \times 10^{-4}$$

The second-order approximation is thus very close to the exact result. The corresponding approximation to the reliability index is

$$\beta_R \approx -\Psi^{-1}(2.63 \times 10^{-4}) = 3.47$$

The transformation (5.22) into a mixed normal space is

$$F_{Z_i}(z_i) = 1 - p_i + p_i\Phi(u_i), \quad i = 1,2$$

where p_i is the solution to the equation

$$p_i = -\frac{u_i^* f_{Z_i}(z_i^*)^2}{\varphi(u_i^*) f_{Z_i}'(z_i^*)}, \quad i = 1,2$$

Due to symmetry, the design point in z-space is on the symmetry line $z_1 = z_2$. The design point has coordinates $(z_1, z_2) = (0.9, 0.9)$. It further follows that $p_1 = p_2 = p$ and $u_1^* = u_2^* = u^*$. From the distribution of Z_i follows

$$F_{Z_i}(0.9) = 0.99, \quad f_{Z_i}(0.9) = 0.2, \quad f_{Z_i}'(0.9) = -2$$

p is therefore obtained as the solution to the equation

$$p = 0.02\,\Phi^{-1}\!\left(\frac{p - 0.01}{p}\right) \Big/ \varphi\left|\Phi^{-1}\!\left(\frac{p - 0.01}{p}\right)\right|$$

Solving this equation gives $p = 0.03703$. The failure surface in u-space has the parametric form

$$(u_1, u_2) = (u_1(t), u_2(t))$$

$$= \left| -\Phi^{-1}\left(\frac{(1-t)^2}{0.03703}\right), -\Phi^{-1}\left(\frac{(t-0.8)^2}{0.03703}\right) \right|, \quad 0.8076 \leqslant t \leqslant 0.9924$$

The point closest to the origin is

$$(u_1^*, u_2^*) = (u_1(0.9), u_2(0.9)) = (0.6123, 0.6123)$$

The failure surface is shown in Fig. 5.10 and the curvature is zero at the point closest to the origin. The first-order (or second-order) approximation to the failure probability is then according to (5.28)

$$P_F \approx p^2 \Phi(-|\mathbf{u}^*|) = 0.03703^2 \Phi(-0.8659) = 2.65 \times 10^{-4}$$

This approximation is also very close to the exact result. The corresponding approximation to the reliability index is

$$\beta_R \approx 3.47$$

5.2 THE PRINCIPLE OF NORMAL TAIL APPROXIMATION

The iteration algorithm (4.32) or (4.36) of Section 4.4.1 for the design point $\mathbf{u}^*$ together with the backward substitution (5.21) can be given a special interpretation known as the *principle of normal tail approximation*. Here this principle is explained for independent basic variables but the principle also applies to dependent variables (Ditlevsen, 1981).

The iteration algorithm is first summarized. From (4.32) follows

$$\mathbf{u}^{(m+1)} = (\mathbf{u}^{(m)^T}\boldsymbol{\alpha}^{(m)})\boldsymbol{\alpha}^{(m)} + \frac{g_u(\mathbf{u}^{(m)})}{|\nabla g_u(\mathbf{u}^{(m)})|}\boldsymbol{\alpha}^{(m)} \qquad (5.31)$$

where $g_u(\mathbf{u}^{(m)})$ is expressed in terms of $\mathbf{z}^{(m)}$ as [(5.10)]

$$g_u(\mathbf{u}^{(m)}) = g(\mathbf{z}^{(m)}) \qquad (5.32)$$

The unit vector $\boldsymbol{\alpha}^{(m)}$ is computed as [(4.29)]

$$\boldsymbol{\alpha}^{(m)} = -\frac{\nabla g_u(\mathbf{u}^{(m)})}{|\nabla g_u(\mathbf{u}^{(m)})|} \qquad (5.33)$$

with the partial derivatives in terms of $\mathbf{z}^{(m)}$ [(5.13)]

$$\frac{\partial g_u}{\partial u_i}(\mathbf{u}^{(m)}) = \frac{\partial g}{\partial z_i}(\mathbf{z}^{(m)}) \frac{\varphi(\Phi^{-1}(F_{Z_i}(z_i^{(m)})))}{f_{Z_i}(z_i^{(m)})} \qquad (5.34)$$

The approximate linear backward substitution of $\mathbf{u}^{(m+1)}$ into $\mathbf{z}^{(m+1)}$ is [(5.21)]

$$z_i^{(m+1)} = z_i^{(m)} + \frac{\varphi(\Phi^{-1}(F_{Z_i}(z_i^{(m)})))}{f_{Z_i}(z_i^{(m)})}(u_i^{(m+1)} - u_i^{(m)}) \qquad (5.35)$$

If the sequence converges to the design point $\mathbf{u}^*$, the approximating tangent hyperplane to the limit state surface has the equation [(5.14)]

$$\sum_{i=1}^{n} \frac{\partial g_u}{\partial u_i}(\mathbf{u}^*)(u_i - u_i^*) = 0 \qquad (5.36)$$

Normal tail approximation consists in changing the distribution function of each basic variable to a substitute normal distribution function $\Phi((z_i - \mu_i)/\sigma_i)$. Parameters (μ_i, σ_i) are determined such that values of the distribution functions and the probability density functions are identical at some point z', i.e., as a solution to

$$\Phi\left(\frac{z_i' - \mu_i}{\sigma_i}\right) = F_{Z_i}(z_i') \qquad (5.37)$$

$$\frac{1}{\sigma_i}\varphi\left(\frac{z_i' - \mu_i}{\sigma_i}\right) = f_{Z_i}(z_i') \qquad (5.38)$$

The solution is

$$\sigma_i = \frac{\varphi(\Phi^{-1}(F_{Z_i}(z_i')))}{f_{Z_i}(z_i')} \qquad (5.39)$$

$$\mu_i = z_i' - \sigma_i \Phi^{-1}(F_{Z_i}(z_i')) \qquad (5.40)$$

Normalized variables x_i are introduced as

$$x_i = \frac{z_i - \mu_i}{\sigma_i} \qquad (5.41)$$

In terms of these variables the failure function is

$$g(\mathbf{z}) = g(x_1\sigma_1 + \mu_1, \ldots, x_n\sigma_n + \mu_n) = g_x(\mathbf{x}) = 0 \qquad (5.42)$$

The failure surface in z- and x-space is shown in Fig. 5.11. The tangent hyperplane at a point $\mathbf{x}'$ related to $\mathbf{z}'$ by (5.41) has the equation

$$\sum_{i=1}^{n} \frac{\partial g_x}{\partial x_i}(\mathbf{x}')(x_i - x_i') = \sum_{i=1}^{n} \frac{\partial g}{\partial z_i}(\mathbf{z}')\sigma_i(x_i - x_i') = 0 \qquad (5.43)$$

It follows from (5.34) and (5.39) to (5.41) that this equation is

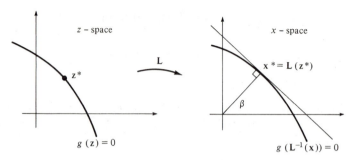

Figure 5.11   Failure surfaces in z- and x-space.

identical to (5.36) in u-space if $z' = z^*$. This suggests an iterative algorithm for determining z^*. A point $z^{(0)}$ is first selected. Parameters μ_i and σ_i are computed from (5.37) and (5.38) with $z' = z^{(0)}$. The normalized variables x_i are defined by (5.41), and the point $x^{(1)}$ closest to the origin in x-space is determined. The corresponding point in z-space is

$$z_i^{(1)} = x_i^{(1)} \sigma_i + \mu_i \tag{5.44}$$

which in general is different from $z^{(0)}$. New normal tail parameters μ_i and σ_i are calculated with $z' = z^{(1)}$, and the point $x^{(2)}$ closest to the origin in the new x-space is found. Continuing this procedure, a sequence $z^{(0)}, z^{(1)}, \ldots, z^{(m)}, \ldots$ is constructed. If the sequence converges, it is shown that the limit corresponds to a stationary value of the function $|T(z)|$ for points z on the limit state surface, (Ditlevsen, 1981). The design point z^* is such a stationary point. If it is the only stationary point, convergence is guaranteed. Otherwise, the sequence may converge toward another stationary point.

Determination of the points closest to the origin in x-space can be done, e.g., by the iteration algorithm in Section 4.4.1. A somewhat faster algorithm is constructed if the iteration in x-space is limited to one iteration by the procedure of Section 4.4.1. $x^{(m+1)}$ is then given by

$$x^{(m+1)} = (x^{(m)^T} \delta^{(m)}) \delta^{(m)} + \frac{g_x(x^{(m)})}{|\nabla g_x(x^{(m)})|} \delta^{(m)} \tag{5.45}$$

where the unit vector $\delta^{(m)}$ is

$$\delta^{(m)} = - \frac{\nabla g_x(x^{(m)})}{|\nabla g_x(x^{(m)})|} \tag{5.46}$$

and the partial derivatives are

$$\frac{\partial g_x}{\partial x_i}(\mathbf{x}^{(m)}) \;=\; \sigma_i\,\frac{\partial g}{\partial z_i}(\mathbf{z}^{(m)}) \;=\; \frac{\partial g_u}{\partial x_i}(\mathbf{x}^{(m)}) \qquad (5.47)$$

The forward iteration (5.45) is thus identical to (5.31). The backward substitution from $\mathbf{x}^{(m+1)}$ to $\mathbf{z}^{(m+1)}$ gives

$$z_i^{(m+1)} \;=\; \sigma_i\,x_i^{(m+1)} + \mu_i \qquad (5.48)$$
$$= \;\sigma_i\,x_i^{(m+1)} + z_i^{(m)} - \sigma_i\,\Phi^{-1}(F_{Z_i}(z_i^{(m)}))$$
$$= \;z_i^{(m)} + \sigma_i\,(x_i^{(m+1)} - x_i^{(m)})$$

where (5.37), (5.40), and (5.41) have been used. This backward substitution is thus identical to (5.35), and the complete iteration algorithm is identical to the one suggested in (5.31) to (5.35). The algorithm is known as the *Rackwitz and Fiessler algorithm* and was proposed by Rackwitz and Fiessler (1978) for a more complex problem of load combination (see also Madsen, 1978). The algorithm is very fast and efficient but it is not guaranteed to converge, and points $\mathbf{z}^{(m)}$ outside the set of the possible values for $\mathbf{Z}$ may occur during the iteration. A flow diagram of the algorithm is shown in Fig. 5.13 and the algorithm is illustrated in Example 5.6.

Example 5.6

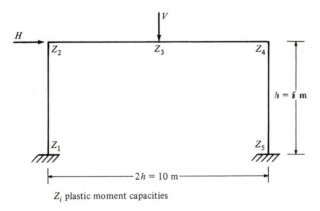

Figure 5.12   Plane frame structure.

Consider the plane frame structure of Fig. 5.12. Plastic hinge mechanisms leading to collapse of the structure are considered and are analyzed by utilizing elastic-plastic stress-strain relations. Hinges are thought to form at the end of elements or at points of load application only. Loads and plastic moment capacities are random variables which are assumed mutually independent and lognormally distributed. There are seven basic variables and their mean values and standard deviations are given in Table 5.3.

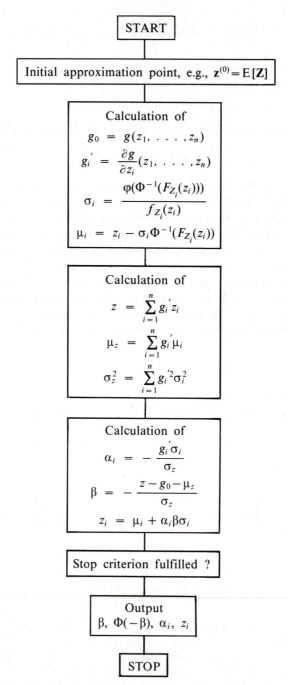

Figure 5.13 Flow diagram for the Rackwitz and Fiessler algorithm.

TABLE 5.3	Mean Values and Standard Deviations of Basic Variables				
Variable	$E[Z_i]$	$D[Z_i]$	$E[\log Z_i]$	$D[\log Z_i]$
$Z_1, \ldots, Z_5$	134.9 kNm	13.49 kNm	4.89956	0.09975
$H = Z_6$	50 kN	15 kN	3.86893	0.29356
$V = Z_7$	40 kN	12 kN	3.64579	0.29356

..l logarithms of the basic variables are normally distributed with
..ues and variances given by

$$E[\log Z] = \log E[Z] - \frac{1}{2}\log\left(\frac{D[Z]^2}{E[Z]^2} + 1\right)$$

$$\text{Var}[\log Z] = \log\left(\frac{D[Z]^2}{E[Z]^2} + 1\right)$$

In this example only the combined mechanism of Fig. 5.14 is considered.
A safety margin M is found by the principle of virtual work:

$$M = Z_1 + 2Z_3 + 2Z_4 + Z_5 - Z_6h - Z_7h$$

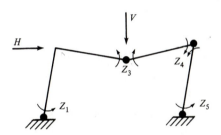

Figure 5.14   Plastic failure mechanism.

An exact mapping of the set of basic variables Z into a set U of
uncorrelated and standardized normal variables is given as

$$U_i = \frac{\log Z_i - E[\log Z_i]}{D[\log Z_i]}, \quad i = 1, 2, \ldots, 7$$

Inserting the values from Table 5.3, the failure function in terms of parameters u_i is

$$g_u(\mathbf{u}) = \exp(0.09975\, u_1 + 4.89956) + 2\exp(0.09975\, u_3 + 4.89956)$$
$$+ 2\exp(0.09975\, u_4 + 4.89956) + \exp(0.09975\, u_5 + 4.89956)$$
$$- 5\exp(0.29356\, u_6 + 3.86893) - 5\exp(0.29356\, u_7 + 3.64579)$$

The minimum distance from the origin to the limit state surface can now
be determined directly by the iteration scheme of Section 4.4.1. The result
is $\beta = 2.883$.

The normal tail approximation technique can also be applied. The distribution function and probability density function for the variables Z_i are

$$F_{Z_i}(z) = \Phi\left|\frac{\log z - E[\log Z_i]}{D[\log Z_i]}\right|$$

$$f_{Z_i}(z) = \frac{1}{z\,D[\log Z_i]}\,\varphi\left|\frac{\log z - E[\log Z_i]}{D[\log Z_i]}\right|$$

The initial approximation point is taken as the mean value point

$$z^{(0)} = E[Z]$$

The values μ_i and σ_i are first determined. From (5.37)

$$\frac{z_1^{(0)} - \mu_1}{\sigma_1} = \Phi^{-1}(F_{Z_1}(z_1^{(0)})) = 0.04986$$

and from (5.39)

$$\sigma_1 = \frac{\varphi\left(\dfrac{z_1^{(0)} - \mu_1}{\sigma_1}\right)}{f_{Z_1}(z_1^{(0)})} = \frac{\varphi(0.04986)}{f_{Z_1}(134.9)} = 13.456$$

From (5.40) then follows

$$\mu_1 = 134.9 - 13.456 \times 0.04986 = 134.229$$

The same result is obtained for $(\mu_2, \sigma_2), \ldots, (\mu_5, \sigma_5)$. For the basic variables Z_6 and Z_7 the results are

$$(\mu_6, \sigma_6) = (47.845, 14.678)$$

$$(\mu_7, \sigma_7) = (38.276, 11.742)$$

The failure surface in x-space is then

$$13.456\,x_1 + 134.229 + 2 \times 13.456\,x_3 + 2 \times 134.229 + 2 \times 13.456\,x_4$$
$$+ 2 \times 134.229 + 13.456\,x_5 + 134.229 - 5 \times 14.678\,x_6$$
$$- 5 \times 47.845 - 5 \times 11.742\,x_7 - 5 \times 38.276 = 0$$

or in normalized form

$$0.130 x_1 + 0.261 x_3 + 0.261 x_4 + 0.130 x_5 - 0.711 x_6 - 0.569 x_7 + 3.633 = 0$$

The point on this hyperplane closest to the origin is

$$\mathbf{x}^{(1)} = 3.633 \begin{vmatrix} -0.130 \\ 0.000 \\ -0.261 \\ -0.261 \\ -0.130 \\ 0.711 \\ 0.569 \end{vmatrix} = \begin{vmatrix} -0.472 \\ 0.000 \\ -0.945 \\ -0.945 \\ -0.472 \\ 2.583 \\ 2.067 \end{vmatrix}$$

The new iteration point $z^{(1)}$ is then, from (5.44),

$$
z^{(1)} = \begin{vmatrix} -0.472 \times 13.456 + 134.229 \\ 0 \times 13.456 + 134.229 \\ -0.945 \times 13.456 + 134.229 \\ -0.945 \times 13.456 + 134.229 \\ -0.472 \times 13.456 + 134.229 \\ 2.583 \times 14.678 + 47.845 \\ 2.067 \times 11.742 + 38.276 \end{vmatrix} = \begin{vmatrix} 127.88 \\ 134.23 \\ 121.51 \\ 121.51 \\ 127.88 \\ 85.76 \\ 62.55 \end{vmatrix}
$$

New values of μ_i and σ_i are

$$
(\mu_i, \sigma_i) = \begin{cases} (134.08, 12.756), & i = 1,5 \\ (134.23, 13.389), & i = 2 \\ (133.61, 12.121), & i = 3,4 \\ (35.79, 25.176), & i = 6 \\ (31.89, 18.362), & i = 7 \end{cases}
$$

and the iteration is continued. The sequence of x points is shown in Table 5.4.

	$x^{(1)}$	$x^{(2)}$	$x^{(3)}$	$\cdots$	$x^{(9)}$	$x^{(10)}$
TABLE 5.4 Iteration in x-Space						
x_1	-0.472	-0.230	-0.233	$\cdots$	-0.221	-0.221
x_2	0.000	0.000	0.000	$\cdots$	0.000	0.000
x_3	-0.945	-0.436	-0.437	$\cdots$	-0.433	-0.433
x_4	-0.945	-0.436	-0.437	$\cdots$	-0.433	-0.433
x_5	-0.472	-0.230	-0.233	$\cdots$	-0.221	-0.221
x_6	2.583	2.267	2.326	$\cdots$	2.393	2.393
x_7	2.067	1.653	1.559	$\cdots$	1.453	1.453
β	3.633	2.891	2.886	$\cdots$	2.883	2.883
$\Phi(-\beta)$	1.40×10^{-4}	1.92×10^{-3}	1.95×10^{-3}	$\cdots$	1.97×10^{-3}	1.97×10^{-3}

The iteration is stopped at $x^{(10)}$. The <u>first-order</u> approximation to the failure probability is

$$
P_F \approx 1.97 \times 10^{-3}
$$

5.3 CLASSIFICATION OF SYSTEMS: STRUCTURE FUNCTIONS

A real structure consists in general of many elements and for each element several failure modes may be relevant. With the exception of the generalized reliability index method, the safety methods dealt with in Chapters 4 and 5 are aimed primarily at assessing the reliability of one element against failing in one particular failure mode.

In this section a short introduction to the mathematical theory of system reliability is given (see also Barlow and Proschan, 1975; Kaufmann et al., 1977). This is followed by a discussion of the classification of systems relevant to structural engineering. The important class of series systems is treated in more detail in the subsequent section of the chapter. Brief introductions to the analysis of parallel systems and series systems of parallel subsystems are also given.

The structure is considered at a fixed point in time, and the state of the structure is assumed to depend — in a unique way — only on the present states of its elements. Each element is assumed to be either in a functioning state or in a failed state. This is also assumed for the structure, and the first step is to express the state of the structure in terms of the states of its elements. It is here of great use to introduce binary *state indicator variables* a_i as

$$a_i = \begin{cases} 1 & \text{if element } i \text{ is functioning} \\ 0 & \text{if element } i \text{ has failed} \end{cases}, \quad i = 1, \ldots, n \quad (5.49)$$

where n is the number of elements in the structure.

A state indicator variable a_S for the system is defined similarly. According to the assumptions, a_S can be given as a function of $a_1, \ldots, a_n$ in a unique way:

$$a_S = \varphi(\mathbf{a}), \quad \mathbf{a} = (a_1, \ldots, a_n) \quad (5.50)$$

The function φ is called the *structure function*. If the structure function is nondecreasing, but not constant, in any argument, the structure is said to be *coherent*.

The structure function is simple for series systems, parallel systems, and k-out-of-n systems. A series system, such as a chain, is in a functioning state if each element is functioning. The structure function is therefore

$$\varphi(\mathbf{a}) = \varphi_S(\mathbf{a}) = \prod_{i=1}^{n} a_i = \min\{a_1, \ldots, a_n\} \quad (5.51)$$

A parallel system is in a functioning state if at least one element is functioning. The structure function is therefore

$$\varphi(\mathbf{a}) = \varphi_P(\mathbf{a}) = 1 - \prod_{i=1}^{n}(1 - a_i) = \max\{a_1, \ldots, a_n\} \quad (5.52)$$

A wire rope is an example of a parallel system. A k-out-of-n system is in a functioning state if at least k out of its n elements are functioning. The structure function is therefore

$$\varphi(\mathbf{a}) \;=\; \varphi_K(\mathbf{a}) \;=\; \begin{cases} 1 & \text{if } \sum_{i=1}^{n} a_i \geqslant k \\ 0 & \text{otherwise} \end{cases} \tag{5.53}$$

The main structure of a bridge is an example of such a system since it functions if at least two of its main girders are functioning. For more complicated systems the determination of the structure function may be difficult, but systematic methods have been developed. These methods are generally based on the characterization of the coherent system by a fault tree, as described in Barlow et al. (1975).

In some cases the structure function can be determined directly by inspection. The structure can then be represented by its path sets or its cut sets. A *path vector* is defined as a set of elements which, if they all function, assures that the system functions. A *minimal path vector* is defined as a path vector for which the failure of any element results in system failure. A *path set* is defined as the set of element numbers in a path vector. To the jth minimal path set R_j the function $\rho_j(\mathbf{a})$ is defined as

$$\rho_j(\mathbf{a}) \;=\; \prod_{i \in R_j} a_i \tag{5.54}$$

$\rho_j(\mathbf{a})$ is clearly the structure function for a series arrangement of the elements of the jth path vector. Such a series arrangement is called a *minimal series path structure*. The whole system functions if at least one of the minimal series path structures functions. The system can therefore be seen as a parallel arrangement of the minimal series path structures and the structure function is

$$\varphi(\mathbf{a}) \;=\; 1 - \prod_{j=1}^{p} [1 - \rho_j(\mathbf{a})] \tag{5.55}$$

where p is the number of minimal path sets.

Similarly, a *cut vector* is defined as a set of elements which, if they all fail, results in system failure. A minimal cut vector, a *cut set*, and a minimal cut set are then defined analogously with the definitions above. To the jth minimal cut set K_j, the function $k_j(\mathbf{a})$ is defined as

$$k_j(\mathbf{a}) \;=\; 1 - \prod_{i \in K_j} (1 - a_i) \tag{5.56}$$

$k_j(\mathbf{a})$ is the structure function for the jth *minimal parallel cut structure*. Since the whole system fails if just one of the minimal parallel cut structures fails, the system can be represented as a series arrangement of the minimal parallel cut structures and the system function is

$$\varphi(\mathbf{a}) = \prod_{j=1}^{k} k_j(\mathbf{a}) \tag{5.57}$$

where k is the number of minimal cut sets.

Example 5.7

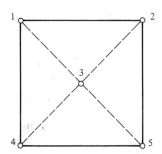

Figure 5.15 Foundation supported by piles.

Figure 5.15 shows a vertically loaded foundation supported by five vertical piles. The piles can loose their bearing capacity, and as a result of this the foundation can become unstable. The foundation is stable if three or more piles can carry a load with the exception of the combinations {1,3,5} and {2,3,4}. The minimum path sets are thus

$$R_1 = \{1,2,3\}, \quad R_2 = \{1,2,4\}, \quad R_3 = \{1,2,5\}, \quad R_4 = \{1,3,4\},$$
$$R_5 = \{1,4,5\}, \quad R_6 = \{2,3,5\}, \quad R_7 = \{2,4,5\}, \quad R_8 = \{3,4,5\}$$

Correspondingly, the minimum cut sets are

$$K_1 = \{1,5\}, \quad K_2 = \{2,4\}, \quad K_3 = \{1,2,3\}$$
$$K_4 = \{1,3,4\}, \quad K_5 = \{2,3,5\}, \quad K_6 = \{3,4,5\}$$

Figure 5.16 shows the minimal path representation and the minimal cut representation for the system. The structure function is

$$\varphi(\mathbf{a}) = 1 - \prod_{j=1}^{8}[1-\rho_j(\mathbf{a})] = \prod_{i=1}^{6} k_i(\mathbf{a})$$

In the preceding the state of the system was related to the state of its elements. Next the reliability of the system is related to the reliabilities of its elements. The state variables are now random variables A_i with expected values P_{R_i}:

$$P_{R_i} = E[A_i] = P(A_i = 1) \tag{5.58}$$

Minimal path set representation for system:

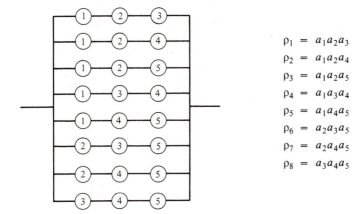

$$p_1 = a_1 a_2 a_3$$
$$p_2 = a_1 a_2 a_4$$
$$p_3 = a_1 a_2 a_5$$
$$p_4 = a_1 a_3 a_4$$
$$p_5 = a_1 a_4 a_5$$
$$p_6 = a_2 a_3 a_5$$
$$p_7 = a_2 a_4 a_5$$
$$p_8 = a_3 a_4 a_5$$

Minimal cut set representation for system:

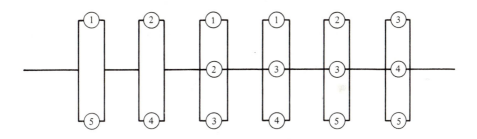

$$k_1 = 1-(1-a_1)(1-a_5), \qquad k_2 = 1-(1-a_2)(1-a_4)$$
$$k_3 = 1-(1-a_1)(1-a_2)(1-a_3), \qquad k_4 = 1-(1-a_1)(1-a_3)(1-a_4)$$
$$k_5 = 1-(1-a_2)(1-a_3)(1-a_5), \qquad k_6 = 1-(1-a_3)(1-a_4)(1-a_5)$$

Figure 5.16 Minimal path and minimal cut representation for foundation system.

The system reliability is similarly P_R, where

$$P_R = E[\varphi(\mathbf{A})] = P(\varphi(\mathbf{A})=1) \qquad (5.59)$$

In the evaluation of P_R in (5.59) there are two different difficulties. First, the random variables A_i are generally not independent due to common loading and load sharing of the elements, common quality

control, etc. The second difficulty is related to the structure .
tion. If the structure function is given as, e.g., in (5.55), the saɯ.
element is often a member of several minimal parallel cut structures, thus introducing dependencies between their state variables.

For independent state variables two general ways of calculating
the reliability can be mentioned. In the first method the expression
for the structure function is expanded into a multinomial expression
in the variables A_i using the fact that A_i^2 and A_i are identical. The
expected value of $\varphi(\mathbf{a})$ is then easily calculated. The second method
uses the identity

$$E[\varphi(\mathbf{A})] = P_{R_i} E[\varphi(\mathbf{A}) \mid A_i = 1] + (1 - P_{R_i})E[\varphi(\mathbf{A}) \mid A_i = 0] \quad (5.60)$$

Repeated application of this identity results in analyses with still
fewer elements. For both methods, however, the computational work
may be considerable.

Example 5.8

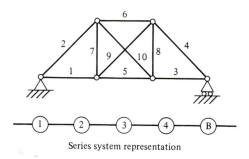

Series system representation

Figure 5.17 Plane truss structure.

Figure 5.17 shows a plane statically indeterminate truss structure. The structure fails if one of the elements 1, 2, 3, or 4 fails or if at least two of the
remaining elements fail. A series system representation of the system is also
given in Fig. 5.17, where system B is a 5-out-of-6 system since at least five
of the six elements 5, 6, . . . , 10 must function for the system to function.
The structure function is

$$\varphi(\mathbf{a}) = A_1 A_2 A_3 A_4 \prod_{i \neq j}[1 - (1 - A_i)(1 - A_j)]$$

where i and j takes the values 5, 6, . . . , 10.

Let the state variables A_i be independent. The typical factor is written as $[A_j + (1 - A_j)A_i]$ and all factors with identical index j are grouped.
The expectation of any expression containing the factor $A_i(1 - A_i)$ vanishes
and therefore, for fixed j,

$$\prod_{i \neq j}[A_j + (1 - A_j)A_i] \sim A_j + (1 - A_j)\prod_{i \neq j}A_i$$

The product of these factors for different values is next formed. The only contributions come from the product of all the first terms and the product of any of the last terms with the first terms of all the first factors. Thus

$$\prod_{i \neq j}[A_j + (1 - A_j)A_i] \sim \prod_j A_j + \sum_j\left((1 - A_j)\prod_{i \neq j}A_i\right)$$

Then the system reliability is

$$P_R = \mathrm{E}[\varphi(\mathbf{A})] = P_{R_1}P_{R_2}P_{R_3}P_{R_4}\left[\prod_{j=5}^{10}P_{R_j} + \sum_{j=5}^{10}\left((1 - P_{R_j})\prod_{i=5, i \neq j}^{10}P_{R_i}\right)\right]$$

In a realistic model the element reliabilities depend on the loading. Consequently, statical calculations should be performed for each of the seven states in which the structure can survive, i.e., with either no elements failed or just one of the elements 5, 6, . . . , 10 failed. There are thus easily created practical problems with the amount of calculation work involved.

For civil engineering structures the element state variables are almost never independent, and a treatment of the structure within the framework given above is therefore practical only for systems with a very simple configuration. However, many civil engineering structures belong to the class of series systems. This is certainly true for all statically determinate structures. It is also true for statically indeterminate structures when perfect plasticity of the elements is assumed and failure is defined as the formation of a plastic mechanism. Moreover, for any structure, if system failure is defined as the first failure of any element, an analysis as a series system will, of course, give the reliability. It appears that the class of series systems is most important and it has been the object of much research. The next section deals with series systems. The analysis of nonseries systems is still under development and in particular, first- and second-order reliability methods for parallel systems (cut vectors) and series systems of parallel subsystems (minimal parallel cut structures). These developments are also described briefly here.

5.4 RELIABILITY OF SERIES SYSTEMS

The structure function for a series system is given in (5.51). To each failure mode arising from the failure of an element, possibly in one of several failure modes, a safety margin M_i can be defined in terms of the basic variables. The system fails if at least one safety margin is negative. Boolean state variables for the system are here denoted A_S and B_S, where

$$A_S = \begin{cases} 1 & \text{if all } M_i > 0, \\ 0 & \text{otherwise} \end{cases} \qquad B_S = 1 - A_S \qquad (5.61)$$

For each failure mode the Boolean state variables A_i and B_i are defined similarly:

$$A_i = \begin{cases} 1 & \text{if } M_i > 0, \\ 0 & \text{otherwise} \end{cases} \qquad B_i = 1 - A_i, \quad i = 1, 2, \ldots, k \qquad (5.62)$$

The system function for the series system gives

$$A_S = A_1 A_2 \cdots A_k \qquad (5.63)$$

First some general bounds on the failure probability of the system in terms of the mode failure probabilities are derived. From (5.63) follows

$$A_S = A_1 A_2 \cdots A_{k-1} - A_1 A_2 \cdots A_{k-1} B_k \qquad (5.64)$$

Repeated application of this formula gives

$$B_S = B_1 + A_1 B_2 + A_1 A_2 B_3 \qquad (5.65)$$
$$+ \cdots + A_1 A_2 \cdots A_{k-1} B_k$$

Since the state variables can only take the values 0 and 1, it follows that

$$\max_i \{B_i\} \leqslant B_S \leqslant \sum_{i=1}^{k} B_i \qquad (5.66)$$

Taking expected values the probability of failure of the system is then bounded by the probabilities of failure in the individual modes as (Cornell, 1967)

$$\max_i \{P(M_i \leqslant 0)\} \leqslant P_F \leqslant \sum_{i=1}^{k} P(M_i \leqslant 0) \qquad (5.67)$$

or in a simpler notation

$$\max_i P_i \leqslant P_F \leqslant \sum_{i=1}^{k} P_i \qquad (5.68)$$

where the notation P_i has been introduced as

$$P_i = P_{F_i} = P(M_i \leqslant 0) \qquad (5.69)$$

Closer bounds on P_F can be given in terms of the failure probabilities in any mode and the joint failure probabilities in any two modes. Since all state variables are nonnegative, it follows that

$$A_1 A_2 \cdots A_i \geqslant \max\{1 - (B_1 + B_2 + \ldots + B_i), 0\} \qquad (5.70)$$

which combined with (5.65) leads to

$$B_S \geq B_1 + \sum_{i=2}^{k} \max\left\{B_i - \sum_{j=1}^{i-1} B_i B_j, 0\right\} \tag{5.71}$$

For any $j \leq i$ it is also true that

$$A_1 A_2 \cdots A_i \leq A_j = 1 - B_j \tag{5.72}$$

Combining (5.72) with (5.65) gives

$$B_S \leq \sum_{i=1}^{k} B_i - \sum_{i=2}^{k} \max_{j<i}\{B_i B_j\} \tag{5.73}$$

Taking expected values in (5.71) and (5.73) gives the bounds on the failure probability suggested and extensively used by Ditlevsen (1979).

$$P_1 + \sum_{i=2}^{k} \max\left\{P_i - \sum_{j=1}^{i-1} P_{ij}, 0\right\} \leq P_F \leq \sum_{i=1}^{k} P_i - \sum_{i=2}^{k} \max_{j<i} P_{ij} \tag{5.74}$$

P_i is defined in (5.69) and the notation P_{ij} has been used for the joint probability

$$P_{ij} = E[B_i B_j] = P(M_i \leq 0 \text{ and } M_j \leq 0) \tag{5.75}$$

The bounds depend on the numbering of the failure modes, and different orderings may correspond to the greatest lower bound and the smallest upper bound. Practical experience suggests that it is good to order the failure modes according to decreasing values P_i. Closer bounds can, of course, be derived if the joint probabilities of failure in any three modes are also introduced. It appears, however, that for practical situations this improvement is small compared to the sometimes substantial improvement in using (5.74) instead of (5.68) (Hohenbichler and Rackwitz, 1983).

Computation of approximations to P_i and P_{ij} is based on the ideas presented in Section 5.1. A transformation of the z-space of basic variables into a space of standardized normal variables is carried out, and the probabilities are computed exactly or approximately in this space. The first step in the calculation is the determination of the design point $\mathbf{u}^*$ for each failure mode.

$$\mathbf{u}_i^* = \beta_i \boldsymbol{\alpha}_i \tag{5.76}$$

In a *first-order system reliability analysis,* the failure set is approximated by the polyhedral set bounded by the tangent hyperplanes at the design points. The corresponding failure probability can then be determined from the formulas given in Section 4.5.1 for probability contents in polyhedral sets. More insight into the importance of the different failure modes is, however, obtained by using the bounds (5.74) and computing the individual and joint failure mode probabilities separately. The individual failure mode probabilities are in the first-order analysis determined as

$$P_i = \Phi(-\beta_i) \tag{5.77}$$

The first-order approximation to P_{ij} is obtained by approximating the joint failure set by the set bounded by the tangent hyperplanes at the design points for the two failure modes. Figure 5.18 shows the projection of the failure surfaces for the two failure modes on the plane spanned by the origin and the two design points $\mathbf{u}_i^*$ and $\mathbf{u}_j^*$. The safety margins $g_i(\mathbf{U})$ and $g_j(\mathbf{U})$ are then replaced by linear safety margins M_i and M_j:

$$M_i = \beta_i - \sum_{r=1}^{n} \alpha_{ir} U_r \tag{5.78}$$

$$M_j = \beta_j - \sum_{s=1}^{n} \alpha_{js} U_s \tag{5.79}$$

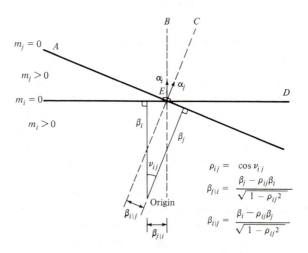

Figure 5.18 Geometrical illustration of multimode failure set, from Ditlevsen (1981).

M_i and M_j are standardized normally distributed with correlation coefficient ρ_{ij}:

$$\rho_{ij} = \rho[M_i, M_j] = \sum_{r=1}^{n} \alpha_{ir} \alpha_{jr} \tag{5.80}$$

Referring to Fig. 5.18, it therefore follows that

$$\rho_{ij} = \cos \nu_{ij} \tag{5.81}$$

The joint failure probability P_{ij} is thus approximated as

$$P_{ij} = \Phi(-\beta_i, -\beta_j; \rho_{ij}) = \int_{-\infty}^{-\beta_i} \int_{-\infty}^{-\beta_j} \varphi(x, y; \rho_{ij}) dx dy \qquad (5.82)$$

where $\varphi(, ; \rho)$ is the probability density function for a bivariate normal vector with zero mean values, unit variances, and correlation coefficient ρ.

$$\varphi(x, y; \rho) = \frac{1}{2\pi \sqrt{1 - \rho^2}} \exp\left| -\frac{1}{2} \frac{x^2 + y^2 - 2\rho xy}{1 - \rho^2} \right| \qquad (5.83)$$

$\Phi(, ; \rho)$ is the corresponding cumulative distribution function, which has the property

$$\frac{\partial^2 \Phi}{\partial x \partial y} = \frac{\partial \Phi}{\partial \rho} \qquad (5.84)$$

Using this identity, P_{ij} can be expressed by a single integral

$$P_{ij} = \Phi(-\beta_i, -\beta_j; 0) + \int_0^{\rho_{ij}} \frac{\partial \Phi}{\partial \rho}(-\beta_i, -\beta_j; z) dz \qquad (5.85)$$

$$= \Phi(-\beta_i)\Phi(-\beta_j) + \int_0^{\rho_{ij}} \varphi(-\beta_i, -\beta_j; z) dz$$

The probability P_{ij} must be evaluated numerically, but simple bounds on P_{ij} can be given, thus avoiding any numerical integration. For practical purposes these bounds will generally be sufficient. Figure 5.18 shows a situation with $\rho_{ij} > 0$. P_{ij} is the probability content in the angle AED. P_{ij} is larger than each of the probability contents in the angles AEC and BED but less than their sum. This gives

$$\max\{\Phi(-\beta_i)\Phi(-\beta_{j\,|\,i}), \Phi(-\beta_j)\Phi(-\beta_{i\,|\,j})\} \leqslant P_{ij} \qquad (5.86)$$

$$\leqslant \Phi(-\beta_i)\Phi(-\beta_{j\,|\,i}) + \Phi(-\beta_j)\Phi(-\beta_{i\,|\,j}), \quad \rho_{ij} > 0$$

For $\rho_{ij} < 0$ similar bounds are

$$0 \leqslant P_{ij} \leqslant \min\{\Phi(-\beta_i)\Phi(-\beta_{j\,|\,i}), \Phi(-\beta_j)\Phi(-\beta_{i\,|\,j})\}, \quad \rho_{ij} < 0 \qquad (5.87)$$

The distances $\beta_{j\,|\,i}$ and $\beta_{i\,|\,j}$ in Fig. 5.18 are found by geometrical considerations and use of (5.81). The distances can be expressed as

$$\beta_{i\,|\,j} = \frac{\beta_i - \rho_{ij}\beta_j}{\sqrt{1 - \rho_{ij}^2}} \qquad (5.88)$$

$$\beta_{j\,|\,i} = \frac{\beta_j - \rho_{ij}\beta_i}{\sqrt{1 - \rho_{ij}^2}} \qquad (5.89)$$

The notion of a conditional reliability index $\beta_{i\,|\,j}$ is used, since it is easily shown that $\beta_{i\,|\,j}$ is the reliability index for failure in mode i given that $M_j = 0$.

$$\beta_{i\,|\,j} \;=\; \frac{E[M_i \mid M_j=0]}{D[M_i \mid M_j=0]} \tag{5.90}$$

The first-order system reliability method is illustrated in an extension of Example 5.6.

Example 5.9

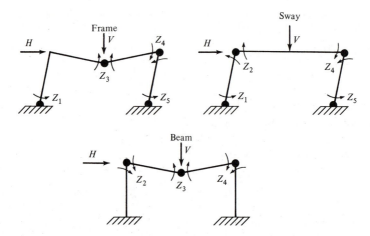

Figure 5.19 Plane frame structure with three plastic failure mechanisms.

The plane frame structure of Example 5.6 is again considered but now three failure modes as shown in Fig. 5.19 are considered. The principle of virtual work gives the three safety margins, which are linear in this case:

$$M_1 = Z_1 + Z_2 + Z_4 + Z_5 - Z_6 h$$
$$M_2 = Z_1 + 2Z_3 + 2Z_4 + Z_5 - Z_6 h - Z_7 h$$
$$M_3 = Z_2 + 2Z_3 + Z_4 - Z_7 h$$

All basic variables are assumed mutually independent and lognormally distributed with mean values and standard deviations as in Example 5.6. The design points for the three failure modes are

$$\mathbf{u}_1^* = 2.71 \begin{vmatrix} -0.084 \\ -0.084 \\ 0.000 \\ -0.084 \\ -0.084 \\ 0.986 \\ 0.000 \end{vmatrix}, \quad \mathbf{u}_2^* = 2.88 \begin{vmatrix} -0.077 \\ 0.000 \\ -0.150 \\ -0.150 \\ -0.077 \\ 0.827 \\ 0.509 \end{vmatrix}, \quad \mathbf{u}_3^* = 3.44 \begin{vmatrix} 0.000 \\ -0.084 \\ -0.164 \\ -0.084 \\ 0.000 \\ 0.000 \\ 0.979 \end{vmatrix}$$

The linearized safety margins in u-space are

$$M_1 = 0.084 U_1 + 0.084 U_2 + 0.084 U_4 + 0.084 U_5 - 0.986 U_6 + 2.71$$
$$M_2 = 0.077 U_1 + 0.150 U_3 + 0.150 U_4 + 0.077 U_5 - 0.827 U_6 - 0.509 U_7 + 2.88$$
$$M_3 = 0.084 U_2 + 0.164 U_3 + 0.084 U_4 - 0.979 U_7 + 3.44$$

The failure modes have been numbered according to increasing values of their first-order reliability indices. These indices are

$$\beta_1 = 2.71, \qquad \beta_2 = 2.88, \qquad \beta_3 = 3.44$$

The correlation matrix of the linearized safety margins is

$$\{\rho_{ij}\} = \begin{vmatrix} 1 & \cdot\ \cdot\ \cdot & \text{sym.} \\ 0.841 & 1 & \cdot\ \cdot\ \cdot \\ 0.014 & 0.536 & 1 \end{vmatrix}$$

A matrix $\mathbf{P}$ of individual and joint failure mode probabilities is now formulated as

$$\mathbf{P} = \begin{vmatrix} P_1 & \cdot\ \cdot\ \cdot & \text{sym.} \\ P_{12} & P_2 & \cdot\ \cdot\ \cdot \\ P_{13} & P_{23} & P_3 \end{vmatrix} = \begin{vmatrix} 3.36 \times 10^{-3} & \cdot\ \cdot\ \cdot & \text{sym.} \\ 9.24 \times 10^{-4} & 1.99 \times 10^{-3} & \cdot\ \cdot\ \cdot \\ 1.14 \times 10^{-6} & 4.25 \times 10^{-5} & 2.91 \times 10^{-4} \end{vmatrix}$$

where (5.77) and (5.85) have been used.

The simple bounds in (5.68) give the following bounds on the first-order system failure probability:

$$3.36 \times 10^{-3} \leqslant P_F^{FO} \leqslant 5.64 \times 10^{-3}$$

corresponding to bounds on a reliability index defined through (5.3):

$$2.53 \leqslant \beta_R^{FO} \leqslant 2.71$$

Use of the closer bounds in (5.74) results in

$$P_F^{FO} \geqslant 3.36 \times 10^{-3} + (1.99 \times 10^{-3} - 9.24 \times 10^{-4})$$
$$+ (2.91 \times 10^{-4} - 1.14 \times 10^{-6} - 4.25 \times 10^{-5}) = 4.67 \times 10^{-3}$$
$$P_F^{FO} \leqslant 3.36 \times 10^{-3} + (1.99 \times 10^{-3} - 9.24 \times 10^{-4})$$
$$+ (2.91 \times 10^{-4} - 4.25 \times 10^{-5}) = 4.67 \times 10^{-3}$$

The corresponding reliability index is

$$\beta_R^{FO} = 2.60$$

Numerical integration to determine the probabilities P_{ij} can be completely avoided by bounding P_{ij} as in (5.86). The matrix of conditional reliability indices is

$$\{\beta_{i\,|\,j}\} = \begin{vmatrix} - & 0.53 & 2.66 \\ 1.11 & - & 1.23 \\ 3.40 & 2.25 & - \end{vmatrix}$$

Upper and lower bounds for $\mathbf{P}$ are then

$$\mathbf{P}^L = \begin{vmatrix} 3.36 \times 10^{-3} & \cdots & & \text{sym.} \\ 5.93 \times 10^{-4} & 1.99 \times 10^{-3} & & \cdots \\ 1.14 \times 10^{-6} & 3.18 \times 10^{-5} & 2.91 \times 10^{-4} \end{vmatrix}$$

$$\mathbf{P}^U = \begin{vmatrix} 3.36 \times 10^{-3} & \cdots & & \text{sym.} \\ 1.04 \times 10^{-3} & 1.99 \times 10^{-3} & & \cdots \\ 2.27 \times 10^{-6} & 5.61 \times 10^{-5} & 2.91 \times 10^{-4} \end{vmatrix}$$

A combination of these bounds with the bounds in (5.74) yields

$$P_F^{FO} \geqslant 3.36 \times 10^{-3} + (1.99 \times 10^{-3} - 1.04 \times 10^{-3})$$
$$+ (2.91 \times 10^{-4} - 2.27 \times 10^{-6} - 5.61 \times 10^{-5}) = 4.54 \times 10^{-3}$$

$$P_F^{FO} \leqslant 3.36 \times 10^{-3} + (1.99 \times 10^{-3} - 5.93 \times 10^{-4})$$
$$+ (2.91 \times 10^{-4} - 3.18 \times 10^{-5}) = 5.02 \times 10^{-3}$$

and the corresponding bounds on the first-order system reliability index are

$$2.57 \leqslant \beta_R^{FO} \leqslant 2.61$$

A *second-order system reliability method* is based on a more accurate approximation of the failure surface than the first-order method. The individual failure mode probabilities are computed by approximating the failure surface by a quadratic surface with the same curvatures at the design point. The probabilities P_i are then determined as described in Section 4.5.2. The second-order approximation to P_{ij} is obtained by approximating the joint failure set by the set bounded by the hyperplanes at the point $\mathbf{u}_{ij}^*$ on the joint failure surface closest to the origin. The difference between the first- and second-order approximations to P_{ij} is illustrated in Fig. 5.20 for a case of two basic variables. In general, $\mathbf{u}_{ij}^*$ will not be in the hyperplane spanned by the origin and the design points $\mathbf{u}_i^*$ and $\mathbf{u}_j^*$. $\mathbf{u}_{ij}^*$ is found as the solution to an optimization problem with two constraints in the same way as each design point was found as the solution to an optimization problem with one constraint. $\mathbf{u}_{ij}^*$ is the solution to

$$\min |\mathbf{u}| \; ; \quad g_i(\mathbf{u}) \leqslant 0, \quad g_j(\mathbf{u}) \leqslant 0 \tag{5.91}$$

Alternatively, the minimization is formulated in z-space as

$$\min |\mathbf{T}(\mathbf{z})| \; ; \quad g_i(\mathbf{z}) \leqslant 0, \quad g_j(\mathbf{z}) \leqslant 0 \tag{5.92}$$

with $\mathbf{u}_{ij}^*$ being determined from the solution $\mathbf{z}_{ij}^*$ by $\mathbf{u}_{ij}^* = \mathbf{T}(\mathbf{z}_{ij}^*)$. The constraints are now formulated as inequalities, and the optimal solution does not necessarily correspond to an equality sign for both constraints. Example 5.9 is extended in the following to include this second-order system reliability analysis.

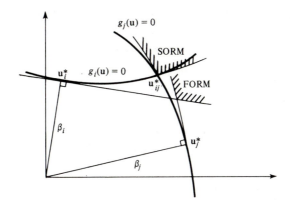

Figure 5.20 Difference between first- and second-order approximation to joint failure set.

Example 5.10

The second-order system reliability analysis is applied to the structure dealt with in Examples 5.6 and 5.9. The second-order approximations to the probabilities P_i are determined from (4.87) as

$$P_1^{SO} = 3.22 \times 10^{-3}, \quad P_2^{SO} = 2.67 \times 10^{-3}, \quad P_3^{SO} = 2.83 \times 10^{-4}$$

The linearization points on the joint failure surfaces are

$$\mathbf{u}_{12}^* = \begin{vmatrix} -0.227 \\ -0.057 \\ -0.338 \\ -0.391 \\ -0.227 \\ 2.674 \\ 0.977 \end{vmatrix}, \quad \mathbf{u}_{13}^* = \begin{vmatrix} -0.226 \\ -0.504 \\ -0.564 \\ -0.304 \\ -0.226 \\ 2.627 \\ 3.329 \end{vmatrix}, \quad \mathbf{u}_{23}^* = \begin{vmatrix} -0.060 \\ -0.233 \\ -0.564 \\ -0.346 \\ -0.060 \\ 0.350 \\ 3.366 \end{vmatrix}$$

The approximating linear safety margins for the joint failure probabilities are:

Mode 1 and 2:
$$M_1 = 0.084 U_1 + 0.085 U_2 + 0.082 U_4 + 0.084 U_5 - 0.986 U_6 + 2.71$$
$$M_2 = 0.074 U_1 + 0.147 U_3 + 0.146 U_4 + 0.074 U_5 - 0.875 U_6 - 0.425 U_7 + 2.90$$
$$P_{12} = \Phi(-2.71, -2.90; 0.887) = 1.09 \times 10^{-3}$$

Mode 1 and 3:
$$M_1 = 0.085 U_1 + 0.083 U_2 + 0.083 U_4 + 0.085 U_5 - 0.986 U_6 + 2.71$$
$$M_3 = 0.083 U_2 + 0.166 U_3 + 0.083 U_4 - 0.979 U_7 + 3.44$$
$$P_{13} = \Phi(-2.71, -3.44; 0.007) = 1.06 \times 10^{-6}$$

Mode 2 and 3:

$$M_2 = 0.076\,U_1 + 0.145\,U_3 + 0.148\,U_4 + 0.076\,U_5 - 0.446\,U_6 - 0.864\,U_7 + 3.21$$
$$M_3 = 0.085\,U_2 + 0.164\,U_3 + 0.084\,U_4 - 0.979\,U_7 + 3.44$$
$$P_{23} = \Phi(-3.21, -3.44; 0.882) = 1.54 \times 10^{-4}$$

The second-order approximation to the matrix $\mathbf{P}$ is

$$\mathbf{P} = \begin{vmatrix} P_1 & \cdots & \text{sym.} \\ P_{12} & P_2 & \cdots \\ P_{13} & P_{23} & P_3 \end{vmatrix} = \begin{vmatrix} 3.22 \times 10^{-3} & \cdots & \text{sym.} \\ 1.09 \times 10^{-3} & 2.67 \times 10^{-3} & \cdots \\ 1.06 \times 10^{-6} & 1.54 \times 10^{-4} & 2.83 \times 10^{-4} \end{vmatrix}$$

Application of the system reliability bounds (5.74) yields coinciding bounds

$$P_F^{SO} = 4.93 \times 10^{-3}$$

corresponding to the reliability index

$$\beta_R^{SO} = 2.58$$

Results which can be considered exact have been obtained by extensive simulation using an importance sampling technique. The results are

$$\mathbf{P} = \begin{vmatrix} P_1 & \cdots & \text{sym.} \\ P_{12} & P_2 & \cdots \\ P_{13} & P_{23} & P_3 \end{vmatrix} = \begin{vmatrix} 3.22 \times 10^{-3} & \cdots & \text{sym.} \\ 1.19 \times 10^{-3} & 2.69 \times 10^{-3} & \cdots \\ 1.07 \times 10^{-6} & 1.57 \times 10^{-4} & 2.83 \times 10^{-4} \end{vmatrix}$$

yielding coinciding bounds on the system reliability

$$P_F = 4.85 \times 10^{-3}$$

corresponding to the reliability index

$$\beta_R = 2.59$$

The results in Examples 5.6, 5.9, and 5.10 illustrate the various levels of refinement which are considered relevant in a system reliability analysis of a series system. The level of refinement must be chosen in each individual case depending on the demand for accuracy. For most cases the simple first-order results are sufficient.

5.5 RELIABILITY OF NONSERIES SYSTEMS

This section deals with the reliability of parallel systems (cut structures), series systems of parallel subsystems (series arrangement of minimal cut structures) and parallel systems of series subsystems (parallel arrangement of minimal path series structures). The analysis is not yet as well developed as the analysis of series systems, and only some main results are given.

The structure function for a *parallel* system is given in (5.52). To each element failure mode a safety margin M_j can be defined in terms of the basic variables $\mathbf{Z}$. The system fails if all safety margins are negative. The failure probability is thus

$$P_F = P\left|\bigcap_{j=1}^{k} \{g_j(\mathbf{Z}) \leq 0\}\right| \tag{5.93}$$

Simple bounds on P_F similar to (5.68) for series systems are

$$0 \leq P_F \leq \min_j P_j \tag{5.94}$$

These bounds are generally of little use, and closer bounds, e.g., similar to (5.74), have not been derived. The computation of P_F is therefore based on (5.93) directly. As for single elements and series systems, the computation of P_F first involves a mapping of the basic variables into a set of mutually independent and standardized normal variables, e.g., by the Rosenblatt transformation.

In a *first-order system reliability analysis,* the failure set is approximated by the intersection of the sets outside the tangent hyperplanes at the design points for the individual element failure modes. Figure 5.20 shows this approximation for $k = 2$. The corresponding approximation to the failure probability is

$$P_F \approx \Phi_k(-\boldsymbol{\beta}; \mathbf{R}) \tag{5.95}$$

where $\boldsymbol{\beta}$ contains the reliability indices and $\mathbf{R}$ is the correlation matrix between the linearized safety margins for the individual element. Example 5.9 and the continuation in Example 5.10 contain a comparison between first-order results and exact results for a case with $k = 2$. A rather large difference is observed for the probability P_{23} in the example. This is not a unique observation, and in general, first-order reliability methods cannot be considered sufficiently accurate for parallel systems.

An asymptotic result for the probability in (5.93) has been derived by Hohenbichler (1984). The result is asymptotic in the same sense as defined by Breitung (1984) and used in (4.72). The *joint design point* $\mathbf{u}^*$ is defined as the point in the failure set closest to the origin and thus the solution to the minimization problem

$$\min |\mathbf{u}|; \quad g_1(\mathbf{u}) \leq 0, \, g_2(\mathbf{u}) \leq 0, \, \ldots, g_k(\mathbf{u}) \leq 0 \tag{5.96}$$

or alternatively, $\mathbf{u}^* = \mathbf{T}(\mathbf{z}^*)$ with $\mathbf{z}^*$ being the solution of

$$\min |\mathbf{T}(\mathbf{u})|; \quad g_1(\mathbf{z}) \leq 0, \, g_2(\mathbf{z}) \leq 0, \, \ldots, g_k(\mathbf{z}) \leq 0 \tag{5.97}$$

For $\mathbf{u} = \mathbf{u}^*$ not all constraints are necessarily active; i.e., $g_j(\mathbf{u}) = 0$ is not necessarily valid for all $j \in \{1, 2, \ldots, k\}$. Let the constraints be numbered such that the constraints $1, \ldots, l$ are active at $\mathbf{u}^*$ while the remaining constraints $l+1, \ldots, k$ are nonactive. Unit vectors $\boldsymbol{\alpha}_i$ are defined as

$$\boldsymbol{\alpha}_i \;=\; -\frac{\nabla g_i(\mathbf{u}^*)}{|\,\nabla g_i(\mathbf{u}^*)\,|}, \quad i = 1, \ldots, l \tag{5.98}$$

For the linearized safety margins at $\mathbf{u}^*$ the reliability indices $\boldsymbol{\beta} = \{\beta_i\}$ and the correlation matrix $\mathbf{R} = \{\rho_{ij}\}$ are

$$\beta_i \;=\; \boldsymbol{\alpha}_i^T \mathbf{u}^*, \quad i = 1, \ldots, l \tag{5.99}$$

$$\rho_{ij} \;=\; \boldsymbol{\alpha}_i^T \boldsymbol{\alpha}_j, \quad i,j = 1, \ldots, l \tag{5.100}$$

It is assumed that the $\boldsymbol{\alpha}$-vectors are linearly independent, implying that $l \leq n$, where n is the dimension of the basic variable vector. It follows from Lagranges theorem that $\mathbf{u}^*$ is given uniquely as a linear combination of the $\boldsymbol{\alpha}$-vectors.

$$\mathbf{u}^* \;=\; \sum_{i=1}^{l} \lambda_i \nabla g_i(\mathbf{u}^*) \;=\; \sum_{i=1}^{l} \lambda_i (-\,|\,\nabla g_i(\mathbf{u}^*)\,|\,\boldsymbol{\alpha}_i) \tag{5.101}$$

with $\lambda_i \leq 0$. A matrix $\mathbf{D}$ containing the second derivatives of the active constraints in the directions perpendicular to the $\boldsymbol{\alpha}$'s is defined as

$$D_{ij} \;=\; \sum_{p=1}^{l} \lambda_p \frac{\partial^2 g_p(\mathbf{u}^*)}{\partial u_i \partial u_j}, \quad i,j = l+1, \ldots, n \tag{5.102}$$

The asymptotic result can now be stated as

$$P_F \;\sim\; \Phi_l(-\boldsymbol{\beta}; \mathbf{R})\,[\det(\mathbf{I} - \mathbf{D})]^{-1/2}, \quad |\mathbf{u}^*| \to \infty \tag{5.103}$$

It is observed that for a system with only one element (5.103) reduces to the asymptotic result in (4.72). If all constraints are active at the joint design point, then P_F is asymptotically as in (5.95) but with linearization of all safety margins at the joint design point and not at the individual design points, and with a correction factor $[\det(\mathbf{I} - \mathbf{D})]^{-1/2}$. In Example 5.10 both constraints are active at the design point in all three cases and the correction factor has been neglected. The experience with the asymptotic result in (5.103) is not yet very large. It is, however, expected that the asymptotic result can provide a close approximation to the exact result in most cases of high-reliability systems.

The structure function for a *series system of parallel subsystems* is given in (5.57). The failure probability is

$$P_F \;=\; P\left| \bigcup_{i=1}^{k} \left(\bigcap_{r=1}^{l_i} \{g_{ir}(\mathbf{Z}) \leq 0\} \right) \right| \tag{5.104}$$

$$=\; P\left| \bigcup_{i=1}^{k} \left(\bigcap_{r=1}^{l_i} \{M_{ir} \leq 0\} \right) \right|$$

where k is the number of subsystems and l_i is the number of elements in the ith parallel subsystem. $M_{ir} = g_{ir}(\mathbf{Z})$ is the safety margin for the rth element in the ith parallel subsystem (ith minimal parallel parallel cut structure). With the notation

$$P_i = P\left| \bigcap_{r=1}^{l_i} \{ M_{ir} \leq 0 \} \right| \tag{5.105}$$

$$P_{ij} = P\left| \left(\bigcap_{r=1}^{l_i} \{ M_{ir} \leq 0 \} \right) \cap \left(\bigcap_{s=1}^{l_j} \{ M_{js} \leq 0 \} \right) \right| \tag{5.106}$$

the simple bounds (5.68) and the bounds (5.74) are still valid. When the subsystem failure probabilities are small, these bounds are generally close. The computation of bounds on P_F in (5.104) can therefore be based on the techniques already developed for pure series and parallel structures.

The structure function for a *parallel system of series subsystems* is given in (5.55). The failure probability is

$$P_F = P\left| \bigcap_{i=1}^{s} \left(\bigcup_{r=1}^{t_i} \{ g_{ir}(\mathbf{Z}) \leq 0 \} \right) \right| \tag{5.107}$$

$$= P\left| \bigcap_{i=1}^{s} \left(\bigcup_{r=1}^{t_i} \{ M_{ir} \leq 0 \} \right) \right|$$

where s is the number of subsystems and t_i is the number of elements in the ith series subsystem. $M_{ir} = g_{ir}(\mathbf{Z})$ is the safety margin for the rth element in the ith series subsystem. The system functions (5.55) and (5.57) are identical and it is thus always possible to find an equivalent description of the system as a series system of parallel subsystems for which the reliability can be computed through the bounds given above. If all elements are different the equivalent system has $\prod_{i=1}^{s} t_i$ subsystems of s elements in parallel. The equivalent system may thus easily be very large and the computation of its reliability not be feasible in practice.

Upper bounds on P_F in (5.107) can be obtained by considering smaller series systems of parallel subsystems. Denoting

$$P_i = P\left| \bigcup_{r=1}^{t_i} M_{ir} \leq 0 \right| \tag{5.108}$$

$$P_{ij} = P\left| \left(\bigcup_{r=1}^{t_i} M_{ir} \leq 0 \right) \cap \left(\bigcup_{s=1}^{t_j} M_{js} \leq 0 \right) \right| \tag{5.109}$$

$$= P\left| \bigcup_{r=1}^{t_i} \bigcup_{s=1}^{t_j} \{ M_{ir} \leq 0 \cap M_{js} \leq 0 \} \right|$$

it follows from (5.94) that

$$P_F \leqslant \min_{i} P_i \tag{5.110}$$

and furthermore, that a closer upper bound on P_F is

$$P_F \leqslant \min_{i,j} P_{ij} \tag{5.111}$$

This procedure can be continued, thus leading to even closer upper bounds on P_F — however, at the cost of larger amounts of calculation. A lower bound on P_F can be obtained by computing the reliability of a system consisting of a subset of the $\prod_{i=1}^{s} t_i$ series system of parallel systems of size s. Due to a lack of experience, nothing definite can be stated about the closeness of these bounds.

5.6 SENSITIVITY MEASURES

It is often of interest to know the sensitivity of the failure probability or the reliability index to variations of parameters in the safety problem. The parameters, denoted by **p**, may include parameters in the distributions of the basic variables and deterministic parameters in the limit state function. Sensitivity measures are here derived for components, for series systems, and for parallel systems.

The *component* reliability is computed as explained in section 5.1. The connection between the basic variable vector **Z** and the vector **U** of independent and standardized normal variables is

$$\mathbf{Z}(\mathbf{p}) = \mathbf{T}^{-1}(\mathbf{U},\mathbf{p}) \tag{5.112}$$

T is generally taken as the Rosenblatt transformation (5.16), and the possible dependence of the distribution of **Z** on elements in **p** is explicitly expressed. The failure set in u-space is $F(\mathbf{p})$,

$$F(\mathbf{p}) = \{\mathbf{u} \mid g(\mathbf{u},\mathbf{p}) \leqslant 0\} \tag{5.113}$$

and the reliability index is $\beta_R(\mathbf{p})$,

$$\beta_R(\mathbf{p}) = -\Phi^{-1}[P(F(\mathbf{p}))] \tag{5.114}$$

The reliability index is computed for $\mathbf{p} = \mathbf{p}_0$, and the sensitivity of the reliability index to changes in parameter values is desired, i.e., the partial derivatives $\partial \beta_R(\mathbf{p}_0)/\partial p_i$ are desired. The partial derivatives can be computed by numerical differentiation, but more efficient methods which do not require a repeated computation of the reliability index are of more interest. This section presents asymptotic results for the partial derivatives following Hohenbichler (1984). The asymptotic results have been found to provide very good approximations.

Breitung (1984) showed the asymptotic results (4.72) and (4.73), which in terms of the failure probability and β_R are,

$$P(F(\mathbf{p}_0)) \sim \Phi(-\beta(\mathbf{p}_0)) \prod_{i=1}^{n-1} [1 - \beta(\mathbf{p}_0)\kappa(\mathbf{p}_0)]^{-1/2}, \quad \beta(\mathbf{p}_0) \to \infty \quad (5.115)$$

$$\beta_R(\mathbf{p}_0) \sim \beta(\mathbf{p}_0), \quad \beta(\mathbf{p}_0) \to \infty \quad (5.116)$$

$\beta(\mathbf{p}_0)$ is the first-order reliability index, and the results are asymptotic in the sense defined in Chapter 4. In the same asymptotic sense Hohenbichler (1984) has shown

$$\frac{\partial}{\partial p_i} \beta_R(\mathbf{p}_0) \sim \frac{\partial}{\partial p_i} \beta(\mathbf{p}_0), \quad \beta(\mathbf{p}_0) \to \infty \quad (5.117)$$

The first-order reliability index is

$$\beta(\mathbf{p}_0) = |\mathbf{u}^*| = (\mathbf{u}^{*T}\mathbf{u}^*)^{1/2} \quad (5.118)$$

where $\mathbf{u}^*$ is the design point which can be expressed as in (5.101),

$$\mathbf{u}^* = -\beta(\mathbf{p}_0) \frac{\nabla g(\mathbf{u}^*,\mathbf{p}_0)}{|\nabla g(\mathbf{u}^*,\mathbf{p}_0)|} = \lambda \nabla g(\mathbf{u}^*,\mathbf{p}_0) \quad (5.119)$$

For a *distribution parameter* follows

$$\frac{\partial}{\partial p_i} \beta(\mathbf{p}_0) = \frac{1}{\beta} \mathbf{u}^{*T} \frac{\partial}{\partial p_i} \mathbf{u}^* = \frac{1}{\beta} \mathbf{u}^{*T} \frac{\partial}{\partial p_i} \mathbf{T}(\mathbf{z}^*,\mathbf{p}_0) \quad (5.120)$$

where $\mathbf{z}^* = \mathbf{T}^{-1}(\mathbf{u}^*,\mathbf{p}_0)$.

For a *limit state function parameter* follows

$$\frac{\partial}{\partial p_i} \beta(\mathbf{p}_0) = \frac{1}{\beta} \mathbf{u}^{*T} \frac{\partial}{\partial p_i} \mathbf{u}^* \quad (5.121)$$

$$= \frac{1}{\beta} \left(-\beta \frac{\nabla g(\mathbf{u}^*,\mathbf{p}_0)^T}{|\nabla g(\mathbf{u}^*,\mathbf{p}_0)|} \right) \frac{\partial}{\partial p_i} \mathbf{u}^*$$

$$= \frac{\frac{\partial}{\partial p_i} g(\mathbf{u}^*,\mathbf{p}_0)}{|\nabla g(\mathbf{u}^*,\mathbf{p}_0)|}$$

Example 5.11

Let the basic variables be mutually independent and let the basic variable Z_1 be normally distributed with mean value μ_1 and standard deviation σ_1. Z_1 and U_1 are related by

$$U_1 = T_1(Z_1) = \frac{Z_1 - \mu_1}{\sigma_1}$$

The sensitivity factors with respect to μ_1 and σ_1 follow from (5.120) as

$$\frac{\partial \beta}{\partial \mu_1} = \frac{1}{\beta} u_1^* \frac{\partial}{\partial \mu_1} \left(\frac{z_1^* - \mu_1}{\sigma_1} \right) = -\frac{u_1^*}{\beta \sigma_1}$$

$$\frac{\partial \beta}{\partial \sigma_1} = \frac{1}{\beta} u_1^* \frac{\partial}{\partial \sigma_1} \left(\frac{z_1^* - \mu_1}{\sigma_1} \right) = -\frac{(u_1^*)^2}{\beta \sigma_1}$$

Let the basic variable Z_2 have a lognormal distribution with mean value μ_2 and coefficient of variation V_2. Z_2 and U_2 are related by

$$U_2 = T_2(Z_2) = \frac{\log Z_2 - E[\log Z_2]}{D[\log Z_2]}$$

where the mean value and standard deviation of $\log Z_2$ are

$$E[\log Z_2] = \log \mu_2 - \frac{1}{2}\log(1 + V_2^2)$$

$$D[\log Z_2] = \sqrt{\log(1 + V_2^2)}$$

The sensitivity factors with respect to μ_2 and V_2 follow from (5.120) as

$$\frac{\partial \beta}{\partial \mu_2} = -\frac{u_2^*}{\beta \mu_2 \sqrt{\log(1 + V_2^2)}}$$

$$\frac{\partial \beta}{\partial V_2} = -u_2^* \frac{V_2}{\beta \sqrt{\log(1 + V_2^2)}(1 + V_2^2)} \left(\frac{u_2^*}{\sqrt{\log(1 + V_2^2)}} - 1 \right)$$

The sensitivity factor results (5.120) and (5.121) are for the reliability index. Sensitivity factors for the failure probability are expressed in terms of sensitivity factors for the reliability index as

$$\frac{\partial}{\partial p_i} P(F(\mathbf{u},\mathbf{p}_0)) = \frac{\partial}{\partial p_i} \Phi(-\beta_R(\mathbf{p}_0)) \tag{5.122}$$

$$= \varphi(-\beta_R(\mathbf{p}_0)) \frac{\partial}{\partial p_i} \beta_R(\mathbf{p}_0)$$

$$\sim \varphi(-\beta(\mathbf{p}_0)) \frac{\partial}{\partial p_i} \beta(\mathbf{p}_0)$$

The failure probability of a *series system* is asymptotically equal to the sum of the component failure probabilities, i.e., it is equal to the upper bound in (5.67). When the minimum component reliability index is unique, the two bounds in (5.67) coincide asymptotically. The asymptotic values of the sensitivity factors for the system failure probability are

$$\frac{\partial}{\partial p_i} P(F(\mathbf{u},\mathbf{p}_0)) \sim \sum_i \frac{\partial}{\partial p_i} P(F_i(\mathbf{u},\mathbf{p}_0)) \tag{5.123}$$

$$\sim \sum_i \varphi(-\beta_i(\mathbf{p}_0)) \frac{\partial}{\partial p_i} \beta_i(\mathbf{p}_0)$$

The joint design point $\mathbf{u}^*$ for a *parallel system* is given in (5.101). Hohenbichler (1984) has shown the asymptotic result for the sensitivity factors of a parallel system

$$\frac{\partial}{\partial p_i}\beta_R(\mathbf{p}_0) \;\sim\; \frac{\partial}{\partial p_i}\beta^*(\mathbf{p}_0) \qquad\qquad (5.124)$$

where β^* is the length of $\mathbf{u}^*$. For a *distribution parameter*, the sensitivity factors are thus as in (5.120)

$$\frac{\partial}{\partial p_i}\beta_R(\mathbf{p}_0) \;\sim\; \frac{1}{\beta^*}\mathbf{u}^{*T}\frac{\partial}{\partial p_i}\mathbf{T}(\mathbf{z}^*,\mathbf{p}_0) \qquad\qquad (5.125)$$

For a *limit state function parameter*, the sensitivity factor is similarly to (5.121)

$$\frac{\partial}{\partial p_i}\beta_R(\mathbf{p}_0) \;\sim\; \frac{1}{\beta^*}\mathbf{u}^{*T}\frac{\partial}{\partial p_i}\mathbf{u}^* \qquad\qquad (5.126)$$

$$= \frac{1}{\beta^*}\sum_{k=1}^{l}\lambda_k\,\nabla g_k(\mathbf{u}^*,\mathbf{p}_0)^T\frac{\partial}{\partial p_i}\mathbf{u}^*$$

$$= \frac{1}{\beta^*}\sum_{k=1}^{l}\lambda_k\frac{\partial}{\partial p_i}g_k(\mathbf{u}^*,\mathbf{p}_0)$$

The sensitivity factors for the failure probability are as in (5.122) in terms of the sensitivity factors for the reliability index.

5.7 SUMMARY

This chapter presents first- and second-order reliability methods. These methods are used to compute approximations to the failure probability for failure modes of structures with parameters described by their joint distribution and with the set of possible values of the parameters divided by a limit state surface into a failure set and a safe set. The methods are based on an exact transformation into a standardized normal space and on an approximation of the limit state surface by its tangent hyperplane or quadratic approximation at a suitably selected design point in this space. Various algorithms are presented for finding the design point and the principle of normal tail approximation is explained.

The mathematical theory of system reliability is briefly outlined and a classification of structures relevant in structural engineering is discussed. The important class of series systems is then analyzed by first- and second-order reliability methods using bounds on the system reliability in terms of individual and joint failure mode probabilities. Parallel systems, series systems of parallel subsystems, and parallel systems of series subsystems are also analyzed but in somewhat less detail. A section is devoted to sensitivity measures. It is described how measures for the sensitivity of the reliability to changes in parameters is determined. The sensitivity measures are computed for parameters in the distributions of the basic variables and for parameters in the limit state functions.

REFERENCES

BARLOW, R. E. and F. PROSCHAN, *Statistical Theory of Reliability and Life Testing,* Holt, Rinehart and Winston, 1975.

BARLOW, R. E., J. B. FUSSEL and N. D. SINGPURWALLA, eds., *Theoretical and Applied Aspects of System Reliability and Safety Assessment,* Society for Industrial and Applied Mathematics, Philadelphia, 1975.

BREITUNG, K., "Asymptotic Approximations for Multinormal Integrals," *Journal of Engineering Mechanics,* ASCE, Vol. 110, 1984, pp. 357-366.

CHEN, X. and N. C. LIND, "Fast Probability Integration by Three Parameter Normal Tail Approximation," *Structural Safety,* Vol. 1, 1983, pp. 269-276.

CORNELL, C. A., "Bounds on the Reliability of Structural Systems," *Journal of the Structural Division,* ASCE, Vol. 93, 1967, pp. 171-200.

DITLEVSEN, O., "Narrow Reliability Bounds for Structural Systems," *Journal of Structural Mechanics,* Vol. 7, 1979, pp. 453-472.

DITLEVSEN, O., "Principle of Normal Tail Approximation," *Journal of Engineering Mechanics,* ASCE, Vol. 107, 1981, pp. 1191-1208.

DITLEVSEN, O., "Gaussian Safety Margins," in *Proceedings,* Fourth International Conference on Application of Statistics and Probability in Soil and Structural Engineering, ICASP4, University of Firenze, Italy, June 1983, pp. 785-824.

HOHENBICHLER, M., "An Asymptotic Formula for the Probability of Intersections," *Berichte zur Zuverlässigkeitstheorie der Bauwerke,* Heft 69, LKI, Technische Universität München, 1984, pp. 21-48.

HOHENBICHLER, M., "Mathematische Grundlagen der Zuverlässigkeitsmethode erster Ordnung, und einige Erweiterungen," Doctoral

Thesis at the Technical University of Munich, Munich, West Germany, 1984.

HOHENBICHLER, M. and R. RACKWITZ, "Nonnormal Dependent Vectors in Structural Reliability, *Journal of the Engineering Mechanics Division,* ASCE, Vol. 107, 1981, pp. 1127-1238.

HOHENBICHLER, M. and R. RACKWITZ, "First-Order Concepts in System Reliability," *Structural Safety,* Vol. 1, 1983, pp. 177-188.

ISSC (International Ship Structures Congress), Report of Committee 1, Proceedings of the Second International Ship Structures Congress, Delft, Netherland, July 20-24, 1964.

KAUFMANN, A., D. GROUCHKO and R. CRUON, *Mathematical Models for the Study of the Reliability of Systems,* Academic Press, New York, 1977.

LINDLEY, D. V., *Probability and Statistics, Vol. 2: Inference,* Cambridge University Press, Cambridge, 1965.

MADSEN, H. O., "Some Experience with the Rackwitz and Fiessler Algorithm for the Calculation of Structural Reliability under Combined Loading," in *DIALOG 77,* Danish Engineering Academy, Lyngby, Denmark, 1978, pp. 73-98.

MADSEN, H. O. and O. BACH-GANSMO, "Design Wave Determination by Fast Integration Technique," in *System Modelling and Optimization,* Proceedings of the 11th FIP Conference, Copenhagen, July 25-29, 1983, Springer-Verlag, Berlin, 1984, pp. 471-477.

RACKWITZ, R. and B. FIESSLER, "Structural Reliability under Combined Random Load Sequences," *Computers & Structures,* Vol. 9, 1978, pp. 489-494.

ROSENBLATT, M., "Remarks on a Multivariate Transformation," *The Annals of Mathematical Statistics,* Vol. 23, 1952, pp. 470-472.

CALIBRATION

6.1 THE PROCESS OF CALIBRATION

Calibration is the process of assigning values to the parameters in a design code. The *code parameters* are all numerical quantities specified in the code, except physical constants. Code parameters are selected primarily with a view to achieve a desired level of reliability in error-free structures. A code may be calibrated by judgment, fitting, optimization, or by a combination of these approaches.

Judgment. Calibration by judgment (or guesswork) was the common approach until quite recently. If a code after some years of service has proved to perform satisfactorily, the parameter values may be accepted as correct or, if there is sufficient economic pressure, some of the code parameters may be changed moderately toward smaller safety margins. Conversely, poor performance may generate a pressure to increase the safety margins; such increases are often more drastic. If there is no change of format, the code should gradually settle down to a steady state by this process, calibrated on an empirical basis and characterized by the absence of sufficient reason to change matters − neither in the direction of improved reliability nor improved economy of construction. The slenderness limitation ($l/r < 200$) in metal construction provides an example of a parameter that has reached a steady state, because it has little economic significance and little influence on performance.

Fitting. A radical change in code format is sometimes made after a code has been functioning for a number of years. Then the problem is to transfer the accumulated practical experience with the performance from the old code to the new code. A simple and conservative method to do this type of calibration is to adjust the parameters of the new code such that it gives the same minimum permissible physical dimensions as the old code. Of course, this is not always feasible, but when it is, it may be advantageous at least as a first step. In a sense, the new code just mimics the old code, which could seem rather futile. Why rewrite a code if neither reliability nor economy can be improved? Any code change is a nuisance for all parties involved. Such rewriting, however, could be advantageous for several reasons, for example if the new format were more correct in philosophy, more in line with standards for other materials or standards in other countries, or if it were simpler to use.

Code optimization. Finally, a code may be calibrated by a more formal process of explicit optimization (Ravindra and Lind, 1983). The first step is to define the *scope,* i.e., the class of structures to which the code is meant to apply. The scope is a parameterized set of structures. For all structures encompassed by this scope a class of relevant *failure modes* is identified. The code is characterized in a mathematical framework as an element of a set called the code format. The code parameters may be considered as variables. As these variables take on various different values, a set of different codes is generated. This set is called the *code format* of the code; the original code is one of many realizations of the format. Each realization is characterized by its particular set of values of the parameters, corresponding to a point in the parameter space for the code format.

The second step in the writing of a structural code is to define the *code objective.* For a Level I code (i.e., a limit states design code with specified partial safety factors) the objective may be stated at any higher level. A Level IV objective, for example, is the maximization of total expected utility; a Level III objective might be to achieve a constant specified reliability over the data set; on Level II one could aim for a constant reliability index or for a *target reliability index* of specified variation over the data set.

The third step is the determination of the frequency of occurrence of a particular safety check. Since a code, in general, cannot be both simple and exactly meet the objective, it is necessary to define the most important structural data for which the objective is to be met as closely as possible. For example, if most structural actions at a cross section are confined within a dead load to live load ratio of 0.5 to 2.0, it is generally possible to meet the objective closer over this range than over the hypothetical range of 0 to ∞. The frequency of occurrence is a scalar point function in data space and is called the *demand function*.

The fourth step in the design of a code is to select a measure of closeness between a code realization and its objective. For example, let β^* denote the target value of the reliability index in a particular safety check, and let β be the actual value produced by the check procedure. The difference $\beta^* - \beta$ varies over the data set; for some structural sections (say, "slender columns"), it may be positive, while for others (say, "intermediate columns"), it may be negative. Then, the criterion of closeness of the code to the objective may be that the expected value of $(\beta^* - \beta)^2$ should not exceed a given value.

The final step is the selection of a sequence of trial code formats, arranged in order of decreasing simplicity. Even the simplest objective cannot be met exactly by a Level I code, except at a practically unacceptable level of complexity. It is necessary to confine the search to a set of formats that lead to sufficiently simple design procedures. In each format, there exists generally an optimal realization which comes closest to the objective. With the criterion of closeness, one can select the best of these realizations as the simplest one that meets the criterion.

A number of structural design codes have been developed with this philosophy, including the National Building Code of Canada (NBCC, 1977; Siu et al., 1975), the Ontario Highway Bridge Code (OHBDC, 1983; Nowak and Lind, 1979), and the American National Standards Institute's Loading Standard (Ellingwood et al., 1980). The parameters in these codes have been selected using the code optimization concepts of data space and measure of closeness, as illustrated in Example 6.1. The rigor of formal code optimization can sometimes stand in the way of developing code provisions that lead to simple design procedures. Example 6.2 shows how the Nordic Committee on Building Regulations calibrated a Level I format to incorporate a number of desirable features to conform with the Level II format using a more informal approach.

Example 6.1

The limit states design criteria in the National Building Code of Canada (NBCC, 1977) for all materials and types of construction has the following format for an ordinary building (importance factor $\gamma = 1.0$) [(3.7)]:

$$\varphi R \geqslant \alpha_D D + \psi(\alpha_L L + \alpha_W W + \alpha_T T)$$

where W is the characteristic wind load. The selection of the code parameters $\varphi, \alpha_D, \alpha_L, \alpha_W, \ldots$, and ψ was described by Siu et al. (1975). It is based on a calibration to existing design standards for cold-formed steel, hot-rolled steel, reinforced concrete, and wood. The objective was to have common load factors for all structural materials. The resistance factors would be different according to material and limit state. The statistical data on loads and limit state resistances were collected. The nominal value of load was the value specified by the code authorities, and the nominal resistance was the value predicted by theoretical or empirical formulas used in conjunction with the design standard.

The data space (consisting of different structures, different locations in the structure, and geographical location) was characterized in this study by two variables: live-load-to-dead-load ratio q_l and wind-load-to-dead-load ratio q_w. The frequencies of occurrence of different design situations were estimated by several engineers individually, reconciled and presented as weighting factors for combinations of q_l and q_w. The data space would vary with the technology (i.e., material) and the individual limit states and should be properly represented in the calibration by the code committee on the basis of usage. However, only one set of design situations and their frequencies of occurrence was assumed for all technologies and limit states in this study. The calibration procedure was:

1. For each limit state for the structural material (e.g., hot-rolled, steel-column compression failure), the reliability index values implied in the existing code NBCC (1975) for all design situations were examined. The value of the reliability index β was determined using the approximate expression of Rosenblueth and Esteva (1972),

$$\beta = \frac{\log \mu_R - \log \mu_Q}{\sqrt{V_R^2 + V_Q^2}}$$

μ ≠ V only

in which μ_R and μ_Q are, respectively, the mean resistance and mean total load effect, and V_R and V_Q are the coefficients of variation. Table 6.1 shows an example of the variation of the implied reliability index. A weighted average value of the reliability index, β_{avg}, was found for the structural material for the limit state under consideration,

$$\beta_{avg} = \frac{\sum f_i \beta_i}{\sum f_i}$$

where f_i is the weighting factor based on the frequency of occurrence of the specific design situation (i.e., combination of q_l and q_w values in Table 6.1).

2. From an analysis of the β_{avg} values for different materials under different limit states, representative target reliability index values β^* were selected for different limit states as constants for all structural members, e.g., $\beta^* = 4.00$ for yielding in tension and flexure, $\beta^* = 4.75$ for compression and buckling failures, and $\beta^* = 4.25$ for shear failures.

TABLE 6.1 Reliability Index Reflected in the Existing Design Standard, NBCC (1975), after Siu et al. (1975)

β-values for Hot-rolled Steel Columns under Compression Failure.

q_w	q_l					
	0.00	0.50	1.00	2.00	3.00	4.00
0.00	5.04	5.32	5.02	5.02	4.30	4.14
0.50	5.21	4.64	5.00	4.55	4.30	4.14
1.00	4.93	4.93	4.83	4.55	4.30	4.14
2.00	4.52	4.52	4.52	4.83	4.69	4.32
3.00	4.29	4.29	4.29	4.80	4.79	4.68

Weighted average: $\beta_{avg} = 4.67$

3.   The evaluation of $\varphi, \alpha_D, \alpha_L, \ldots$ was carried out by the code optimization procedure. For a selected set of $\varphi, \alpha_D, \alpha_L, \ldots$, and for a given material under a particular limit state, the implied value of the reliability index was determined for the new code at a chosen data point. An objective function reflecting the measure of closeness was formulated as

$$\Delta = \sum_{materials} \sum_{limit\ states} \sum_{data\ space} f_i\,(\beta_i^* - \beta_i)^2$$

where β_i^* is the target reliability index for a particular limit state and material. The minimization of Δ and round-off resulted in the following optimal calibration of the format:

$$\varphi R \;\geqslant\; \begin{cases} 1.25D + 1.50L \\ 1.25D + 0.70\,(\,1.50L + 1.40W) \end{cases}$$

The resistance factors for different technologies were obtained as, for example:

Cold-formed steel	Yielding	$\varphi_1 = 0.90$
Hot-rolled steel	Yielding	$\varphi_2 = 0.85$
	Compression	$\varphi_3 = 0.74$
Reinforced concrete	Flexure	$\varphi_4 = 0.83$
	Compression	$\varphi_5 = 0.68$
	Shear	$\varphi_6 = 0.64$

For details of procedure, reference is made to Siu et al. (1975) or more recent examples in the literature (Schwartz et al., 1981; Thoft-Christensen and Baker, 1982; Ravindra and Lind, 1983).

Example 6.2

The Nordic Committee on Building Regulations considered the typical calculation model (NKB, 1978)

$$cmI_X = g_1(Y_1 + I_{Y_1}) + g_2(Y_2 + I_{Y_2})$$

in which c, g_1, and g_2 are constants that include geometrical data; m is a material property (yield strength, for example); Y_1 is the permanent load and Y_2 is a variable load; and I_X, I_{Y_1}, and I_{Y_2}, called *judgment variables*, are random variables that provide an allowance for model uncertainties. I_X and m are assumed to be lognormally distributed; Y_1, I_{Y_1}, and I_{Y_2} are assumed to be normally distributed; and the variable load Y_2 is assumed distributed as the largest value in a sample of size 1, 10, or 100 drawn from a normal population. The procedure described in Chapter 4 gives the mean value $c\mathrm{E}[m]$ for a prescribed safety index β as

$$c\mathrm{E}[m]\exp(-\beta\alpha_X V_X) = g_1\mathrm{E}[Y_1](1+\beta\alpha_1 V_1) + g_2\mathrm{E}[Y_2](1+\beta\alpha_2 V_2)$$

in which the sensitivity factors α_i ($i=X,1,2$) are functions of β and the standard deviations of all the random variables. The coefficient of variation V_X is determined from

$$V_X^2 = V_m^2 + V_{I_X}^2 + V_m^2 V_{I_X}^2$$

and V_1 and V_2 are determined in a similar manner. Figure 6.1 shows $\mathrm{E}[m]$ as a function of the characteristic live-load-to-dead-load-effect ratio for several parameter assumptions and for a normally distributed live load of varying coefficient of variation. The characteristic values were judiciously chosen as the 98% fractile of the loads; as a result, the coefficient of variation of the live load has only little influence on the mean resistance. NKB (1978) further gives similar curves for a live load distributed as the largest in a sample of size 10 and 100. These curves lie within the range of those shown in Fig. 6.1, and it was judged acceptable in calibration to represent all live loads by the particular case of a normal distribution with the coefficient of variation V_{Y_2} equal to 0.3.

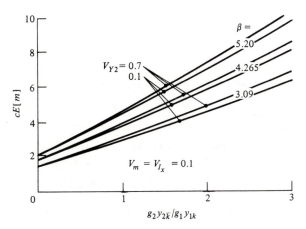

Figure 6.1 Mean resistance versus live-load-to-dead-load-effect ratio, Level II design.

The sensitivity factors α_i do not vary very much in practical applications. As in previous studies (Ravindra et al., 1974) the α_i were next assigned a set of constant values α_i^0 for a particular value of the reliability index, subject to $\sum(\alpha_i^0)^2 = a^2 \geqslant 1$. The Nordic recommendations prescribe values of β ranging from 3.1 (less serious failure consequences and type I failure) to 5.2 (very serious failure consequences and type III failure), with the β-value 4.265 representing the average or common case. For this value of β, the α_i^0-vector was "determined so that a design performed by this fixed vector on an average yields the same dimensions as a design performed by application of the varying α_i-vector . . . ." Moreover, it was decided that α_1^0 should equal zero, so that a convenient dead load factor value of 1.00 would result. This gave $\alpha_X^0 = 0.850$ and $\alpha_2^0 = 0.585$; Fig. 6.2 shows the value of $cE[m]$ compared with the approximation.

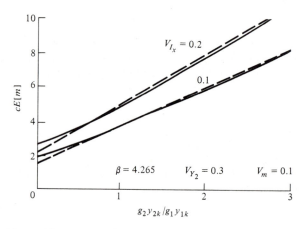

Figure 6.2 Mean resistance versus live-load-to-dead-load-effect ratio, comparison of Level I and Level II design.

Denoting partial safety factors by γ and characteristic values by a subscript c, a Level I design according to this model is expressed as

$$cm_c/\gamma_m = g_1 y_{1c}\gamma_1 + g_2 y_{2c}\gamma_2$$

The characteristic values are

$$m_c = E[m]\exp(-1.645 V_m) \quad 5\% \text{ fractile}$$
$$y_{1c} = E[Y_1] \quad\quad\quad\quad\quad\quad \text{dead load}$$
$$y_{2c} = E[Y_2](1 + 2.054 V_{Y_2}) \quad 98\% \text{ fractile}$$

Inserting the values $\beta = 4.265$, $V_m = V_{I_X} = 0.1$, $V_{Y_1} = V_{I_{Y_1}} = 0.05$, $V_{Y_2} = 0.3$, and $V_{I_{Y_2}} = 0.323$ gives the load factors $\gamma_1 = 1.00$ for the dead load and $\gamma_2 = 1.30$ for the live load. On the observation that curves such as those in Fig. 6.1 are close to the set of straight lines concurrent at a point on the abcissa, it is reasonable to decide that the load factors shall be constant, independent of β. After some algebra, it is found that this implies that $\beta\alpha_2^0$ is almost constant, equal to 2.5 (actually varying from 2.42 to 2.54 for V_{Y_2} varying from 0.2 to 0.7). This gives $\alpha_X^0 = (a^2 - 6.25/\beta^2)^{1/2}$, and the strength partial safety factor

$$\gamma_m = \exp(\beta V_X \alpha_X^0) \exp(-1.645 V_m)$$

Finally, the value of a is assigned so as to give Level I designs that closely approximate Level II designs; the following approximation was chosen (NKB, 1978),

$$a = 1 + 2\left(\frac{\beta - 2.5}{100 V_X}\right)^2$$

The NKB also recommended that α_X^0 should not be less than 0.7 and never greater than 1.0.

6.2 ESTIMATION OF DISTRIBUTION PARAMETERS

The calibration of a structural design code also presents some formidable problems of a different kind: How are the observed data on loads to be translated into the design loads to be tabulated in the code? Similarly, how are the observations of material strength that are generally obtainable to be translated into nominal strength values specified by the designer? And how is the supplier's compliance with the strength specifications to be reckoned? These questions are ultimately statistical problems, concerned with the representation of empirical data and testing of hypotheses. In this and the following section some practical approaches to these statistical problems are examined.

Consider, to be specific, the problem of selecting the design snow load in a locality where there are n observations $(x_1, x_2, ..., x_n)$ of an annual maximum snow load. The classical approach is to choose a family F of mathematical distributions, such as the normal distributions. This family is parameterized; for the class of normal distributions the problem is reduced to estimating the mean and variance, which is done routinely by calculating the sample mean $\bar{x}$, (3.11), and the sample variance $s_{\bar{x}}^2$, (3.12). These are known to be *unbiased estimators* of the parameters, meaning that the expected value of the error, considered as a random variable over the set of all such samples, is zero. Remembering that sampling uncertainty contributes to the total uncertainty in design, the value of the variance of the snow load is increased, calculated as the sum of the intrinsic variance of the distribution of the snow load and the sampling variance.

This classical approach may be unsatisfactory for at least two reasons. First, unbiased estimators are not necessarily optimal, as

one can understand from an example of a single-parameter family
such as the exponential. If one wishes to fit an exponential distribution to the snow data, the sample mean $\bar{x}$ could be used as an
unbiased estimator for the distribution parameter μ. But the sample
standard deviation s_X could equally well be used as an estimator
for μ — and this would generally lead to a different result. Figure
6.3 shows the situation; all exponential distributions fall on the
locus $\mu = \sigma$. A particular set of sample statistics is represented by
point A. Several plausible estimators are represented by points B,
C, and D in the figure.

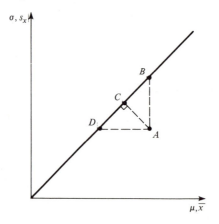

Figure 6.3 Illustration of estimators for exponential distribution
parameter.

One can similarly ask whether it is best to estimate the two
parameters of the normal distribution by the first two sample
moments only — or whether higher order moments should also be
taken into account. Second, the distribution assumption can have a
significant influence on the final design because of "tail sensitivity." The probability of failure is influenced particularly by
higher-valued deviations in the loads and lower-valued deviations in
the strength; the most probable failure situation occurs near the
design point. The failure probability ought to be very sensitive to
the highest sample values and insensitive to the lowest sample
values of the load. Now, consider a structure with normally distributed resistance with mean value μ_R and standard deviation σ_R,
and with normally distributed load. The mean value and standard
deviation of the load distribution are taken as the sample statistics,
thus ignoring the statistical uncertainty in these estimates. The
estimated probability of failure becomes

$$P_F = \Phi\left(-\frac{\mu_R - \bar{x}}{\sqrt{\sigma_R^2 + s_X^2}}\right) \tag{6.1}$$

It is observed that this expression is symmetric in the observations x_i; all carry equal weight. Indeed, if a sample shows moderate dispersion, except that it contains one very low load value, this value tends to increase σ_X relatively more than it lowers $\bar{x}$. As a result, this occurrence may actually raise the calculated failure probability P_F in (6.1), although by all reason it ought to lower P_F slightly.

The weakness of the classical approach is that a particular distribution type is *assumed*. It is quite a different matter to *approximate* the unknown load distribution by a particular distribution. Indeed, for the purpose of evaluating reliability, many distributions can be represented effectively by a normal distribution fitted at the design point as shown in Section 5.2. No procedure is generally accepted for a "best" normal tail approximation of a distribution from a sample.

Present practice mostly applies the classical approach and includes statistical uncertainty in the estimates. When the samples do not contain the above mentioned "outliers", it is a practical experience that this procedure works satisfactorily.

6.3 SELECTION OF CHARACTERISTIC STRENGTH LEVEL

In Level I analysis, the information about any basic random variable is represented by only one parameter, called the *characteristic value* of the random variable. The characteristic value r_c of a material strength R is usually chosen to equal a certain fractile r_α of the distribution of R. Alternatively, some codes prescribe a specified multiple, k, of standard deviations below the mean strength. If the strength is normally distributed, the two approaches are equivalent when k and α satisfy $\alpha = \Phi(-k)$. If the strength deviates from the normal distribution the fractile for a given k, and hence the reliability, will generally be different.

The design value r_d was defined in Chapter 3 as the characteristic value multiplied by a safety factor, φ, generally smaller than one. The design value r_d therefore corresponds to a lower fractile in the distribution for R than r_c. Let the design value be a multiple, k_d, of standard deviations below the mean strength. For a normally distributed strength the design value is thus the $\Phi(-k_d)$-fractile, and it follows that k_d is related to k, φ, and the coefficient of variation V_R as

$$\mu_R - k_d \sigma_R \;=\; \varphi(\mu_R - k\sigma_R) \tag{6.2}$$

leading to

$$k_d \;=\; \frac{1 - \varphi(1 - kV_R)}{V_R} \tag{6.3}$$

It is possible to select a particular fractile for the characteristic value such that the reliability is insensitive to distribution assumptions and parameters. This circumstance is of fundamental importance in the justification of Level I design. Two different materials, A and B, may have different strength distributions and yet have identical design strength. If so, they are necessarily treated as identical in Level I design. If the code admits design in both materials, it is necessary that substitution of one material for the other does not change the reliability.

To determine the design strength fractile such that designs in the two different materials have the same reliability under the same loading, let the loading S be represented by a normal distribution (or normal tail approximation) with mean value μ_S and standard deviation σ_S. Moreover, let the strength of material A be represented by a normally distributed variable with mean value μ_R and standard deviation σ_R, while material B has a normal representation with mean value $\mu_R + \Delta\mu_R$ and standard deviation $\sigma_R + \Delta\sigma_R$. Since the design strengths are equal, $r_d = \mu_R - k_d\sigma_R = \mu_R + \Delta\mu_R - k_d(\sigma_R + \Delta\sigma_R)$, from which

$$\Delta\mu_R \;=\; k_d \Delta\sigma_R \tag{6.4}$$

The reliability of a design in material A under the loading S equals $\Phi(\beta)$, with

$$\beta \;=\; \frac{\mu_R - \mu_S}{\sqrt{\sigma_R^2 + \sigma_S^2}} \tag{6.5}$$

while a design in material B has reliability $\Phi(\beta + \Delta\beta)$, with

$$\Delta\beta \;=\; \frac{\Delta\mu_R}{\sqrt{\sigma_R^2 + \sigma_S^2}} - \frac{(\mu_R - \mu_S)\sigma_R}{(\sigma_R^2 + \sigma_S^2)^{3/2}}\Delta\sigma_R \tag{6.6}$$

neglecting terms of higher order in $\Delta\sigma_R$. Using (6.4) and setting $\Delta\beta = 0$ gives

$$k_d \;=\; \frac{\beta}{\sqrt{1 + \sigma_S^2/\sigma_R^2}} \;=\; \alpha_R\beta \;=\; \frac{\beta}{\sqrt{1 + V_S^2/(\Theta V_R)^2}} \tag{6.7}$$

in which

$$\alpha_R \;=\; \frac{\sigma_R}{\sqrt{\sigma_R^2 + \sigma_S^2}} \tag{6.8}$$

$$\Theta = \frac{\mu_R}{\mu_S} = \frac{1 + \sqrt{\beta^2 V_R^2 + \beta^2 V_S^2 - \beta^4 V_R^2 V_S^2}}{1 - \beta^2 V_R^2} \qquad (6.9)$$

Θ is called the *central safety factor*. One can conclude that the design value r_d must satisfy

$$r_d = \mu_R - k_d \sigma_R = \mu_R - \beta \alpha_R \sigma_R \qquad (6.10)$$

if the reliability is to be constant, equal to $\Phi(\beta)$, for small variations of the parameters of the normal distribution (or normal tail approximation) of the strength distribution. Equation (6.10) shows that r_d equals the value of R at the design point of a Level II reliability analysis with failure function $g(r,s) = r - s$. r_d is the $\Phi(-k_d)$-fractile of the normal tail approximation of R. Since the R-distribution coincides with its normal tail approximation at the design point, r_d is also the $\Phi(-k_d)$-fractile of the R-distribution.

Now, it has been established in the context of the principle of normal tail approximation that the reliability is practically invariant under moderate changes in distribution shape, as long as the distribution and its density are held constant at the design point. Together with the present result, this means that the reliability for a given load distribution is constant to a first order of approximation under small changes of the strength distribution, as long as the $\Phi(-\alpha_R \beta)$-fractile remains at the design point.

In practice, the load distribution varies from member to member according to application, since, for example, the dead-load-to-live-load ratio varies. A change in σ_S influences α_R and hence k_d and the design point fractile. Fortunately, the practical range of variation of V_S is limited and the specific standardized safety margin k_d is quite insensitive to such variation (see Table 6.2). The table shows, for example, that the 1% fractile is optimum for a material with $V_R = 0.15$ when $V_S = 0.20$ and $\beta = 2.5$. For some materials μ_R can vary widely but V_R is almost constant. Then, any value of k close to k_d can be used as characteristic value, combined with a φ-value to satisfy (6.3). If V_R varies appreciably $k = k_d$ should be used with $\varphi = 1$.

TABLE 6.2	Optimal $\Phi(-k_d)$	Fractiles of Strength	R for $\beta = 2.5$	
V_S	0.10	0.20	0.30	0.40
$V_R = 0.1$	0.012	0.022	0.028	0.030
0.15	0.008	0.010	0.012	0.012
0.20	0.007	0.007	0.008	0.008
0.25	0.006	0.006	0.007	0.007
0.30	0.006	0.006	0.006	0.006

Example 6.3

A material is used in a context of varying loadings; the governing load is represented by a normal tail approximation with coefficient of variation V_S that varies between 0.1 and 0.5, with a mode of about 0.3. These values and their assumed relative frequency, f, are shown in the left-hand column of Table 6.3. They include the effects of model uncertainty and variations in influence factors, while fluctuations in strength due to fabrication, strength analysis, etc., are included in the strength model. The material occurs in several qualities and types, with varying mean strength μ_R and coefficient of variation varying between 0.15 and 0.30, a typical value being 0.25. It is desired to achieve a uniform reliability corresponding to the target reliability index $\beta^* = 3.00$. The exact central safety factor Θ by (6.9) for $\beta = 3.00$ is shown in Table 6.3.

TABLE 6.3	Separation of Safety Factors					
				Strength		
Loading			V_R	0.15	0.20	0.25	0.30
V_S	f	Θ_S	Θ_R	1.954	2.456	3.631	8.460
0.1	0.2	1.118	Θ	1.911	2.572	4.059	10.050
			Θ_p	2.185	2.746	4.059	9.463
			β	3.46	3.13	3.00	2.98
			P_F	2.7×10^{-4}	8.8×10^{-4}	13.5×10^{-4}	14.5×10^{-4}
0.3	0.6	1.233	Θ	2.409	3.027	4.474	10.430
			Θ_p	2.409	3.027	4.475	10.431
			β	3.00	3.00	3.00	3.00
			P_F	13.5×10^{-4}	13.5×10^{-4}	13.5×10^{-4}	13.5×10^{-4}
0.5	0.2	1.413	Θ	3.026	3.659	5.129	11.118
			Θ_p	2.761	3.470	5.129	11.956
			β	2.71	2.89	3.00	3.03
			P_F	33.5×10^{-4}	19.5×10^{-4}	13.5×10^{-4}	12.5×10^{-4}
Global β				2.939	2.992	3.000	3.001
Total P_F				16.6×10^{-4}	13.9×10^{-4}	13.5×10^{-4}	13.5×10^{-4}

If it is chosen to calculate the design value of R on the basis of $\beta = 3.00$ and an assumed "average" value of $V_S = 0.3$, (6.7), (6.9) and (6.10), give the central resistance factors $\Theta_R = \mu_R / r_d$ as tabulated. A similar procedure, assuming a constant value $V_R = 0.25$, gives the central load factors $\Theta_S = s_d / \mu_S$ as shown. In design these assumptions together give the central safety factor $\Theta_p = \Theta_R \Theta_S$ shown for comparison in the table. Using this value instead of Θ gives the reliability index β shown for comparison in Table 6.3. The table shows indeed that a systematic error is introduced by ignoring the influence of V_S on Θ_R and of V_R on Θ_S. As a result, the safety margin is exaggerated for a combination of low V_R and low V_S, and too small for low V_R and high V_S. For a material with $V_R = 0.15$, for example, this would lead to systematic overdesign for structures governed by load combinations of small dispersion and underdesign for large V_S.

But the net effect of these errors is very small, as shown by the last two lines in the table. The probability of failure and the corresponding β are practically independent of the material parameter V_R. It can be concluded that judicious factoring of Θ in the form

$$\Theta(V_R,V_S) \; = \; \Theta_R(V_R)\Theta_S(V_S)$$

is a permissible approximation in Level I design.

6.4 CONTROL OF CHARACTERISTIC STRENGTH

Except for natural soils, the material strength parameters are generally determined after the design of the structure, during the construction phase. While load parameters are prescribed quantities, strength parameter values are presumed by the designer and subsequently verified. And, unlike loads, materials are furnished by a supplier who has an economic interest partially conflicting with the owner. Also, unlike the case of loads, the outcome of the statistical process is a decision: whether or not the material has acceptable strength.

Using the terminology of elementary statistical quality control, consider, following Ofverbeck (1980), a population of elements, each characterized by the value of a relevant variable, R, such as cylinder strength or percent elongation. A certain value of R is specified. The specified value is most often taken to be the same as the characteristic value r_c assumed in design. This is also assumed here, but they may be different. For example, in the testing of wood it is common standard practice to use a specified 5% fractile regardless of application; but the characteristic value depends on the desired reliability in a particular design application. Now, it is not feasible to test all elements; instead, the population is partitioned into *lots* and a sample of elements is taken at random from each lot and tested. The test results are inserted into the *control rule,* which is a two-valued function that tells whether to *accept* or *reject* the lot.

It is the aim of the control that no more than a prescribed fraction, p_0, of the accepted elements are to be defective, i.e., have the outcome of R falling below r_c. For a particular lot, let the true fraction of elements with strength less than r_c be p, the *fraction defective* of the lot. Ideally, the control procedure should lead to acceptance of all lots with $p \leqslant p_0$, and to rejection of all lots with $p > p_0$, but no amount of sample control can accomplish that. Because p is unknown, there is a probability of a *type 1 error,* i.e., the error of accepting a lot that should have been rejected. At the same time, there is also a probability of a *type 2 error,* namely, rejecting a lot that should have been accepted.

The probability of acceptance P_a depends on the fraction defective. It may also depend on the parameters of the distribution of R. If P_a is a function only of the fraction defective, $P_a = P_a(p)$, this function is called the *operating characteristic* function (OC function) of the control rule. This will be the case if the class of distributions of R is a single-parameter family of functions. It may also be the case for multiparameter classes of distributions, and the control rule is then said to be *parameter-free.*

Since p is unknown, one cannot guarantee that all accepted lots have $p \leqslant p_0$. However, one can demand that the expected value of p is at most equal to p_0 in accepted lots. This means that in the new population of all elements in all accepted lots there are no more than the proportion p_0 of defective elements,

$$E[p \mid \text{accept}] \leqslant p_0 \qquad (6.11)$$

The proportion of defectives p varies from lot to lot. The probability density function of the fraction defective, p, is denoted by $f_p(p)$ and the probability density function of p for accepted lots is proportional to $f_p(p) P_a(p)$. Thus, the expected value of p for accepted lots is

$$E[p \mid \text{accept}] = \frac{\int_0^1 p\, f_p(p)\, P_a(p)\, dp}{\int_0^1 f_p(p)\, P_a(p)\, dp} \qquad (6.12)$$

Equation (6.12) shows that (6.11) does not completely characterize the OC function, but the control rule may be optimized with (6.11) as a constraint. The probability density function of the fraction defective is usually unknown. In some cases it may be estimated from production control records. Since the probability of acceptance of a randomly chosen lot is close to 1, i.e., the nominator in (6.12) is close to 1, a sufficient condition for (6.11) is

$$P_a(p) \leqslant \frac{p_0}{p}, \quad 0 < p < 1 \qquad (6.13)$$

As Ofverbeck (1980) points out, (6.13) can be taken as a "safe region" for operating characteristic functions, in the sense that any control rule whose OC function satisfies (6.13) is severe enough to maintain safety in the long run regardless of the quality level of the production. It should be recognized that there may be other, equally important safety considerations besides this "safety in the long run". In the following, three control rules are presented.

6.4.1 Control by Attributes

Suppose that the only information used for a sample of n elements is the number, k, of elements for which the strength is less than the specified value r_c. The control process is classified as *control by attributes*. The control rule is of the form

$$\text{accept iff} \quad r_{(q)} > r_c \qquad (6.14)$$

in which $r_{(q)}$ is the qth smallest value in the sample. Since k is binomially distributed, one gets the OC function

$$P_a(p) = \sum_{i=0}^{q-1} \binom{n}{i} p^i (1-p)^{n-i} \qquad (6.15)$$

showing that the control rule is distribution-free.

The principal advantage of this method is that it is distribution-free, requiring no prior assumption about the distribution. It is also often an advantage that it is unnecessary to determine the value of the strength for all elements in the sample. Instead, one may test for a fixed load r_c and merely count the failures. The main disadvantage can be inferred from the form of the OC function. Since q in (6.14) must be one of the integers $1, 2, \ldots, n$, it is only possible to obtain n different OC curves, and for small sample size n, none of these may resemble the desired one. Also, for small n, not even the choice $q = 1$ may be strict enough, thus forcing an increased sample size.

6.4.2 Control by Variables

If the control rule depends on all variables $x_1, \ldots, x_n$ in a sample, the process is classified as *control by variables*. It is generally necessary to assume that R is restricted to a parametric family of distributions, but the computations tend to be tedious in the general case; in practice, it is commonly assumed that the variable follows the normal distribution, or that some simple monotonic transformation does. Often, $\log X$ is used, reflecting a lognormal distribution assumption. If the distribution is assumed normal with mean and variance unknown, a p_0-fractile with confidence level K is given by, [(3.15)]

$$r_{sp_0} = \bar{r} + k(n)s_R \qquad (6.16)$$

in terms of the usual sample statistics and $k(n)$, that depends on p_0 and K. Chapter 3 gives some values of $k(n)$, and an approximation to $k(n)$ is shown in (3.22). Acceptance occurs if r_{sp_0} is greater than the specified value r_c

$$\text{accept iff} \quad \bar{r} + k(n)s_R > r_c \tag{6.17}$$

As explained in Chapter 3 the random variable $\bar{R} + k(n)\,S_R$ is approximately normally distributed with mean value and standard deviation as in (3.19) and (3.20). The control rule may be shown to be parameter free, and the OC function can be approximated as (Hald, 1952)

$$P_a(p) \approx 1 - \Phi\left|\frac{u_p - k(n)}{\left(\dfrac{1}{n} + \dfrac{k(n)^2}{2(n-1)}\right)^{1/2}}\right| \tag{6.18}$$

where $u_p = (r_c - \mu)/\sigma = \Phi^{-1}(p)$. For $p = p_0$ this probability is $1 - K$ [(3.21)]. Fig. 6.4 shows two examples of the OC function computed from (6.18). As the sample size n increases the OC function approaches a step function.

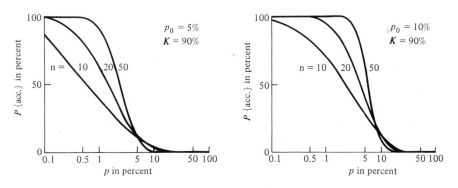

Figure 6.4 Examples of OC function for normal distribution.

If the distribution is assumed to be lognormal, $Y = \log X$ is normally distributed. Thus with sample statistics $\bar{y}$ and s_Y,

$$\bar{y} = \frac{1}{n}\sum_{i=1}^{n}\log x_i \tag{6.19}$$

$$s_Y^2 = \frac{1}{n-1}\sum_{i=1}^{n}(\log x_i - \bar{y})^2 \tag{6.20}$$

a p_0 fractile of R with confidence level K is

$$r_{sp_0} = \exp(\bar{y} + k(n)s_Y) \tag{6.21}$$

with $k(n)$ as in (6.16). Equations (6.19) to (6.21) are called the lognormal limits; it is, unfortunately, common practice to use other expressions based on sample mean and sample coefficient of variation of the sample values directly which, at best, are bad approximations (Ofverbeck, 1980).

The normal and exact lognormal limits and the corresponding control rules make the most effective use of information contained in the example. This, however, entails a high sensitivity to any deviation from the assumed class of distributions. If the sample size is small, such deviation is particularly influential because $k(n)$ is large (see Table 3.1) and s_R is highly dispersed. Ofverbeck (1980) studied the effect of misuse of the normal and lognormal limits by taking many samples of size 10 from different distributions and applying (6.16) and (6.21) to estimate the 75% confidence limits of the 5% fractile $(k_{0.05}(10) = -2.1)$. Simulation of 1000 samples gave the results shown in Table 6.4, expressed conveniently in terms of the true confidence level attained, i.e., the frequency with which the computed characteristic value underestimates the true 5% fractile. This value should be 75% ideally. If it is lower, it means that there is a tendency to give characteristic values that are too high. The so-called "approximate lognormal" limit $\bar{r} \exp(k(n)s_R /\bar{r})$ is seen to be unacceptable.

TABLE 6.4 Confidence Levels in Percent					
Limit ($n = 10$, $K = 75\%$, $p_0 = 0.05$)		Normal Limit	Exact Lognormal	Approx. Lognormal	Power Limit
Equation		(6.16)	(6.21)		(6.22)
Actual Distribution	V_R				
Normal	0.1	75*	66	64	-
	0.3	75*	-	36	-
	0.5	75*	-	8	-
Lognormal	0.1	83	75*	73	83
	0.3	91	75*	67	82
	0.5	95	75*	61	81
Weibull	0.1	57	51	46	77
	0.3	72	52	36	77
	0.5	89	53	21	76
Truncated	0.006	-	-	-	73
normal	0.077	-	-	-	72
	0.16	-	-	-	73
Rectangular		-	-	-	77
Exponential		-	-	-	74
ST1312 yield strength (8946 obs.)		83	79	-	80
HT36 yield strength (6259 obs.)		87	83	-	80
C15 concrete compr. str. (5267 obs.)		91	83	-	80
C40 concrete compr. str. (63878 obs.)		83	76	-	79

Source: Data extracted from Ofverbeck (1980).

* Exact

6.4.3 The Power Limit

Control by attributes employs only a minimum of information in the sample. Control by variables employs a maximum of the information (most of which is relevant to the quantile only through the detailed assumptions in the model) but is sensitive to deviations from the distribution assumptions. As a practical compromise Ofverbeck (1980) suggested retaining only the q lowest values $r_{(1)}, \ldots, r_{(q)}$ in the sample of size n and defined the *power limit* r_p as

$$r_p = r_{(q)}^{1-\xi} \prod_{i=1}^{q-1} r_{(i)}^{\xi/(q-1)} \qquad (6.22)$$

where ξ depends on the fractile, the confidence level, the sample size and the value of q. The choice of q reflects a compromise between utilization of sample information and minimization of distribution sensitivity. Suggested values of q and corresponding value of ξ are given in Table 6.5.

TABLE 6.5 Parameters of the Power Limit (6.22) for the 5% Fractile 75% Confidence Level

n	5	6	7	8	9	10	12	14
q	2	2	2	2	2	3	3	3
ξ	5.93	5.35	4.85	4.42	4.03	3.31	2.96	2.66

n	16	18	20	25	30	40	50
q	3	3	4	4	5	6	7
ξ	2.41	2.19	2.22	1.86	1.80	1.58	1.44

Source: Data extracted from Ofverbeck (1980).

The power limit is the p_0-fractile for confidence level K for the class of power distributions $F(r) = (r/r_0)^k$, $0 \leq r \leq r_0$, $k > 0$. Power distributions are of little intrinsic interest, but they are excellent approximations to the tails of a very large class of distributions as k ranges from zero to infinity. Table 6.4 shows the general good performance of the power limit for $n = 10$ for samples from various theoretical distributions and for real strength data.

Example 6.4

In a sample of size $n = 10$ of the 28-day compressive cylinder strength of a batch of concrete, the three lowest values were 18.7, 19.8, and 20.1 MPa. By Table 6.5, $q = 3$ and $\xi = 3.31$. The 5% power limit for 75% confidence is thus

$$r_p = (r_{(1)} r_{(2)} \cdots)^{\xi/(q-1)} r_{(q)}^{1-\xi} = (18.7 \times 19.8)^{3.31/(3-1)} 20.1^{-2.31} = 17.4 \text{ MPa}$$

6.5 SUMMARY

Assigning values to the parameters in a code format to achieve a target reliability of structures (free of gross error) is called calibration. A code may be calibrated by judgment, by successive corrections based on experience, or by more formal methods. Code optimization is a formal method of calibration, in which an average deviation of the reliability measure from a target is minimized for the class of structures within the scope of the code.

The distributions of the strength and load parameters are of unknown type, but modeled by standard mathematical distributions selected as a compromise between good fit and simple calculation. The straightforward approach to fit a normal or a lognormal distribution to a set of sample values by using the sample mean and variance is discussed.

In Level I analysis the information about any basic random variable is represented by a single characteristic value. This value must be chosen at a fractile appropriate for the dispersions and the target reliabilities as given in Section 6.3, in order that the resultant designs be insensitive to deviations from the distributions assumed.

The statistical problem of estimating a characteristic strength or load value for a particular population on the basis of a sample is considered in Section 6.4. Some of the conventional approaches are shown to have shortcomings and the power limit is introduced as an alternative.

REFERENCES

ELLINGWOOD, B. et al., "Development of a Probability Based Load Criterion for American National Standard A58," *National Bureau of Standards Publication 577,* Washington, D. C., 1980.

HALD, A., *Statistical Theory with Engineering Applications,* John Wiley, New York, 1952.

NBCC (National Building Code of Canada), National Research Council of Canada, Ottawa, Ontario, 1975, 1977, 1980.

NKB (The Nordic Committee on Building Regulations), "Recommendations for Loading and Safety Regulations for Structural Design," *NKB-Report,* No. 36, Copenhagen, November 1978.

NOWAK, A. S. and N. C. LIND, "Practical Bridge Code Calibration," *Journal of the Structural Division,* ASCE, Vol. 105, 1979, pp. 2497-2510.

OFVERBECK, P., "Small Sample Control and Structural Safety," Report TVBK-3009, Department of Structural Engineering, Lund Institute of Technology, Lund, Sweden, 1980.

OHBDC (Ontario Highway Bridge Design Code), Ontario Ministry of Transportation and Communication, Downsview, Ontario, 1983.

RAVINDRA, M. K. and N. C. LIND, "Theory of Structural Code Optimization," *Journal of the Structural Division,* ASCE, Vol. 99, 1973, pp. 541-553.

RAVINDRA, M. K. and N. C. LIND, "Trends in Safety Factor Optimization," *Beams and Beam Columns,* ed. R. Narayanan, Applied Science Publishers, Barking, Essex, UK, 1983, pp. 207-236.

RAVINDRA, M. K., N. C. LIND and W. W. SIU, "Illustrations of Reliability-Based Design," *Journal of the Structural Division,* ASCE, Vol. 100, 1974, pp. 1789-1811.

ROSENBLUETH, E. and L. ESTEVA, "Reliability Basis for Some Mexican Codes," *ACI Publication SP-31,* 1972, pp. 1-41.

SCHWARTZ, M. W., et al., "Load Combination Methodology Development, Load Combination Program: Project II Final Report," NUREG/CR-2087, UCRL-53025, Lawrence Livermore Laboratory, Cal., July 1981.

SIU, W. W., S. R. PARIMI and N. C. LIND, "Practical Approach to Code Calibration," *Journal of the Structural Division,* ASCE, Vol. 101, 1975, pp. 1469-1480.

THOFT-CHRISTENSEN, P. and M. BAKER, *Structural Reliability Theory and Its Applications,* Springer-Verlag, Berlin, 1982.

EXTREME VALUE DISTRIBUTION AND STOCHASTIC PROCESSES

7.1 TWO DISTRIBUTION-FREE RESULTS

In many practical problems involving statistical information, the underlying distribution functions are not known, and it is therefore of interest to base conclusions on results that do not depend on the distribution functions. Here only two such distribution-free results are discussed, namely one relating to the problem of empirical determination of the "tail" of a distribution function and another to the probability of exceeding the mth largest test observation in a given number of future trials. These results are discussed in more detail by Gumbel (1958) and Hogg and Craig (1966), who also give further references.

7.1.1 Order Statistics

Let the continuous stochastic variable X have the distribution function $F(x)$ and the probability density function $f(x)$. Now consider n independent observations $x_1 \leqslant x_2 \leqslant \cdots \leqslant x_n$ taken in order of increasing magnitude. The probability density function $f_m(x)$ of the stochastic variable X_m, where m indicates the *order*, follows from the observation that there are $m-1$ independent stochastic variables less than and $n-m$ larger than X_m.

$$f_m(x) = \frac{n!}{(m-1)!(n-m)!} F(x)^{m-1}[1 - F(x)]^{n-m} f(x) \qquad (7.1)$$

For most of the commonly used distributions, (7.1) does not give useful closed-form results for the statistics of X_m, such as the mean value and variance of X_m. However, if the function $F(X_m)$ is considered as a stochastic variable instead of X_m, simple formulas are found for the expectation and variance. With this in mind, (7.1) can be written in the form

$$f_m(x_m)dx_m = m \begin{vmatrix} n \\ m \end{vmatrix} F(x_m)^{m-1}[1 - F(x_m)]^{n-m} dF(x_m) \qquad (7.2)$$

Multiplication of (7.2) with $F(x_m)$ and integration over the probability mass then gives the expectation of $F(X_m)$:

$$E[F(X_m)] = m \begin{vmatrix} n \\ m \end{vmatrix} \int_0^1 F^m[1-F]^{n-m} dF \qquad (7.3)$$

$$= m \begin{vmatrix} n \\ m \end{vmatrix} \frac{m!(n-m)!}{(n+1)!} = \frac{m}{n+1}$$

Similarly, the variance of $F(X_m)$ is found to be

$$Var[F(X_m)] = E[F(X_m)^2] - E[F(X_m)]^2 \qquad (7.4)$$

$$= \frac{m(m+1)}{(n+1)(n+2)} - \frac{m^2}{(n+1)^2}$$

$$= \frac{1}{n+2} E[F(X_m)](1 - E[F(X_m)])$$

The results (7.3) and (7.4) are useful when a number of independent observations $x_1, x_2, \ldots, x_n$ are plotted in order to estimate the distribution function $F(x)$ empirically. Equation (7.3) says that the expected value of the distribution function evaluated at the observation of order m is equal to $m/(n+1)$ and thereby suggests plotting the points $(x_m, m/(n+1))$, $m = 1, \ldots, n$ (Fig. 7.1). The result (7.4) says that the variance of $F(X_m)$, i.e., the ordinate of the plotting point, attains the maximum value $1/4(n+2)$ at the median and decreases symmetrically toward the ends of the interval $[0,1]$.

7.1.2 Quantile Predictions

Another simple distribution-free result concerns the probability of exceeding the observation of order m from a sample of n independent observations M times in N future trials. This problem is fundamental for estimating strength and load parameters directly from observations without any assumptions regarding the underlying

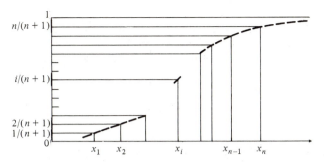

Figure 7.1 Plotting positions $(x_i, i/(n+1))$ for n observations.

distribution.

In terms of the distribution function $F(x)$, the conditional probability of exceeding X_m, given that $X_m = x_m$, in one trial is $1 - F(x_m)$. According to the binomial formula the probability of exceeding x_m exactly M times in N trials is then

$$P(N,M) = \binom{N}{M}[1 - F(x_m)]^M F(x_m)^{N-M} \tag{7.5}$$

The unconditional probability can be found by use of the probability density function $f_m(x_m)$ from (7.2).

$$p(n,m;N,M) = E[P(N,M)] \tag{7.6}$$

$$= m\binom{n}{m}\binom{N}{M}\int_0^1 F^{m+N-M-1}(1-F)^{n-m+M}\,dF$$

$$= m\binom{n}{m}\binom{N}{M}\frac{(m+N-M-1)!(n-m+M)!}{(n+N)!}$$

$$= \frac{\binom{N-M+m-1}{m-1}\binom{n-m+M}{M}}{\binom{n+N}{n}}$$

Given n observations, it follows in particular from (7.6) that an additional observation, $N=1$, is less than X_m with probability

$$p(n,m;1,0) = \frac{m}{n+1} \tag{7.7}$$

This result states that a new observation of a continuous stochastic variable X has equal probability of falling in any of the $n+1$ intervals in which the previous n observations divide the interval of definition of X. This further supports the use of the plotting position of Fig. 7.1.

Another result of immediate interest in safety considerations is the probability that the largest observations in a test series with n observations will not be exceeded in N future trials. This probability follows from (7.6) as

$$p(n,n;N,0) = \frac{n}{N+n} \tag{7.8}$$

Usually, the size of the test series is not large compared with the desired number of future observations, and it then follows from (7.8) that there is a considerable probability of exceeding the largest test observation. Due to the symmetry of formula (7.6), the probability of future observations less than x_1 is the same as the probability of observations larger than x_n.

It is seen from the simple result (7.8) that it would require a very extensive test series to determine limits that are only exceeded with very low probability without making any assumptions about an underlying distribution function. Often such assumptions about the distribution function will be based on smoothing and interpolation of the data, if possible in combination with knowledge of an underlying mechanism. The next section considers distribution functions with special relevance to extremes.

Example 7.1

Design wind speeds are often based on a quantile of the distribution of the maximum wind speed in one year. The quantile exceeded with probability $1/n$ is usually called the n-year wind. This wind speed is not directly related to the maximum wind in an n-year period. In fact, the probability of exceeding the n-year wind in a k-year period is

$$P_{n,k} = 1 - \left(1 - \frac{1}{n}\right)^k$$

if the annual maximum wind speeds are considered to be independent. For $1/n \ll 1$, $P_{n,k}$ can be written in the form

$$P_{n,k} = 1 - \exp\left|k \log\left(1 - \frac{1}{n}\right)\right| \approx 1 - \exp\left(-\frac{k}{n}\right)$$

Thus, in particular, the probability of exceeding the n-year wind in n years asymptotically approaches the value $1 - e^{-1} = 0.63$ for large n.

7.2 EXTREME-VALUE DISTRIBUTIONS

Let $F(x)$ be the distribution function of a stochastic variable X, and consider n independent trials. The distribution function $F_n(x_n)$ of the largest observation X_n is

$$F_n(x) \; = \; F(x)^n \tag{7.9}$$

For differentiable distribution functions the corresponding density function is

$$f_n(x) \; = \; nF(x)^{n-1}f(x) \tag{7.10}$$

This is recognized as a special case of (7.1).

Formulas (7.9) and (7.10) are illustrated in Fig. 7.2 for the standard normal distribution $\Phi(x)$. It follows from (7.9) that for any value of x, where $0 < F(x) < 1$, the distribution function $F_n(x)$ decreases with increasing sample size n. This implies that the expected value of X_n increases with increasing sample size n. In the particular example of Fig. 7.2 the steepness of $F_n(x)$ increases with n leading to decreasing variance of X_n, but this is not generally true.

Some distribution functions only change position and length scale under the transformation (7.9), and the distributions of the

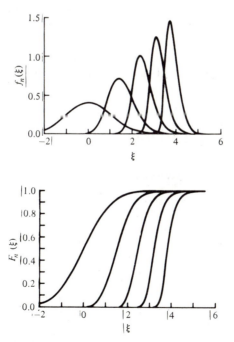

Figure 7.2 Distribution of the largest of n independent normal variables, $n = 1, 10, \ldots, 10^4$.

maximum value drawn from the corresponding populations therefore have a number of simple properties. Furthermore, the distributions are approached asymptotically with increasing sample size under rather weak restrictions on the initial distributions. These distribution functions are therefore important in describing maximum values, even when the individual observations follow a different distribution. The specific results as well as the general conclusions are easily applied to minimum values by use of the identity $\min\{x_i\} = \max\{-x_i\}$.

The extreme-value distributions are characterized by being invariant with respect to sample size apart from changes in position and length-scale parameters. The distribution of the maximum of a sample of size n is given by (7.9). Thus, if $F(x)$ is an extreme-value distribution, $F_n(x)$ can be obtained from $F(x)$ by translation and scaling of the argument. Extreme-value distributions therefore satisfy a relation of the form

$$F\left(\frac{x - \alpha_n}{\beta_n}\right) = F(x)^n, \quad \beta_n > 0 \qquad (7.11)$$

where α_n and β_n may depend on n. This relation was investigated by Fisher and Tippett (1928), who first observed that when the two arguments in (7.11) are equal, either $F(x) = 0$ or $F(x) = 1$. This leads to three types of distributions. The first is characterized by $\beta_n = 1$. In this case the arguments in (7.11) are never identical, and the interval of definition of X is infinite. If $\beta_n \neq 1$, the point of identical argument can correspond to either $F(x) = 0$ or $F(x) = 1$, i.e., either a finite lower or upper limit. The three distributions for maximum values are given in Table 7.1 for easy reference. Table 7.2 gives the corresponding functions for minimum values. μ_n is the expected value, and σ_n^2 is the variance.

TABLE 7.1	Asymptotic Maximum Distributions		
Type	1	2	3
$F(x)$	$\exp[-\exp(-x)]$ $-\infty < x < \infty$	$\exp(-x^{-k})$ $0 < x < \infty$	$\exp[-(-x)^k]$ $-\infty < x < 0$
α_n β_n	$\log n$ 1	0 $n^{1/k}$	0 $n^{-1/k}$
$\mu_n - \alpha_n$ σ_n^2	$\gamma = 0.5772$ $\pi^2/6$	$n^{1/k}\Gamma(1 - \frac{1}{k})$ $n^{2/k}[\Gamma(1 - \frac{2}{k}) - \Gamma^2(1 - \frac{1}{k})]$	$-n^{-1/k}\Gamma(1 + \frac{1}{k})$ $n^{-2/k}[\Gamma(1 + \frac{2}{k}) - \Gamma^2(1 + \frac{1}{k})]$

The type 1 distribution, also called the exponential type, has a constant length scale, while the position parameter α_n increases with $\log n$ in the case of maxima. The expected value μ_n follows

TABLE 7.2	Asymptotic Minimum Distributions		
Type	1	2	3
$F(x)$	$1 - \exp(-\exp x)$ $-\infty < x < \infty$	$1 - \exp[-(-x)^{-k}]$ $-\infty < x < 0$	$1 - \exp(-x^k)$ $0 < x < \infty$
α_n β_n	$\log n$ 1	0 $n^{1/k}$	0 $n^{-1/k}$
$\mu_n - \alpha_n$ σ_n^2	$-\gamma$ $\pi^2/6$	$-n^{1/k}\Gamma(1 - \frac{1}{k})$ $n^{2/k}[\Gamma(1 - \frac{2}{k}) - \Gamma^2(1 - \frac{1}{k})]$	$n^{-1/k}\Gamma(1 + \frac{1}{k})$ $n^{-2/k}[\Gamma(1 + \frac{2}{k}) - \Gamma^2(1 + \frac{1}{k})]$

the position parameter and is equal to Euler's constant γ for $n = 1$. This distribution of the maximum value is approached asymptotically when the stochastic variable X is unlimited, and $1 - F(x)$ approaches zero exponentially for large values of x. The normal distribution satisfies these conditions, but the convergence to the asymptotic distribution is rather slow. The slow convergence appears in Fig. 7.2 as a change of shape of distribution function with sample size n. The type 1 extreme distribution is often used for the maximum wind speed or wind pressure. It is also related to the gust factor, discussed in Section 10.1.

The asymptotic extreme distribution of type 2 follows from that of type 1 by the logarithmic transformation $x = k \log y$. For a given value of k only moments of order $n < k$ are finite. Although this by itself is not enough to disqualify the distribution, it turns out to be of less practical interest (Johnson, 1953).

The distribution of type 3 concerns upward bounded stochastic variables. The length scale is proportional to $n^{-1/k}$, and when the limit is taken to be zero, the expected value μ_n and the standard deviation σ_n are proportional to $n^{-1/k}$. This implies that the coefficient of variation is independent of n. The parameter k is determined by the coefficient of variation. From Tables 7.1 and 7.2 follows

$$\left(\frac{\sigma_n}{\mu_n}\right)^2 = \frac{\Gamma(1 + 2/k)}{\Gamma^2(1 + 1/k)} - 1 \tag{7.12}$$

When this distribution is used for the minimum value, it is sometimes called the *Weibull distribution,* and it contains the well-known exponential and Rayleigh distributions as special cases for $k = 1$ and $k = 2$, respectively. It is often used to model phenomena that can be produced by a "weakest link" mechanism such as brittle failure and fatigue failure (Weibull, 1951).

In Tables 7.1 and 7.2 the asymptotic extreme distributions are given in standardized form. Evidently, the asymptotic properties are not changed when a linear function of x is used as argument of the original distribution function. This generalization as well as the relation between the extreme distribution types have been discussed by Gumbel (1958).

Example 7.2

The hazard function $\rho(t)$ was introduced in Section 1.2 as the probability of failure at time t conditional on survival until time t. In terms of the distribution function $F(t)$ for the lifetime of any of the members of the original population,

$$\rho(t) = \frac{f(t)}{1 - F(t)} = -\frac{d}{dt}\log[1 - F(t)]$$

Alternatively, $F(t)$ can be expressed in terms of the hazard function

$$1 - F(t) = \exp\left[-\int_0^t \rho(\tau)d\tau\right]$$

If $\rho(\tau)$ is a power function, e.g., $\rho(\tau)$ is proportional to τ^{k-1}, then $F(t)$ is the Weibull distribution function as seen from Table 7.2. In fact, Weibull offered the power function representation of the hazard function as an explanation of the usefulness of this distribution (Weibull, 1951).

7.3 DISCRETE PROCESSES

Many phenomena can be described by events taking place at discrete points in time generated by some stochastic mechanism. They are conveniently described by one model generating the points and, if necessary, another model for the event. The simplest case is that of points generated independent of previous history, the *Poisson process* (Parzen, 1962). For this widely studied process some results will be given concerning waiting times, interarrival times, correlation properties, and extremes. These results find direct application when the Poisson process is used to model loads. They also find application to extremes of continuous processes encountered in connection with structural response to earthquake, wave, and wind loads. The concept of a renewal process, i.e., a discrete process that retains memory of the latest event, is also introduced.

7.3.1 The Simple Poisson Process

Let $N(t)$ be a stochastic variable denoting the number of points in the interval $(0,t]$. $N(t)$ may be considered as a stochastic function of the deterministic variable t. A particular member of the sample space is called a realization, in the present case a step-function as shown in Fig. 7.3. When $N(t)$ is considered as a function of t, it is called a *stochastic process*.

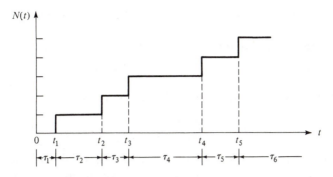

Figure 7.3 Poisson process with waiting times t_n and interarrival times τ_n.

The stochastic process $N(t)$ is called a *Poisson process* when the points t_i are generated according to the following mechanism. The probability of exactly one event occurring in the interval $(t, t + \Delta t]$ is asymptotically proportional to the length Δt of the interval, while the probability of more than one event in the interval is a higher-order term for $\Delta t \to 0$. Furthermore, events in disjoint intervals are stochastically independent. The parameter $v(t)$, determined by

$$v(t) = \lim_{\Delta t \to 0} \frac{1}{\Delta t} P(\text{one event in } (t, t + \Delta t])$$

is called the *intensity,* and it defines the Poisson process completely. If v is a constant, the corresponding process is called *homogeneous,* while the process is called nonhomogeneous if v is a function of t.

Let $P_n(t)$ be the probability of exactly n events having occurred in the interval $(0, t]$. $P_n(t)$ is determined by observing that if n events have occurred at t, either n or $n-1$ events must have occurred at $t - \Delta t$.

$$P_n(t) = P_n(t - \Delta t)(1 - v\Delta t) + P_{n-1}(t - \Delta t)v\Delta t + o(\Delta t) \quad (7.13)$$

By dividing by Δt and taking the limit $\Delta t \to 0$, the following differential equation is obtained:

$$\frac{d}{dt} P_n(t) + v\, P_n(t) = v\, P_{n-1}(t) \quad (7.14)$$

The probability of no events, $P_0(t)$, follows from (7.14) with $P_{-1}(t) = 0$,

$$P_0(t) = \exp\left[-\int_0^t v(\tau)d\tau\right] \quad (7.15)$$

This formula also determines the distribution of the waiting time T_1 until the first event, because the probability of $T_1 > t$ is equal to $P_0(t)$.

$$F_{T_1}(t) = 1 - \exp\left[-\int_0^t v(\tau)d\tau\right] \quad (7.16)$$

As discussed in Example 7.2, $v(t)$ can be interpreted as a hazard function.

It is convenient to introduce the notation

$$m(t) = \int_0^t v(\tau)\,d\tau, \quad t > 0 \tag{7.17}$$

and it is easily verified that the solution to (7.14) is

$$P_n(t) = \frac{m(t)^n}{n!} e^{-m(t)}, \quad t > 0 \tag{7.18}$$

Thus the number of events in the interval $(0,t]$ follows the Poisson distribution with expected value and variance given by $m(t)$.

$$E[N(t)] = \text{Var}[N(t)] = m(t) \tag{7.19}$$

The fact that only the integral of $v(t)$ appears in (7.18) is a consequence of the independence of the events.

7.3.2 Waiting Time and Interarrival Time

In safety and reliability considerations waiting times and interarrival times are of considerable interest. The *waiting time* T_n is defined as the time of the nth event (Fig. 7.3). The distribution of T_n can be found by a simple extension of the argument leading to (7.16). The event $T_n > t$ is characterized by $N(t) < n$. For the Poisson process it follows from (7.18) that

$$1 - F_{T_n}(t) = \sum_{k=0}^{n-1} \frac{m(t)^k}{k!} e^{-m(t)} = \int_{m(t)}^{\infty} \frac{z^{n-1}}{(n-1)!} e^{-z}\,dz \tag{7.20}$$

Thus the probability density function for T_n is

$$f_{T_n}(t) = \frac{v(t)m(t)^{n-1}}{(n-1)!} e^{-m(t)}, \quad t \geq 0 \tag{7.21}$$

For *constant* v, T_n is gamma-distributed with parameters n and $1/v$. An alternative statement of this is that $2vT_n$ is chi-square-distributed with $2n$ degrees of freedom (Hogg and Craig, 1966). The expectation and variance are

$$E[T_n] = n/v, \quad \text{Var}[T_n] = n/v^2 \tag{7.22}$$

The time differences between consecutive events are measured by the interarrival times $\tau_n = T_n - T_{n-1}$, where $\tau_1 = T_1$. The distribution function $F_{\tau_n}(s)$ for τ_n is found by first considering the event $\tau_n \leq s$ conditional on $T_{n-1} = t$ and then integrating over t using (7.21). In the Poisson process the events are independent of previous history, and the conditional distribution function $F_{\tau_n \mid T_{n-1}}(s \mid t)$ therefore follows from (7.16) by simple change of integration limits,

$$1 - F_{\tau_n \mid T_{n-1}}(s \mid t) = \exp\left[-\int_t^{t+s} v(\tau)d\tau\right] \tag{7.23}$$

$$= \exp[-m(t+s) + m(t)]$$

Use of (7.21) then gives the unconditional distribution function.

$$1 - F_{\tau_n}(s) = \int_0^\infty [1 - F_{\tau_n \mid T_{n-1}}(s \mid t)]\, f_{T_{n-1}}(t)dt \tag{7.24}$$

$$= \int_0^\infty \frac{v(t)m(t)^{n-2}}{(n-2)!}\, e^{-m(t+s)}dt$$

For constant intensity v the conditional distribution function (7.23) is independent of t, and (7.24) reduces to

$$F_{\tau_n}(s) = 1 - e^{-vs} \tag{7.25}$$

i.e., all the interarrival times τ_n are exponentially distributed with expectation $E[\tau_n] = 1/v$. This is the same as the waiting time T_1 in accordance with the lack of memory of the Poisson process. The modification of this result in the case of a renewal process is discussed in Section 7.3.5.

Example 7.3

Assume that changes of occupancy in a particular type of building occur according to a homogeneous Poisson process. The intensity v of changes for a single building can be estimated from a survey covering k buildings over the time t by the maximum likelihood method (Hogg and Craig, 1966). The idea is to express the probability of the observed event $L(v)$ with v as a free parameter and then maximize this probability with respect to v. Let the observed number of changes be n. The intensity of the total number of changes is kv, and the *likelihood function* $L(v)$ then follows from (7.18) with $m(t) = kvt$

$$L(v) = \frac{(kvt)^n}{n!}\, e^{-kvt}$$

Maximizing $L(v)$ with respect to v yields the *maximum likelihood estimate* $v = n/kt$. In this case the same estimate would have resulted from use of the expectation (7.19).

7.3.3 Filtered Poisson Processes

A useful extension of the simple Poisson process is obtained by attributing stochastic properties to the events. Following Parzen (1962), the stochastic process

$$X(t) = \sum_{k=1}^{N(t)} w(t, t_k, Y_k) \tag{7.26}$$

is called a *filtered Poisson process* when $N(t)$ is a simple Poisson process generating the points t_k. $w(t,\tau,Y)$ is a given response function and Y_k are independent stochastic variables. The stochastic variables Y_k may be vectors. $w(t,\tau,Y)$ is defined to be zero for $t < \tau$, and the event described by $w(t,\tau,Y)$ is thus initiated at $t = \tau$. The sum (7.26) can then be written in the alternative form

$$X(t) = \int_0^\infty w(t,\tau,Y)\, dN(\tau) \tag{7.27}$$

General results can be obtained for the expectation $E[X(t)]$ and the covariance $\text{Cov}[X(t),X(s)]$. They follow from similar results for the simple Poisson process (Cox and Miller, 1977). With the notation $\Delta N(t) = N(t+\Delta t) - N(t)$ it follows from the definition of the intensity $v(t)$ that

$$E[\Delta N(t)] \approx v(t)\Delta t \tag{7.28}$$

and from the independence of previous events

$$E[\Delta N(t)\Delta N(s)] = P(\Delta N(t) = \Delta N(s) = 1) \tag{7.29}$$

$$= P(\Delta N(t) = 1 \mid \Delta N(s) = 1)\, v(s)\Delta s$$

$$= [\delta(t-s) + v(t)]\Delta t\, v(s)\Delta s$$

Here $\delta(t-s)$ is the Dirac delta function. Combinations of these results give the covariance function of the Poisson process

$$\text{Cov}[\Delta N(s)\Delta N(t)] = \delta(t-s)v(t)\Delta t\,\Delta s \tag{7.30}$$

As could be expected from the independence of previous events, the covariance function is zero for $t \neq s$.

The expectation of the filtered Poisson process $X(t)$ follows directly from (7.27) and (7.28),

$$E[X(t)] = \int_0^\infty v(\tau)\, E[w(t,\tau,Y)]\, d\tau \tag{7.31}$$

The covariance function follows similarly from (7.27) to (7.30),

$$\text{Cov}[X(t),X(s)] = \int_0^\infty v(t)\, E[w(t,\tau,Y)\, w(s,\tau,Y)]\, d\tau \tag{7.32}$$

In (7.31) and (7.32) the upper limit can be replaced by t and $\min(t,s)$, respectively.

Provided that the form of the response function $w(t,\tau,Y)$ allows the lower limit of (7.27) to be taken as $-\infty$, similar results are valid for that case. When, furthermore, the response function depends on t and τ only through the difference $t - \tau$ and v is constant, the process $X(t)$ is homogeneous and stationary. More generally a stochastic process is called *strongly stationary* when it is invariant under a translation of the time parameter. A stochastic process is called *weakly stationary* if the expectation and the covariance function are independent of a time translation.

While the results (7.31) and (7.32) cover a rather general class of processes, they are restricted to the first two moments. Extensions to higher-order moments and even the distribution function of $X(t)$ are possible (Parzen, 1962; Cox and Miller, 1977), but the general result is quite complicated and does not give the distribution of the extreme value of $X(t)$ within a finite interval of t. Before returning to this important question, the results (7.31) and (7.32) are illustrated by an example.

Example 7.4

Consider a filtered Poisson process with rectangular pulses of length a and height Y_k (Fig. 7.4). The response function $w(t,\tau,Y)$ is

$$w(t,\tau,Y) = \begin{cases} Y, & 0 < t - \tau < a \\ 0, & t < \tau \text{ or } \tau + a < t \end{cases}$$

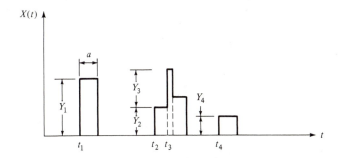

Figure 7.4 Filtered Poisson process with pulses of equal length.

Let the intensity ν be a constant. The expectation now follows from (7.31). For $0 < t < a$ the process is still influenced by the start at $t = 0$, while for $t > a$ the expectation becomes independent of t. With the notation $T = \min(t,a)$,

$$E[X(t)] = E[Y]\nu T$$

In (7.32) for the covariance function the initiation time τ for the observed events at t and s is the same, and the formula contains only one stochastic variable, Y. The covariance function is symmetric in t and s, and it is therefore no restriction to assume that $t \geqslant s$. The evaluation of the integral is illustrated in Fig. 7.5. Contributions to the integral are obtained when $0 < \tau < s$ and $\tau > t - a$. The result is then found to be $E[Y^2]\nu$ times the length of the interval $[\max(0, t - a), s]$

$$\text{Cov}[X(t),X(s)] = \begin{cases} 0, & t - s > T \\ E[Y^2]\nu(T + s - t), & t - s \leqslant T \end{cases}$$

The covariance is extended symmetrically for $s > t$.

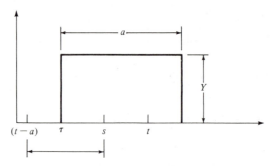

Figure 7.5 The integral (7.32) receives contributions from the indicated interval, $\max(0, t-a) < \tau < s$.

In this particular case the distribution function $F_X(x)$ can be obtained directly. The event $X(t) \leqslant x$ can also be identified by $\sum Y_k \leqslant x$, where the summation extends to all responses initiated in the time interval $(\max(0, t-a), t]$. The probability of the event $X(t) \leqslant x$ can then be evaluated by summation of the similar probabilities under the condition of n initiated responses

$$F_X(x) = \sum_{n=0}^{\infty} P\left(\sum_{k=1}^{n} Y_k \leqslant x \right) \frac{(vT)^n}{n!} e^{-vT}$$

where again $T = \min(t, a)$. Clearly, the possibility of obtaining a closed-form expression depends on the distribution of Y. Here only two cases are considered: exponentially distributed stochastic variables Y and the deterministic value $Y = c$. In both these cases it is convenient to find the finite probability of $X(t) = 0$, and then to evaluate the probability density function of $f_X(x)$. From the first term

$$P(X(t) = 0) = e^{-vT}$$

and from the remaining terms

$$f_X(x) = \sum_{n=1}^{\infty} f_{X \mid N}(x \mid n) \frac{(vT)^n}{n!} e^{-vT}, \quad 0 < x$$

In the case of exponentially distributed stochastic variables Y_k with expectation $1/\lambda$, the sum of n of these is gamma-distributed [see (7.21)], with $m = vt$. Thus

$$f_{X \mid N}(x \mid n) = \frac{\lambda^n x^{n-1}}{(n-1)!} e^{-\lambda x}$$

and upon substitution with $k = n - 1$

$$f_X(x) = \lambda v T \exp(-vT - \lambda x) \sum_{k=0}^{\infty} \frac{(vT\lambda x)^k}{k!(k+1)!}$$

$$= \lambda \sqrt{vT/\lambda x} \, \exp(-vT - \lambda x) \, I_1(2\sqrt{vT\lambda x})$$

$I_1(\ )$ is the modified Bessel function of order 1, and the summation follows from the series representation given, e.g., by Abramowitz and Stegun (1965). When $Y = c$, X/c can only take integer values, and the probability density function $f_X(x)$ giving the probability of $X = x$ is defined only for $x/c = 0, 1, 2, \ldots$:

$$f_X(x) = \frac{(vT)^{x/c}}{(x/c)!} e^{-vT}, \quad x/c = 0, 1, 2, \ldots$$

Thus, in this case, where $P(Y = c) = 1$, X/c follows a Poisson distribution.

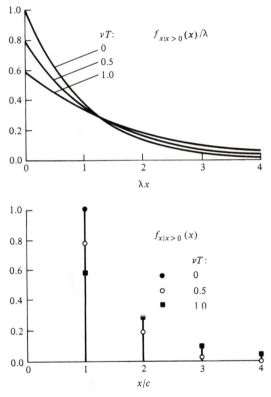

Figure 7.6 Conditional probability density as function of normalized levels $x\lambda$, x/c for normalized duration $vT = 0, 0.5, 1.0$.

Although the expectation and covariance of the process $X(t)$ will be the same in the two cases, when $c = 1/\lambda$ the distributions of $X(t)$ are quite different. This is illustrated in Fig. 7.6, which shows the probability density function for $X(t)$ under the condition $X(t) > 0$ for various intensities of the Poisson process. The limiting form for $vT \to 0$ is the probability density function for Y. It is seen that the influence of interaction is greatest for the concentrated distribution and that the tails of the distributions are more sensitive to interaction than the main probability mass.

7.3.4 Two Distributions of Extremes

A problem of great importance in safety considerations is that of the first crossing by a stochastic process of a given curve. The stochastic process may represent a load, and the curve may represent a strength. The intersection then represents the event of failure. The special case of the first intersection of a constant level is equivalent to the problem of the distribution of the maximum of a stochastic process within a given interval. This is seen by observing that the probability of the maximum remaining below the level is equal to the probability of being below the level initially and not crossing within the interval.

While the filtered Poisson process treated in Example 7.4 readily allows the determination of the correlation function and the distribution function for $X(t)$ for any particular value of t, no solution to the first-passage problem is available. The complication lies in describing mathematically that a passage is indeed the first passage within a given interval.

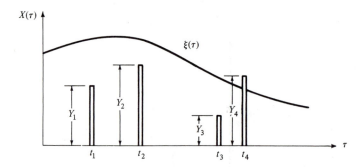

Figure 7.7 First passage of a Poisson "spike" process.

The complication of correlation disappears in the limit of "spikes," i.e., $a = 0$. Then the problem consists in determining the probability that all the spikes of height Y_k in the interval $0 < \tau < t$ are below a given curve $\xi(\tau)$ (Fig. 7.7). This probability follows as a special result from the observation that the spikes exceeding $\xi(\tau)$ may be considered as events in a new Poisson process with intensity

$$\nu_\xi(\tau) = \nu(\tau)[1 - F_Y(\xi)] \qquad (7.33)$$

The probability of no exceedences in the interval $(0,t]$ is then given by (7.15) with $\nu(\tau)$ replaced by $\nu_\xi(\tau)$. When ξ is a constant level, this probability equals the distribution function for the maximum value of the stochastic process $X(t)$ in the interval $(0,t]$

$$F_{max}(\xi) = \exp\left[-\int_0^t (1 - F_Y(\xi))v(\tau)d\tau\right] \qquad (7.34)$$

This formula allows the distribution function $F_Y(\ )$ to depend on τ.

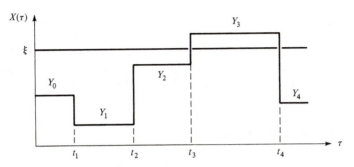

Figure 7.8   First passage of a Poisson square-wave process.

Another variant of the Poisson process that allows a closed-form expression for the extreme value within a given interval is the *Poisson square-wave process*, illustrated in Fig. 7.8. This process is generated by a simple Poisson process with intensity $v(t)$. At each event t_k the value of $X(t)$ is redefined as $X(t_k+) = Y_k$, where Y_k are independent, identically distributed stochastic variables. The expectation is

$$E[X(t)] = E[Y] \qquad (7.35)$$

and the covariance function follows from observing that $X(t)$ and $X(s)$ are identical if there are no points t_k between t and s and are independent if there are one or more points generated. Thus, by use of (7.17) and (7.18),

$$Cov[X(t),X(s)] = Var[Y]\exp\left[-\int_s^t v(\tau)d\tau\right] \qquad (7.36)$$

where it has been assumed that $s \leq t$.

The distribution function $F_{max}(\xi)$ for $\max X(t)$, $0 \leq \tau \leq t$, is found from the distribution function for the maximum under the condition that exactly n changes have taken place in the interval $0 < \tau \leq t$. Because the stochastic variables Y_k are independent and identically distributed, this distribution function is $F(\xi)^n$, and $F_{max}(\xi)$ follows by summation and use of (7.18).

$$F_{\max}(\xi) = F_Y(\xi)\, e^{-m(t)} \sum_{n=0}^{\infty} \frac{1}{n!}[m(t)F_Y(\xi)]^n \tag{7.37}$$

$$= F_Y(\xi)\exp\left[-(1-F_Y(\xi))\int_0^t \nu(\tau)\,d\tau\right]$$

The factor $F_Y(\xi)$ is identified as the probability of $Y_0 \leqslant \xi$, i.e., a kind of initial condition.

Both of the results (7.34) and (7.37) have been used to approximate the distribution of the maximum value of a continuous process in a given interval. Note that in (7.37) it was assumed that all the exceedence probabilities are equal, while this assumption is not necessary in (7.34). This makes it difficult to generalize (7.37) to nonstationary processes.

7.3.5 Renewal Processes

In some cases a sequence of events occur with independent, identically distributed interarrival times. Such a stochastic process is called a *renewal process*. The stationary Poisson process has independent, exponentially distributed interarrival times and is therefore a special type of renewal process. However, the simplicity of the Poisson process is intimately connected with the complete lack of memory implied by the exponential distribution, and the relation between interarrival times and waiting times must be reconsidered in the general case.

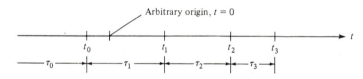

Figure 7.9 Introduction of an arbitrary origin in the interval τ_1.

Let $t_j,\ j = \ldots, -1, 0, 1, \ldots$ denote a sequence of events with interarrival times $\tau_j = t_j - t_{j-1}$, as shown in Fig. 7.9. The stochastic variables τ_j are independent and identically distributed with distribution function $F(\tau)$. In order to obtain a stationary renewal process the time origin must be selected "at random." As all the stochastic variables τ_j are identically distributed, it is no restriction to denote the interval containing the time origin τ_1, but it is important to note that in placing the origin the longer intervals are given preference, and the resultant distribution $F_1(\tau_1)$ of τ_1 is therefore in general different from $F(\tau)$. The relation between $F_1(\tau_1)$ and $F(\tau)$

is most conveniently expressed in terms of the corresponding density functions $f_1(\tau)$ and $f(\tau)$ (Rice and Beer, 1966). The probability of selecting an interval of length τ is proportional with τ, and therefore

$$f_1(\tau) = \frac{\tau}{\mu_\tau} f(\tau) \tag{7.38}$$

The normalizing factor μ_τ is determined by integrating both sides from zero to infinity, whereby

$$\mu_\tau = E[\tau] = \int_0^\infty \tau f(\tau)\, d\tau \tag{7.39}$$

Thus μ_τ is the expectation of τ_j, $j \neq 1$.

The probability density function $f_{T_1}(t)$ of the waiting time T_1 can now be determined via its conditional counterpart $f_{T_1 \mid \tau_1}(t \mid \tau_1)$. For any given length τ_1 of the interval containing the origin its position is arbitrary, and therefore

$$f_{T_1 \mid \tau_1}(t \mid \tau_1) = \frac{1}{\tau_1}, \quad 0 < t < \tau_1 \tag{7.40}$$

The probability density function of τ_1 is given by (7.38), and $f_{T_1}(t)$ then follows by integration.

$$f_{T_1}(t) = \int_0^\infty f_{T_1 \mid \tau_1}(t \mid \tau) f_1(\tau)\, d\tau = \frac{1}{\mu_\tau} \int_t^\infty f(\tau)\, d\tau \tag{7.41}$$

$$= \frac{1}{\mu_\tau} [1 - F(t)]$$

This formula gives the probability density function of the waiting time T_1 to the first event in terms of the distribution function $F(t)$ of the interarrival times. It is an immediate consequence of this formula that $f_{T_1}(t)$ is a nonincreasing function for all values of t.

A simple relation between the moments of T_1 and τ is obtained from (7.41) by multiplication with t^n and integration.

$$E[T_1^n] = \frac{1}{\mu_\tau} \int_0^\infty t^n \int_t^\infty f(\tau)\, d\tau\, dt = \frac{1}{\mu_\tau} \int_0^\infty \frac{t^{n+1}}{n+1} f(t)\, dt \tag{7.42}$$

$$= \frac{1}{n+1} \frac{E[\tau^{n+1}]}{E[\tau]}, \quad n = 1,2,\dots$$

In particular the expectation of T_1 is

$$\mu_{T_1} = \frac{1}{2} \mu_\tau \left[1 + \left(\frac{\sigma_\tau}{\mu_\tau} \right)^2 \right] \tag{7.43}$$

where σ_τ is the standard deviation of τ. Thus the relative magnitude of μ_{T_1} and μ_τ depends on the coefficient of variation of τ, and the Poisson process represents a particular case in which $\mu_{T_1} = \mu_\tau$.

These relations between the waiting time T_1 and the interarrival time τ serve to place the Poisson process within a wider context and are used in connection with a traffic load model in Section 10.4. The results (7.41) and (7.42) remain valid also when the original interval lengths τ_j are not independent, and they have been used to evaluate first-passage times from recurrence times in simulations of continuous stationary processes (Crandall et al., 1966). Renewal processes have also been used to calculate approximate first-passage times of continuous stationary processes (Rice and Beer, 1966). This problem is discussed in Section 7.4.5.

7.4 THE NORMAL PROCESS

The preceding section dealt with discrete events taking place at specific points in time and with some associated discrete stochastic processes. If the phenomenon under consideration, e.g., a wind speed or the deflection of a building member, develops continuously in time, the need arises for a more general type of stochastic process that can describe stochastic properties of continuous functions. The discussion of such processes will be restricted to the so-called "normal process."

A *normal process* $X(t)$ is described by the statistical properties of any set of selected stochastic variables $X(t_1), X(t_2),...,X(t_i),...$. The stochastic process $X(t)$ is said to be normal if any set of stochastic variables $X(t_i)$, $i = 1,2, \ldots$, is of joint normal distribution. This seemingly abstract definition turns out to be quite workable once some properties of joint normal stochastic variables are understood.

7.4.1 The Multidimensional Normal Distribution

The stochastic variable Z is normally distributed if its probability density function is of the form

$$f(z) = \frac{1}{\sigma\sqrt{2\pi}} \exp\left[-\frac{1}{2}\left(\frac{z-\mu}{\sigma}\right)^2\right] \qquad (7.44)$$

Straightforward calculations show that the parameters μ and σ are the expected value and the standard deviation, respectively. It is convenient to introduce the probability density function $\varphi(x)$ and the cumulative distribution function $\Phi(x)$ for a normalized normal variable X, i.e., a normal variable with $\mu = 0$ and $\sigma = 1$.

$$\varphi(y) \;=\; \frac{1}{\sqrt{2\pi}} \exp\!\left(-\frac{1}{2}y^2\right) \tag{7.45}$$

$$\Phi(x) \;=\; \int_{-\infty}^{x} \varphi(y)\,dy \tag{7.46}$$

In the case of independent stochastic variables, the normal distribution is easily generalized to any dimension, but without the assumption of independence a slightly more elaborate procedure is necessary. The approach taken here is similar to the introduction of a linear transformation in Section 4.4.

Consider n independent normalized normal variables X_i, $i = 1, \ldots , n$, and define n new stochastic variables Z_j, $j = 1, \ldots , n$, by the linear transformation

$$Z_j \;=\; \sum_{i=1}^{n} a_{ji} X_i + \mu_j \tag{7.47}$$

From the independence and normalization of the variables X_i, it follows that the expectation of Z_j is

$$E[Z_j] \;=\; \mu_j \tag{7.48}$$

and the covariance between Z_i and Z_k is

$$\mathrm{Cov}[Z_j, Z_k] \;=\; c_{jk} \;=\; \sum_{i=1}^{n} a_{ji} a_{ki} \tag{7.49}$$

It is a simple matter to show that the sum of two independent normal variables is normal, and as a consequence each of the stochastic variables Z_j is normally distributed. The joint probability density function of the variables Z_j, $j = 1, \ldots , n$, follows from that of the variables X_i, $i = 1, \ldots , n$, by a direct variable transformation including the volume element, i.e.,

$$f_{z_1 \ldots z_n}(\mathbf{z})\, dz_1 \cdots dz_n \;=\; \varphi(x_1) \cdots \varphi(x_n)\, dx_1 \cdots dx_n \tag{7.50}$$

$$= (2\pi)^{-n/2} \exp\!\left|-\frac{1}{2}\sum_{j=1}^{n} x_j^2\right| dx_1 \cdots dx_n$$

Upon inversion of (7.47) and introduction of the determinant $|\,c_{jk}\,|$, this relation gives the joint probability density function of the n-dimensional normal distribution

$$f_{z_1 \ldots z_n}(\mathbf{z}) \;=\; \frac{(2\pi)^{-n/2}}{|\,c_{jk}\,|^{1/2}} \exp\!\left|-\frac{1}{2}\sum_{j=1}^{n}\sum_{k=1}^{n}(z_j - \mu_j)c_{jk}^{-1}(z_k - \mu_k)\right| \tag{7.51}$$

c_{jk}^{-1} is the inverse of the covariance matrix c_{jk}.

It is seen that the joint probability density function (7.51) is expressed directly in terms of the expectation vector μ_j and the covariance matrix c_{jk}, and the linear transformation (7.47) is needed only if a normalized decomposition is desired. This implies that a normal stochastic process $X(t)$ is fully determined by its expected value,

$$\mu(t) = E[X(t)] \tag{7.52}$$

and its covariance function,

$$c(t,s) = \text{Cov}[X(t),X(s)] \tag{7.53}$$

7.4.2 The Crossing Problem

A problem of considerable interest in connection with stochastic processes is the crossing of a certain level or barrier by the process. Two simple examples for Poisson processes have already been considered in Section 7.3.4. A similar problem arises in connection with the normal process, e.g., for evaluating wind speeds or deformations and stresses in structures under stochastic load. In the case of discrete processes the actual crossing of the barrier is connected with one of the discrete events in the process, while the crossings of a normal process warrants a closer analysis.

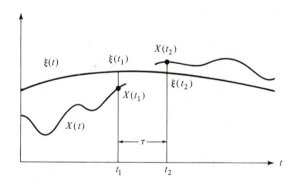

Figure 7.10   The upcrossing of a continuous process.

Consider the situation in Fig. 7.10, where $x(t)$ is a sample function of a stochastic process $X(t)$, and $\xi(t)$ is a specified barrier. Let N_+ denote the number of upcrossings in the interval $[t_1,t_2)$. The probability of an odd number of upcrossings between t_1 and t_2 then is

$$P(N_+ \text{ odd}) = P(X(t_1) < \xi(t_1) \text{ and } X(t_2) > \xi(t_2)) \tag{7.54}$$

The upcrossing rate is now defined as the limit

$$v_\xi^X(t) = \lim_{t_1,t_2 \to t} \frac{1}{|t_2 - t_1|} P(N_+ \text{ odd}) \tag{7.55}$$

if it exists. In the case of continuous processes the correlation between $X(t_1)$ and $X(t_2)$ increases as $\tau = t_2 - t_1 \to 0$, and it is often convenient to express $v_\xi^X(t)$ in terms of the variables

$$X = \frac{1}{2}(X(t_1) + X(t_2)), \quad \widetilde{X} = \frac{1}{\tau}(X(t_2) - X(t_1)) \tag{7.56}$$

$$\xi = \frac{1}{2}(\xi(t_1) + \xi(t_2)), \quad \widetilde{\xi} = \frac{1}{\tau}(\xi(t_2) - \xi(t_1)) \tag{7.57}$$

In terms of the joint probability density function $f_{X\widetilde{X}}(x,\widetilde{x})$ of X and $\widetilde{X}$, the upcrossing rate is

$$v_\xi^X(t) = \lim_{t_1,t_2 \to t} \int_A f_{X\widetilde{X}}(x,\widetilde{x}) \, dx \, d\widetilde{x} \tag{7.58}$$

where the area of integration is shown in Fig. 7.11. In the limit process this area becomes increasingly narrow, and if the mean-value theorem of integral calculus can be used, and the derivative $\dot{X}$ exists, (7.58) takes the form

$$v_\xi^X(t) = \int_\xi^\infty f_{X\dot{X}}(\xi,\dot{x})(\dot{x} - \dot{\xi}) \, d\dot{x} \tag{7.59}$$

This result is known as Rice's formula (Rice, 1944).

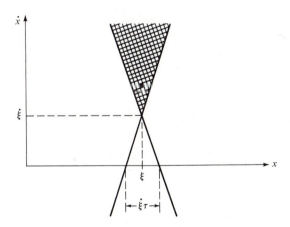

Figure 7.11   Area of integration in (7.58).

In the general case the joint probability function of $X(t)$ and $\dot{X}(t)$ must be obtained from that of $X(t_1)$ and $X(t_2)$ by use of the variable transformation (7.56) followed by a limit process. However, the result for a normal process follows directly from the observation that for any times t_1 and t_2, $X(t_1)$ and $\dot{X}(t_2)$ are jointly normal by virtue of the definition of $\dot{X}(t_2)$. The expectation is

$$E[\dot{X}(t)] = E\left|\lim_{\tau \to 0} \frac{X(t+\tau) - X(t)}{\tau}\right| = \frac{d}{dt}\mu(t) \qquad (7.60)$$

and covariances are by a similar process

$$\text{Cov}[X(t_1),\dot{X}(t_2)] = \frac{\partial}{\partial t_2} c(t_1,t_2) \qquad (7.61)$$

$$\text{Cov}[\dot{X}(t_1),\dot{X}(t_2)] = \frac{\partial^2}{\partial t_1 \partial t_2} c(t_1,t_2) \qquad (7.62)$$

In calculations it is often convenient to replace $X(t)$ and $\xi(t)$ with their normalized equivalents

$$Y(t) = \frac{X(t) - \mu(t)}{\sigma(t)}, \qquad \eta(t) = \frac{\xi(t) - \mu(t)}{\sigma(t)} \qquad (7.63)$$

where $\mu(t)$ and $\sigma(t)$ are the expectation and standard deviation of $X(t)$. By normalization $X(t)$ and $\dot{X}(t)$ become uncorrelated, and this also implies stochastic independence for normal processes. Crossing rates and several other properties of stochastic processes are therefore described more directly by the correlation function

$$\rho(t_1,t_2) = \frac{c(t_1,t_2)}{\sigma(t_1)\sigma(t_2)} \qquad (7.64)$$

Like the covariance function the correlation function is symmetric in its arguments, and it then follows from (7.62) that the variance of the derivative of the normalized process is

$$\omega_0(t)^2 = \frac{\partial^2}{\partial t_1 \partial t_2} \rho(t,t) = \frac{1}{\sigma(t)^2}\left|\frac{\partial^2}{\partial t_1 \partial t_2} c(t_1,t_2) + \dot{\sigma}(t)^2\right| \qquad (7.65)$$

In the particular case of a normal process $X(t)$ the upcrossing frequency is

$$v_\xi^X(t) = v_\eta^Y(t) = \varphi(\eta)\int_{\dot{\eta}}^{\infty} \omega_0^{-1}\varphi(\dot{y}/\omega_0)(\dot{y} - \dot{\eta})\,d\dot{y} \qquad (7.66)$$

$$= \omega_0\,\varphi(\eta)\,\Psi(\dot{\eta}/\omega_0)$$

where the function $\Psi(\zeta)$ is defined as

$$\Psi(\zeta) = \int_{\zeta}^{\infty} (z - \zeta)\varphi(z)\,dz = \varphi(\zeta) - \zeta\Phi(-\zeta) \qquad (7.67)$$

In (7.66) the first factor, $\omega_0(t)$, provides a time scale, the second gives the dependence of the normalized barrier level, and the last gives the dependence on the slope of the normalized level. The function $\Psi(\zeta)$ is shown in Fig. 7.12.

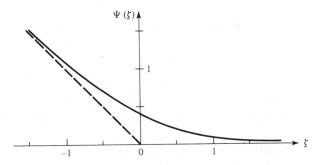

Figure 7.12   The function $\Psi(\zeta) = \varphi(\zeta) - \zeta\Phi(-\zeta)$.

7.4.3 Distribution of Local Extremes

The distribution of the local extremes, e.g., the local maxima, can be found by a procedure similar to that of finding upcrossing rates. The problem will be confined to the determination of the local maxima of the normalized process $Y(t)$ given by (7.63). The occurrence of a maximum implies that $\dot{Y}(t)$ changes from positive to negative, i.e., a downcrossing by $\dot{Y}(t)$ of the zero level. If instead of $\dot{Y}(t)$ the normalized process $\dot{Y}(t)/\omega_0(t)$ is considered, the rate of local maxima follows from analogy with (7.66).

$$\nu_m(t) = \frac{1}{2\pi}\omega_m(t) \tag{7.68}$$

where

$$\omega_m(t)^2 = \frac{\partial^2}{\partial t_1 \partial t_2}\left| \frac{1}{\omega_0(t_1)\omega_0(t_2)} \frac{\partial^2 \rho(t_1,t_2)}{\partial t_1 \partial t_2} \right|_{t_1=t_2=t} \tag{7.69}$$

The occurrence of $\omega_0(t)$ may be considered as a rescaling of time.

The ratio between the rate of zero-upcrossings and the rate of local maxima of the normalized process $Y(t)$ is an important parameter containing information about the regularity of the process. This ratio,

$$\alpha = \frac{\omega_0(t)}{\omega_m(t)} \tag{7.70}$$

is called the regularity factor. It has a particularly simple interpretation for a stationary process. A time interval containing N upcrossings then on average contains N/α local maxima. Let M of these be negative. For symmetric processes there will then also be M positive local minima. The number of zero-upcrossings equals the number of positive maxima minus the number of positive minima, thus giving the equation

$$N = (N/\alpha - M) - M \tag{7.71}$$

The proportion of local maxima that are negative therefore is

$$\alpha \frac{M}{N} = \frac{1-\alpha}{2} \tag{7.72}$$

In the extreme case $\alpha \approx 1$ nearly all local maxima are positive, and the process exhibits a certain regularity associated with a dominance of the frequency v_0^Y. This is discussed further in Section 7.4.4.

The full distribution of the local maxima can be obtained by proceeding as in the case of upcrossings. The occurrence of a local maximum of the stochastic process $Y(t)$ in the interval $[\xi, \xi + d\xi]$ in the time interval $[t, t+dt]$ is described as the joint occurrence of $\xi \leqslant Y(t) < \xi + d\xi$ and a zero-downcrossing of $\dot{Y}(t)/\omega_0(t)$ in the time interval $[t, t+dt]$. The probability density function is then found by normalizing with $v_m(t)dt$, i.e., the probability of the occurrence of a local maximum. The result follows immediately from (7.59) in the form

$$f_m(\xi) = \frac{1}{v_m} \int_{-\infty}^{0} f_{Y,\dot{Y},\ddot{Y}}(\xi,0,z) \mid z \mid dz \tag{7.73}$$

This result is also due to Rice (1944). Computationally, it is again convenient to replace $\ddot{Y}$ with $d(\dot{Y}(t)/\omega_0(t))/dt$ and $\dot{Y}(t)$ with $\dot{Y}(y)/\omega_0(t)$. The variances are

$$\text{Var}[Y] = \text{Var}[\dot{Y}/\omega_0] = 1, \quad \text{Var}\left|\frac{d}{dt}(\dot{Y}/\omega_0)\right| = \omega_m^2 \tag{7.74}$$

and the only nonvanishing covariance is

$$\text{Cov}\left|Y, \frac{d}{dt}(\dot{Y}/\omega_0)\right| = \frac{\partial}{\partial t_2} \frac{1}{\omega_0(t_2)} \frac{\partial}{\partial t_2} \rho(t_1,t_2)_{t_1=t_2=t} \tag{7.75}$$

$$= \frac{1}{\omega_0(t)} \frac{\partial^2}{\partial t_2^2} \rho(t_1,t_2)_{t_1=t_2=t} = -\omega_0(t)$$

In obtaining this result (7.65) is used in connection with $\rho(t,t) \equiv 1$. The correlation coefficient then is $-\alpha$, and for a normal process (7.73) can be expressed by use of (7.51) as

$$f_m(\xi) = \varphi(0) \frac{1}{v_m} \int_{-\infty}^{0} \frac{\mid z \mid}{2\pi \omega_m \sqrt{1-\alpha^2}} \tag{7.76}$$

$$\times \exp\left|-\frac{(\xi^2 + 2\alpha\xi z/\omega_m + (z/\omega_m)^2)}{2(1-\alpha^2)}\right| dz$$

By a change of integration variable to $\eta = z/\omega_m$,

$$f_m(\xi) = \varphi(0)\int_0^{\infty} \frac{1}{\sqrt{1-\alpha^2}} \exp\left| -\frac{1}{2}\xi^2 - \frac{1}{2}\frac{(\eta-\alpha\xi)^2}{1-\alpha^2}\right| \eta\,d\eta \qquad (7.77)$$

$$= \sqrt{1-\alpha^2}\exp\left(-\frac{1}{2}\xi^2\right)\int_{-\alpha\xi/\sqrt{1-\alpha^2}}^{\infty} \varphi(\zeta)\left(\zeta+\frac{\alpha\xi}{\sqrt{1-\alpha^2}}\right)d\zeta$$

The integral has been evaluated in (7.67), and the final result is

$$f_m(\xi) = \sqrt{1-\alpha^2}\,e^{-\xi^2/2}\Psi\left(-\frac{\alpha\xi}{\sqrt{1-\alpha^2}}\right) \qquad (7.78)$$

$$= \sqrt{1-\alpha^2}\,\varphi\left(\frac{\xi}{\sqrt{1-\alpha^2}}\right) + \alpha\xi\,e^{-\xi^2/2}\Phi\left(\frac{\alpha\xi}{\sqrt{1-\alpha^2}}\right)$$

The function $f_m(\xi)$ is shown in Fig. 7.13. It is completely determined by the regularity factor α. The limiting forms are the normal density $\varphi(\xi)$ for $\alpha=0$ and the Rayleigh density for $\alpha=1$. Note that the result (7.72) concerning the fraction of negative maxima can be restated as

$$\int_{-\infty}^{0} f_m(\xi)d\xi = \frac{1-\alpha}{2} \qquad (7.79)$$

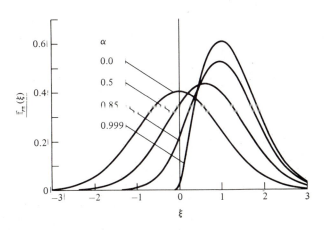

Figure 7.13 Probability density $f_m(\xi)$ of the maxima in a normal process for different values of the regularity factor α.

While the result (7.78) concerns a single local maximum, it is often of more interest to have information about the largest of these in a time interval of length T. In this interval the expected number of local maxima is $N_m = \nu_m T = \nu_0 T/\alpha$, where ν_0 is the zero-upcrossing frequency. Now the procedure is that of Section 7.2

for the extreme-value distribution of N_m trials. With reference to the time interval T, the extreme-value distribution follows, [(7.9)]

$$F_T(\xi) = F_m(\xi)^{N_m} = [1-(1-F_m(\xi))]^{N_m} \qquad (7.80)$$

Integration by parts of (7.78) gives

$$1 - F_m(\xi) = \int_\xi^\infty f_m(\eta)\, d\eta \qquad (7.81)$$

$$= 1 - \Phi\left(\frac{\xi}{\sqrt{1-\alpha^2}}\right) + \alpha\, e^{-\xi^2/2}\, \Phi\left(\frac{\alpha\,\xi}{\sqrt{1-\alpha^2}}\right)$$

By use of the asymptotic result (4.56)

$$\Phi(x) \sim 1 - \varphi(x)(x^{-1} - x^{-3} + \cdots) \qquad (7.82)$$

the asymptotic behavior of (7.81) is found to be

$$1 - F_m(\xi) = \alpha\, e^{-\xi^2/2} + o\left|\xi^{-3}\varphi\left(\frac{\xi}{\sqrt{1-\alpha^2}}\right)\right| \qquad (7.83)$$

The following calculation of asymptotic expressions for the expectation and variance of the extreme value in N_m trials involves simultaneous asymptotics with respect to ξ and N_m, both of which are assumed to be large. Therefore, the variable ξ is replaced by a new variable z, suitably normalized with respect to N_m. The largest of N_m observations is likely to be located around the $1/N_m$ quantile, and the variable

$$z = N_m(1 - F_m(\xi)) = N\, e^{-\xi^2/2} \qquad (7.84)$$

is therefore of order unity for increasing N_m. It is noted that in (7.84) N_m has been estimated by use of the identity $N = \alpha N_m = \nu_0 T$.

An asymptotic expression for $F_T(\xi)$ now follows from (7.80) and (7.84):

$$F_T(\xi) = (1 - z/N_m)^{N_m} = \exp[N_m \log(1 - z/N_m)] \qquad (7.85)$$

$$\approx \exp(-z) = \exp(-N\, e^{-\xi^2/2})$$

Thus asymptotically $F_T(\xi)$ is the extreme-value distribution type 1 in the variable

$$y = \frac{1}{2}\xi^2 - \log N \qquad (7.86)$$

The asymptotic expectation is found by using the expansion

$$\xi \sim \sqrt{2\log N}\left(1 + \frac{y}{2\log N} + \cdots\right) \qquad (7.87)$$

and the extreme type 1 expectation of Table 7.1.

$$\mu_{max} \approx \int_0^1 \xi \, dF_T \approx \sqrt{2 \log N} \int_0^1 (1 + \frac{y}{2 \log N}) \, dF_T \qquad (7.88)$$

$$= \sqrt{2 \log N} + \frac{\gamma}{\sqrt{2 \log N}}$$

where $\gamma = 0.577$ is Euler's constant. The variance also follows from that of the extreme 1 distribution given in Table 7.1.

$$\sigma_{max}^2 \approx \int_0^1 (\xi - \mu_{max})^2 \, dF_T \approx \frac{1}{2 \log N} \int_0^1 (y - \gamma)^2 dF_T \qquad (7.89)$$

$$= \frac{\pi^2}{6} \frac{1}{2 \log N}$$

These results are due to Cartwright and Longuet-Higgins (1956) and Davenport (1964).

The results given in this section have been concerned with a single local extremum or a large sample of these. It is much more difficult to relate one local extremum to the next. There are two aspects of this problem: the time interval between consecutive local extrema and their difference in magnitude, the wave height H. The problem of time distance is closely related to the first-passage problem treated in Section 7.4.5. It can be solved only approximately, and accordingly, correlation between the extreme values is not known explicitly. There is, however, a single explicit expression for the expected wave height of a stationary process (Rice and Beer, 1965). The expected number of extremes in the time T is $2v_m T$, whereby

$$E[H] = \lim_{T \to \infty} \frac{1}{2 v_m T} \int_0^T |\dot{x}(t)| \, dt = \frac{E[|\dot{X}|]}{2 v_m} \qquad (7.90)$$

In particular for a normal process

$$E[H] = \sqrt{2\pi} \, \alpha \, \sigma_X \qquad (7.91)$$

i.e., the expected wave height is proportional to the regularity factor $\alpha = \omega_0/\omega_m$. This dependence is of interest, e.g., in connection with the accumulation of fatigue damage under random load as discussed in Section 9.3.

7.4.4 Narrow-Band Processes and Envelopes

If the regularity factor α is close to unity, nearly all local maxima are greater than the mean of the process. Thus a certain structure of the process may be expected. This is best illustrated by considering a stationary process $X(t)$ and its covariance function $c(\tau)$. Let $c(\tau)$ be given as a Fourier integral

$$c(\tau) = \int_{-\infty}^{\infty} S(\omega) e^{i\omega\tau} d\omega \tag{7.92}$$

$S(\omega)$ is called the spectral density. As $c(\tau)$ is real and even, $S(\omega)$ is also real and even. Furthermore, it can be shown that the spectral properties of $c(\tau)$ as a covariance function imply that $S(\omega)$ is nonnegative (see Miller, 1974, p. 42). This has far-reaching consequences.

Introduce the spectral moments

$$\lambda_j = 2\int_0^{\infty} \omega^j S(\omega) d\omega \tag{7.93}$$

The angular frequencies ω_0 and ω_m then follow from (7.65) and (7.69) in the form

$$\omega_0^2 = \lambda_2/\lambda_0, \qquad \omega_m^2 = \lambda_4/\lambda_2 \tag{7.94}$$

and the regularity factor is

$$\alpha = \frac{\lambda_2}{\sqrt{\lambda_0\lambda_4}} \tag{7.95}$$

The relation between the regularity factor and the spectral density is better illustrated by the relation

$$1 - \alpha^2 = \frac{2}{\lambda_0\lambda_4} \int_0^{\infty} \int_0^{\infty} (\omega_1^2 - \omega_2^2)^2 S(\omega_1) S(\omega_2) d\omega_1 d\omega_2 \tag{7.96}$$

obtained from (7.95) and (7.93) by rearranging the integrals. If $\alpha \approx 1$, it follows from (7.96) that the integrand must be close to zero everywhere; i.e., $S(\omega)$ can only take large values around a particular frequency, ω_0. A stochastic process with $\alpha \approx 1$ is called a *narrow-band process*. By virtue of the representation (7.92), the local character of $S(\omega)$ in turn implies that the covariance function $c(\tau)$ is oscillating with an angular frequency around ω_0.

Due to the dominance of a single frequency in a narrow-band process, the local behavior resembles a harmonic oscillation, and it may be convenient to introduce a representation in terms of an amplitude $R(t)$ and a phase angle $\theta(t)$. The amplitude $R(t)$ is often called the *envelope*. The definition of a representation in terms of an amplitude and a phase angle requires the introduction of an additional process, as illustrated in Fig. 7.14.

Ideally, there should be no directional preference in the phase plane of Fig. 7.14; i.e., the statistical properties of the cartesian coordinates should be independent and identical. However, a simple envelope with a direct physical interpretation results from use of the normalized cartesian coordinates $(Y, \dot{Y}/\omega_0)$, where $Y(t)$ is the

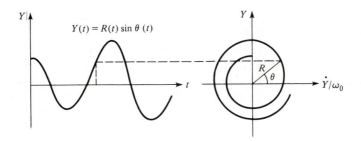

Figure 7.14 Phase-plane representation.

normalized process and $\dot{Y}(t)/\omega_0(t)$ is its normalized time derivative. In terms of the polar coordinates (R,θ),

$$Y(t) = R(t)\sin[\theta(t)] \tag{7.97}$$

$$\dot{Y}(t) = \omega_0 R(t)\cos[\theta(t)] \tag{7.98}$$

whereby

$$R(t)^2 = Y(t)^2 + \dot{Y}(t)^2/\omega_0^2 \tag{7.99}$$

As $Y(t)$ and $\dot{Y}(t)/\omega_0$ are normalized normal variables, it follows from (7.99) that $R(t)$ is Rayleigh distributed,

$$f_R(r) = r\,e^{-r^2/2} \tag{7.100}$$

A more comprehensive description of the envelope requires the determination of a correlation function. It is convenient to consider the covariance of $R(t_1)^2$ and $R(t_2)^2$, which is in the form of expectations of even powers of correlated normal variables for which simple results are available (see Section 8.4.1). In terms of the correlation function $\rho(t_1,t_2)$ of the process $Y(t)$, the correlation function of $R(t)^2$ is

$$k(t_1,t_2)^2 = \frac{\text{Cov}[R(t_1)^2,R(t_2)^2]}{\text{Var}[R(t_1)^2]} \tag{7.101}$$

$$= \frac{1}{2}\left|\rho^2 + \left(\frac{1}{\omega_1}\frac{\partial\rho}{\partial t_1}\right)^2 + \left(\frac{1}{\omega_2}\frac{\partial\rho}{\partial t_2}\right)^2 + \left(\frac{1}{\omega_1\omega_2}\frac{\partial^2\rho}{\partial t_1\partial t_2}\right)^2\right|$$

In the case of a narrow-band stationary process, $\rho'' \approx -\omega_0^2\rho$ and (7.101) takes the form

$$k(\tau)^2 \approx \rho(\tau)^2 + (\rho'(\tau)/\omega_0)^2 \tag{7.102}$$

For a narrow-band process $k(\tau)$ is seen to be the amplitude of the correlation function $\rho(\tau)$, and this implies that $R(t)^2$ is highly correlated over several periods $T = 2\pi/\omega_0$.

Two results concerning the envelope $R(t)$ can be expressed in terms of the function $k(t_1, t_2)$: the joint probability density function of $R(t_1)$ and $R(t_2)$ and the expected rate of upcrossings of a given curve $\eta(t)$ by $R(t)$. In the case of a narrow-band process the variation of $\rho(t_1, t_2)$ may be viewed as an oscillation with the argument $\omega(t_1 - t_2)$ and a slow modulation depending on $\omega(t_1 + t_2)$. When the rate of modulation is considered negligible, the joint probability density of $R(t_1)$ and $R(t_2)$ takes the following form (Rice, 1944):

$$f_{RR}(r_1, r_2) = \frac{r_1 r_2}{1 - k^2} I_0 \left| \frac{k r_1 r_2}{1 - k^2} \right| \exp \left| -\frac{1}{2} \frac{r_1^2 + r_2^2}{1 - k^2} \right| \qquad (7.103)$$

where $I_0(\ )$ is the modified Bessel function of zero order, defined as

$$I_0(z) = \frac{1}{\pi} \int_0^\pi e^{z \cos \theta} \, d\theta \qquad (7.104)$$

The joint probability density function $f_{RR}(r_1, r_2)$ finds application in an approximate method of calculating the first-passage probability of narrow-band processes discussed in Section 7.4.5.

The upcrossing rate of the envelope $R(t)$ plays a major role in another approximate method of calculating first-passage probabilities. It is found by a change of variables followed by use of Rice's formula as discussed in Section 7.4.2. Define the stochastic variables $R(t)$ and $\tilde{R}(t)$ such that

$$R(t_1) = R(t) - \frac{1}{2} \tau \tilde{R}(t), \quad R(t_2) = R(t) + \frac{1}{2} \tau \tilde{R}(t) \qquad (7.105)$$

In the limit $\tau = t_2 - t_1 \to 0$, $k(t_1, t_2) \to 1$. The Bessel function in (7.103) can therefore be replaced by its leading term for large arguments:

$$I_0(z) \sim \frac{e^z}{\sqrt{2\pi z}} [1 + O(z^{-1})] \qquad (7.106)$$

The Jacobi determinant of the transformation (7.105) is τ, and the joint probability density of $R(t)$ and $\dot{R}(t)$ then is

$$f_{R\dot{R}}(r, \dot{r}) = \lim_{\tau \to 0} \tau f_{RR}(r_1, r_2) \qquad (7.107)$$

$$= \lim_{\tau \to 0} \frac{\tau}{\sqrt{2\pi}} \frac{\sqrt{r_1 r_2}}{\sqrt{1 - k^2}} \exp \left| -\frac{1}{4} \frac{(r_1 + r_2)^2}{1 + k} - \frac{1}{4} \frac{(r_1 - r_2)^2}{1 - k} \right|$$

At this point the asymptotic behavior of $k(t_1, t_2)$ for small values of $\tau = t_2 - t_1$ is needed. It follows from (7.101) that

$$k \approx 1 - \frac{1}{2}(\omega_R \tau)^2 \tag{7.108}$$

where

$$\omega_R^2 = \omega_0^2 \frac{1-\alpha^2}{2\alpha^2} \tag{7.109}$$

The limit of (7.107) then is

$$f_{R\dot{R}}(r,\dot{r}) = \frac{r}{\omega_R} e^{-r^2/2} \frac{1}{\sqrt{2\pi}} e^{-(\dot{r}/\omega_R)^2/2} \tag{7.110}$$

$$= \frac{1}{\omega_R} f_R(r) \varphi(\dot{r}/\omega_R)$$

Thus $R(t)$ and $\dot{R}(t)$ are statistically independent and $\dot{R}(t)$ is normally distributed with standard deviation ω_R. The upcrossing frequency of the envelope $R(t)$ thus follows immediately from (7.59).

$$\nu_\eta^R(t) = \omega_R f_R(\eta) \Psi(\dot{\eta}/\omega_R) \tag{7.111}$$

This formula is similar to (7.66) for $\nu_\eta^Y(t)$ but replaces the angular frequency ω_0 with ω_R. For a narrow-band process it follows from (7.109) that $\omega_R \ll \omega_0$; i.e., the time scale of the envelope is much larger than that of the original process.

The definition (7.97) and (7.98) of the envelope in terms of the derivative of the normalized process does not satisfy the ideal requirement that the two cartesian coordinates in the phase plane should be identically distributed. In the case of a stationary process, the spectral density of $\dot{Y}/\omega_0$ is $(\omega/\omega_0)^2 S(\omega)$; i.e., the higher frequencies are amplified. It is possible to define a new process $\hat{X}(t)$ with statistical properties identical to $X(t)$ and such that $\text{Cov}[X(t),\hat{X}(t)]=0$. For a stationary process $X(t)$ the so-called "conjugate process" is given by the Hilbert transformation (Papoulis, 1965, p. 356),

$$\hat{X}(t) = \frac{1}{\pi}\int_{-\infty}^{\infty} \frac{X(s)}{t-s} \, ds \tag{7.112}$$

The effect of a Hilbert transformation is to replace $\cos(\omega t + \varphi)$ with $\sin(\omega t + \varphi)$. This does not change the spectral representation, and an envelope defined by

$$R(t)^2 = X(t)^2 + \hat{X}(t)^2 \tag{7.113}$$

is therefore more smooth than that following from derivatives.

While the general procedure and the results (7.103), (7.110), and (7.111) remain the same, derivatives must be replaced by conjugate functions. Thus the function $k(\ )$ is replaced by a new function given by

$$k(\tau)^2 = \rho(\tau)^2 + \hat{\rho}(\tau)^2 \tag{7.114}$$

and by use of the expansions

$$\rho(\tau) \sim 1 - \frac{1}{2}(\omega_0\tau)^2, \qquad \hat{\rho}(\tau) \sim \lambda\tau \tag{7.115}$$

The angular frequency ω_R in (7.108) is now given by

$$\omega_R^2 = \omega_0^2(1-(\lambda/\omega_0)^2) \tag{7.116}$$

For a stationary process $\lambda=\lambda_1/\lambda_0$, and the role of the regularity factor α is then taken by the new dimensionless parameter

$$\delta = \lambda/\omega_0 = \frac{\lambda_1}{\sqrt{\lambda_0\lambda_2}} \tag{7.117}$$

Comparison with (7.95) shows the appearance of the higher moment λ_4 in the definition of α. This distinction is important in connection with structural response to excitation by a wide-band process.

While the symmetric envelopes defined by (7.99) and (7.113) serve a number of purposes adequately, they do not account for the fraction $(1-\alpha)/2$ of the local extrema placed on the "wrong" side of the mean. It is also clear from (7.100) that $2E[R]$ does not include the expected bandwidth dependence present in the expected wave height $E[H]$. Both of these shortcomings can be overcome by considering the envelope as a local harmonic oscillation around a local equilibrium position (Krenk, 1978). For a stationary process $Y(t)$ the local equilibrium position is

$$Y_0(t) = Y(t) + \ddot{Y}(t)/\omega^2 \tag{7.118}$$

whereby the amplitude is determined by

$$R(t)^2 = (Y(t)-Y_0(t))^2 + (\dot{Y}(t)/\omega)^2 \tag{7.119}$$
$$= (\ddot{Y}(t)/\omega^2)^2 + (\dot{Y}(t)/\omega)^2$$

It follows from (7.74) that if the angular frequency of oscillation is chosen as $\omega=\omega_m$, the two terms in (7.119) are identically distributed, whereby $R(t)$ is Rayleigh distributed

$$f_R(t) = \frac{r}{\alpha^2}\exp\left|-\frac{1}{2}\left(\frac{r}{\alpha}\right)^2\right| \tag{7.120}$$

$Y_0(t)$ is statistically independent of $R(t)$ and normally distributed

$$f_{Y_0}(y) = \frac{1}{\sqrt{1-\alpha^2}}\,\varphi\left(\frac{y}{\sqrt{1-\alpha^2}}\right) \tag{7.121}$$

$Y_0(t)$ and $R(t)$ define the upper and lower envelopes $E_\pm(t)=Y_0(t)\pm R(t)$, and it can be shown that $E_+(t)$ is distributed like the local maxima of $Y(t)$, i.e., as (7.78), and $E_-(t)$ is distributed like the local minima. Furthermore, the use of $2R(t)$ as a measure for the wave height leads to the correct expected value (7.91). Thus (7.119) appears to give an improved estimate of the local amplitude for narrow-band processes.

7.4.5 The First-Passage Problem

A central problem in structural safety analysis is the evaluation of the probability that various forms of structural response, e.g., selected stresses or deformations, remain within some specified limits for a certain time. Alternatively, this may be considered as a problem of finding the probability density function of the first-passage time of the limits. This problem is much more difficult than that associated with mean crossing rates due to the need to account for the history between the starting time and the time of observation.

Let failure be the first upcrossing of the curve $\xi(t)$ by the process $X(t)$ for $t \geqslant 0$. The probability of failure in the time interval $[0,t]$ is called $P_f(t)$. The probability of survival at time t then is

$$1 - P_f(t) = P(N(t)=0 \mid X(0)<\xi(0)) P(X(0)<\xi(0)) \qquad (7.122)$$

where $N(t)$ is the number of barrier crossings in the interval $(0,t]$. The last factor represents the probability of surviving $t=0$.

The relation between the failure probability $P_f(t)$ and the probability density function $f_1(t)$ of the time to the first barrier crossing conditional on $X(0)<\xi(0)$ follows from consideration of a time increment in (7.122).

$$\frac{d}{dt} P_f(t) = f_1(t) P(X(0) < \xi(0)) \qquad (7.123)$$

The problem of finding the failure probability $P_f(t)$ is therefore equivalent to finding the probability density function $f_1(t)$ of the first-passage time conditional on $X(0)<\xi(0)$.

A convenient starting point for a discussion of the function $f_1(t)$ is an integral identity expressing the fact that a barrier crossing at time t must either be the first, or the first crossing must have occurred at some previous time τ. Thus the crossing rate can be given as the sum of two contributions

$$\nu_\xi^X(t) = f_1(t) + \int_0^t K(t \mid \tau) f_1(\tau) \, d\tau \qquad (7.124)$$

where $K(t \mid \tau)$ is the crossing rate at t conditional on the first crossing having occurred at τ.

Although (7.124) is merely a restatement of the problem in terms of the new unknown function $K(t \mid \tau)$, the identity is useful in obtaining approximations and bounds for $f_1(t)$. As the integral is nonnegative,

$$f_1(t) \leqslant v_\xi^X(t) \tag{7.125}$$

and $v_\xi^X(t)$ is seen to provide an upper bound on $f_1(\tau)$. A lower bound is obtained by considering the joint upcrossing frequency $v_{\xi\xi}^X(t,\tau_1)$ and observing that when $\tau_1 < t$ the crossing at τ_1 is the first, or the first has occurred at some previous time τ_2. This gives the identity

$$v_{\xi\xi}^X(t,\tau_1) = K(t \mid \tau_1) f_1(\tau_1) + \int_0^{\tau_1} K(t \mid \tau_1,\tau_2) f_1(\tau_2) \, d\tau_2 \tag{7.126}$$

similar to (7.124). Elimination of the integrand in (7.124) by (7.126) then gives

$$f_1(t) = v_\xi^X(t) - \int_0^t v_{\xi\xi}^X(t,\tau_1) \, d\tau_1 \tag{7.127}$$
$$+ \int_0^t \int_0^{\tau_1} K(t \mid \tau_1,\tau_2) f_1(\tau_2) \, d\tau_2 \, d\tau_1$$

The last integral is nonnegative, providing a lower bound

$$f_1(t) \geqslant v_\xi^X(t) - \int_0^t v_{\xi\xi}^X(t,\tau) \, d\tau \tag{7.128}$$

Continuation of this procedure leads to the formal series representation

$$f_1(t) = v_\xi^X(t) + \sum_{j=1}^\infty (-1)^j \int_0^t \cdots \tag{7.129}$$
$$\int_0^{\tau_{j-1}} v_{\xi\ldots\xi}^X(t,\tau_1, \ldots, \tau_j) \, d\tau_1 \cdots d\tau_j$$

This is the inclusion-exclusion series of Rice (1944). It provides upper and lower bounds upon truncation after an odd or even number of terms. Series representations of this type have been used by Longuet-Higgins (1962) to study intervals between zero crossings and by Roberts (1975) to study first passage of transient structural response.

The series (7.129) suffers the disadvantage that if n crossings are likely to occur at least n terms must be included. This problem is avoided if (7.124) is considered as an integral equation, and an adequate approximation for the kernel $K(t \mid \tau)$ is provided. Special cases of this procedure were considered by Shipley and Bernard (1972), and Bernard and Shipley (1972), while the general nature of the procedure was pointed out by Nielsen (1980).

The simplest approximation of this type follows from the approximation $K(t \mid \tau) = v_\xi^X(t)$; i.e., the crossing rate at t is assumed to be independent of a prior first crossing at τ. In this case (7.121) has the explicit solution

$$f_1(t) = v_\xi^X(t) \exp\left[-\int_0^t v_\xi^X(\tau) d\tau \right] \tag{7.130}$$

The corresponding survival probability is

$$1 - P_f(t) = F(\xi(0)) \exp\left[-\int_0^t v_\xi^X(\tau)d\tau\right] \qquad (7.131)$$

where $F(x)$ is the distribution function of the initial value $X(0)$. The result (7.131) is similar to (7.37) for a Poisson square-wave process, and it is based on assumed independence of barrier crossings.

The damping of most structures is sufficiently small to cause resonance at specific frequencies, and structural response therefore often exhibits correlation over long time intervals, i.e., many vibration periods. In that case the Poisson hypothesis about independent barrier crossings is invalid, and more refined methods must be used. A direct extension of the integral equation method consists in providing an improved approximation for the kernel $K(t \mid \tau)$. If the requirement that the barrier crossing at τ should be the first is dropped,

$$K(t \mid \tau) = \frac{v_{\xi\xi}^X(t,\tau)}{v_\xi^X(\tau)} \qquad (7.132)$$

The main problem in the solution of the integral equation (7.124) then is the evaluation of the kernel (7.132). Details are now given for the upcrossing of a single barrier $\xi(t)$ by a normal process $X(t)$.

First the process and the barrier level are replaced by their normalized equivalents $Y(t)$ and $\eta(t)$ as in (7.63). By a straightforward generalization of Rice's formula (7.59)

$$v_{\eta_1\eta_2}^Y = \int_{\dot\eta_1}^\infty \int_{\dot\eta_2}^\infty f_{\dot{\mathbf{Y}}\mathbf{Y}}(\mathbf{z},\boldsymbol{\eta})(z-\dot\eta_1)(z-\dot\eta_2)\,dz_1 dz_2 \qquad (7.133)$$

$f_{\dot{\mathbf{Y}}\mathbf{Y}}(\dot{\mathbf{y}},\mathbf{y})$ is the normal probability density function (7.51) with the variables $\dot{\mathbf{Y}} = (\dot{Y}(t_1),\dot{Y}(t_2))$ and $\mathbf{Y} = (Y(t_1),Y(t_2))$ and the covariance matrix given in terms of the correlation function $\rho(t_1,t_2)$.

$$\mathbf{c} = \begin{vmatrix} \mathbf{c}_{\dot{\mathbf{Y}}\dot{\mathbf{Y}}} & \mathbf{c}_{\dot{\mathbf{Y}}\mathbf{Y}} \\ \mathbf{c}_{\mathbf{Y}\dot{\mathbf{Y}}} & \mathbf{c}_{\mathbf{Y}\mathbf{Y}} \end{vmatrix} \qquad (7.134)$$

$$= \begin{vmatrix} \omega_1^2 & \rho_{,12} & 0 & \rho_{,1} \\ \rho_{,12} & \omega_2^2 & \rho_{,2} & 0 \\ 0 & \rho_{,2} & 1 & \rho \\ \rho_{,1} & 0 & \rho & 1 \end{vmatrix}$$

Here $\rho_{,1} = \partial\rho(t_1,t_2)/\partial t_1$, etc.

It is advantageous to write (7.133) in terms of the conditional probability density function $f_{\dot{Y}\,|\,Y}(\dot{y}\,|\,\eta)$ for $\dot{Y}$ given $Y = \eta$.

$$v^Y_{\eta_1\eta_2} = f_Y(\eta)\int_{\dot{\eta}_1}^{\infty}\int_{\dot{\eta}_2}^{\infty} f_{\dot{Y}\,|\,Y}(z\,|\,\eta)(z-\dot{\eta}_1)(z-\dot{\eta}_2)\,dz_1\,dz_2 \quad (7.135)$$

The first factor follows directly from (7.51),

$$f_Y(y) = \frac{1}{2\pi\sqrt{1-\rho^2}}\exp\left|-\frac{1}{2(1-\rho^2)}(y_1^2 - 2\rho y_1 y_2 + y_2^2)\right| \quad (7.136)$$

The conditional probability density $f_{\dot{Y}\,|\,Y}(z,\eta)$ is also joint normal and it can be shown that the conditional mean and covariance of $\dot{Y}$ are given by

$$\mu = \mu_{\dot{Y}\,|\,\eta} = c_{\dot{Y}Y}c_{YY}^{-1}\eta \quad (7.137)$$

$$c_{\dot{Y}\,|\,\eta} = c_{\dot{Y}\dot{Y}} - c_{\dot{Y}Y}c_{YY}^{-1}c_{Y\dot{Y}} \quad (7.138)$$

The proof is conveniently carried out by use of Fourier transforms and by means of the so-called "characteristic function" (see, e.g., Stratonovich, 1963, p. 44). Substitutions from (7.134) give the conditional mean

$$\left|\begin{matrix}\mu_1\\ \mu_2\end{matrix}\right| = \frac{1}{1-\rho^2}\left|\begin{matrix}(\eta_2-\rho\eta_1)\rho_{,1}\\ (\eta_1-\rho\eta_2)\rho_{,2}\end{matrix}\right| \quad (7.139)$$

and the conditional covariance matrix

$$c_{\dot{Y}\,|\,\eta} = \left|\begin{matrix}\lambda_1^2 & \kappa\lambda_1\lambda_2\\ \kappa\lambda_1\lambda_2 & \lambda_2^2\end{matrix}\right| \quad (7.140)$$

with

$$\lambda_1^2 = \omega_1^2 - \frac{(\rho_{,1})^2}{1-\rho^2} \quad (7.141)$$

$$\lambda_2^2 = \omega_2^2 - \frac{(\rho_{,2})^2}{1-\rho^2} \quad (7.142)$$

$$\kappa\lambda_1\lambda_2 = \rho_{,12} + \frac{\rho\,\rho_{,1}\,\rho_{,2}}{1-\rho^2} \quad (7.143)$$

In terms of these coefficients,

$$f_{\dot{Y}\,|\,\eta}(z\,|\,\eta) = \frac{1}{2\pi\lambda_1\lambda_2\sqrt{1-\kappa^2}}\exp\left\{-\frac{1}{2(1-\kappa^2)}\left|\left(\frac{z_1-\mu_1}{\lambda_1}\right)^2\right.\right. \quad (7.144)$$

$$\left.\left.-2\kappa\left(\frac{z_1-\mu_1}{\lambda_1}\right)\left(\frac{z_2-\mu_2}{\lambda_2}\right)+\left(\frac{z_2-\mu_2}{\lambda_2}\right)^2\right|\right\}$$

The integral in (7.135) cannot generally be evaluated in closed form, but a very good approximation can be obtained by using the easily verified identity

$$\frac{\partial}{\partial \kappa} f_{\dot{Y}|\eta}(\mathbf{z}\mid\boldsymbol{\eta}) = \lambda_1\lambda_2 \frac{\partial^2}{\partial z_1\partial z_2} f_{\dot{Y}|\eta}(\mathbf{z}\mid\boldsymbol{\eta}) \qquad (7.145)$$

to write $f_{\dot{Y}|\eta}(\mathbf{z}\mid\boldsymbol{\eta})$ in the form

$$f_{\dot{Y}|\eta}(\mathbf{z}\mid\boldsymbol{\eta};\kappa) = f_{\dot{Y}|\eta}(\mathbf{z}\mid\boldsymbol{\eta};0) + \int_0^\kappa \frac{\partial}{\partial k} f_{\dot{Y}|\eta}(\mathbf{z}\mid\boldsymbol{\eta};k)\,dk \qquad (7.146)$$

$$= \frac{1}{\lambda_1}\varphi\left(\frac{z_1-\mu_1}{\lambda_1}\right)\frac{1}{\lambda_2}\varphi\left(\frac{z_2-\mu_2}{\lambda_2}\right)$$

$$+ \lambda_1\lambda_2\int_0^\kappa \frac{\partial^2}{\partial z_1\partial z_2} f_{\dot{Y}|\eta}(\mathbf{z}\mid\boldsymbol{\eta};k)\,dk$$

Introduction of this representation into (7.135) and integration twice by parts lead to the desired result:

$$v_{\eta_1\eta_2}^Y = \lambda_1\lambda_2 f_Y(\boldsymbol{\eta})\left|\Psi\left(-\frac{\mu_1-\dot{\eta}_1}{\lambda_1}\right)\Psi\left(-\frac{\mu_2-\dot{\eta}_2}{\lambda_2}\right)\right. \qquad (7.147)$$

$$+ \kappa\,\Phi\left(\frac{\mu_1-\dot{\eta}_1}{\lambda_1}\right)\Phi\left(\frac{\mu_2-\dot{\eta}_2}{\lambda_2}\right)$$

$$\left.+ \lambda_1\lambda_2\int_0^\kappa (\kappa-k)f_{\dot{Y}|\eta}(\boldsymbol{\mu}-\dot{\boldsymbol{\eta}}\mid\boldsymbol{\eta};k)\,dk\right|$$

Due to the factor $(\kappa-k)$ in the integrand, the main contribution to the integral must come from small values of k. By setting $k=0$ in the exponential in (7.147), the following approximation is obtained:

$$v_{\eta_1\eta_2}^Y \approx \lambda_1\lambda_2 f_Y(\boldsymbol{\eta})\left|\Psi\left(-\frac{\mu_1-\dot{\eta}_1}{\lambda_1}\right)\Psi\left(-\frac{\mu_2-\dot{\eta}_2}{\lambda_2}\right)\right. \qquad (7.148)$$

$$+ \kappa\Phi\left(\frac{\mu_1-\dot{\eta}_1}{\lambda_1}\right)\Phi\left(\frac{\mu_2-\dot{\eta}_2}{\lambda_2}\right)$$

$$\left.+ \varphi\left(\frac{\mu_1-\dot{\eta}_1}{\lambda_1}\right)\varphi\left(\frac{\mu_2-\dot{\eta}_2}{\lambda_2}\right)[\sqrt{1-\kappa^2}-1+\kappa\arcsin(\kappa)]\right|$$

This formula was derived by Krenk and Madsen (1983).

Figure 7.15 shows the joint upcrossing frequency $v_{\eta\eta}^Y(t)$ for a constant barrier level $\eta=2.0$ and the correlation function

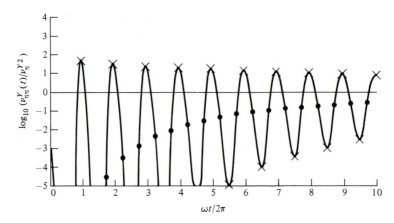

Figure 7.15 Normalized crossing frequency $v_{\xi\xi}^{x}(t)$ from (7.147) and (7.148).

$$\rho(t) = e^{-\zeta\omega_0 |t|} \left| \cos(\sqrt{1-\zeta^2}\omega_0 t) \right. \tag{7.149}$$

$$\left. + \frac{\zeta}{\sqrt{1-\zeta^2}} \sin(\sqrt{1-\zeta^2}\omega_0 |t|) \right|$$

with the parameter value $\zeta = 0.02$. $\rho(t)$ in (7.149) is the correlation function of the stationary response of a linear single-degree-of-freedom structure to white noise. The derivation of (7.149) and related problems of structural response are dealt with in Chapter 8. For stationary response and a constant barrier level, the approximate formula (7.148) becomes exact four times in each interval, namely, for $\rho(t) = 0$ and $\dot{\rho}(t) \approx 0$. These points include the maxima and minima, but also the general agreement is so good that no difference is seen between (7.147) and (7.148) in the figure.

It is seen that the joint upcrossing frequency exhibits marked peaks at each full period, i.e., $t_n = 2\pi n/\omega_0$. The integral equation (7.124) then implies that $f_1(t)$ behaves approximately as a step function in the stationary case. This is illustrated in Fig. 7.16, where $f_1(t)$ obtained by solution of the integral equation (7.124) is compared with simulated results of Christensen and Sørensen (1980). The first step of the calculated $f_1(t)$ is seen to be too wide. This is connected with the fact that the initial condition $X(0) < \zeta(0)$ has not been accounted for in the kernel $K(t \mid \tau)$, and this is of minor importance in applications with homogeneous initial conditions on the process. Apart from this the agreement is good.

In the integral equation (7.124) with the kernel given by (7.132) and (7.148), rather detailed information about the correlation function $\rho(t)$ is used and therefore has to be calculated. For

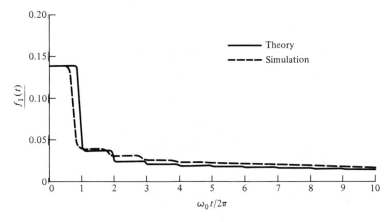

Figure 7.16 First-passage probability density from the integral equation with kernel (7.132) and simulation results of Christensen and Sørensen (1980).

narrow-band processes, i.e., processes dominated by a single frequency, simpler approximations to the first-passage probability density $f_1(t)$ have been developed. Two will be treated here, and they both use that the local extremes lie on the envelope. In both cases the idea is to treat barrier crossings by the envelope instead of upcrossings by the original process. By virtue of (7.101) the envelope will not have narrow-band character, and simpler approximation methods can then be used.

Consider the upcrossing of the normalized barrier $\eta(t)$. The upcrossing rate of the (symmetric) envelope is given by (7.111) in terms of the angular frequency ω_R, and an assumption of independent envelope crossings will give a survival probability in the form (7.131).

$$1 - P_f(t) = F_R(\eta(0)) \exp\left[-\int_0^t v_\eta^R(\tau)\,d\tau\right] \qquad (7.150)$$

This formula is based on the assumption that every excursion of the envelope is followed by an excursion of the normalized process $Y(t)$. However, for normal processes it follows from (7.66) and (7.111) that

$$\frac{v_\eta^R}{v_\eta^Y} = \frac{\omega_R}{\omega_0}\,\frac{\Psi(\dot\eta/\omega_R)}{\Psi(\dot\eta/\omega_0)}\,\sqrt{2\pi}\,\eta \qquad (7.151)$$

Thus for sufficiently high levels, $v_\eta^R > v_\eta^Y$ and some of the excursions of $R(t)$ are not accompanied by excursions of $Y(t)$.

A semiempirical reduction of the envelope crossing frequency was introduced by Vanmarcke (1975). For a single barrier the argument is as follows. If the length of the excursions of $R(t)$ is greater than $1/v_0^Y$, an upcrossing of $Y(t)$ is assumed to be certain. For a smaller length τ of excursion of $R(t)$, the probability of an excursion of $Y(t)$ is assumed to be proportional with τ, i.e., $v_0^Y\tau$. Thus the proportion of unqualified envelope crossings is

$$1 - q = \int_0^{1/v_0^Y} (1 - v_0^Y\tau) f_\tau(\tau) \, d\tau \qquad (7.152)$$

where $f_\tau(\tau)$ is the probability density of τ. The expectation of τ is found from the distance between upcrossings and the fraction of time spent above the barrier level η.

$$E[\tau] = \frac{(1 - F_R(\eta))}{v_\eta^R} = \frac{1}{v_\eta^R} e^{-\eta^2/2} \qquad (7.153)$$

The reduction factor q then follows from (7.152) by the ad hoc assumption that τ is exponentially distributed.

$$q = v_\eta^Y/v_\eta^R \, [1 - \exp(-v_\eta^R/v_\eta^Y)] \qquad (7.154)$$

The rate of qualified upcrossings to be used in (7.150) then is

$$q \, v_\eta^R = v_\eta^Y \, [1 - \exp(-v_\eta^R/v_\eta^Y)] \qquad (7.155)$$

It is seen by reference to (7.151) that qv_η^R represents an interpolation between v_{η_R} for low levels and v_{η_Y} for high levels. The important reference frequency ω_R may be determined from either (7.109) or (7.116).

The ratio qv_η^R/v_η^Y given in the brackets in (7.155) is the relative magnitude of the exponent compared with the Poisson assumption. The ratio is shown in Fig. 7.17 as a function of the level η with $\dot{\eta}=0$ for different values of the parameter ζ. ζ is the damping ratio for a single-degree-of-freedom structure excited by white noise (see Table 8.2). It is seen that the exponent decreases with decreasing damping ζ.

The problems connected with the irregularity of the envelope at high levels can be avoided by considering the local maxima as discrete points on the envelope. For a narrow-band process a typical distance between the local maxima is $T=2\pi/\omega_0$, and failure is then associated with the first occurrence of a point above the barrier level, i.e., $R(nT) \geqslant \eta(nT)$ (Yang and Shinozuka, 1971, 1972; Yang, 1973, 1975).

Let $F_1(nT)$ denote the probability of at least one of the first n points exceeding the barrier. $F_1(nT)$ is then a discretized approximation to the first-passage distribution function at $t = nT$. The probability of survival at $t = nT$ is

$$1 - F_1(nT) = [1 - F_1((n-1)T)][1 - p(nT)] \qquad (7.156)$$

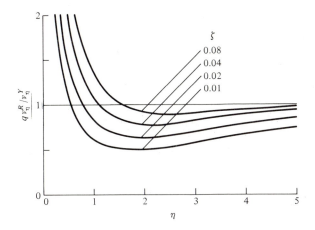

Figure 7.17 Normalized qualified envelope crossing rate for different damping ratios ζ.

where $p(nT)$ is the probability of $R(nT) \geqslant \eta(nT)$ conditional on $R(jT) < \eta(jT)$, $j = 0, \ldots, n-1$. From (7.156)

$$F_1(nT) = 1 - \prod_{j=1}^{n} [1 - p(jT)], \quad n = 1, 2, \ldots \qquad (7.157)$$

Ideally, the probability $p(n)$ should account for all the previous points $j = 0, \ldots, n-1$, but if the point process $R(jT)$ is assumed to be Markovian, $p(nT)$ depends only on the preceding point and reduces to

$$p(nT) = \frac{1}{F_R(\eta_{n-1})} \int_0^{\eta_n} \int_{\eta_n}^{\infty} f_{RR}(r_{n-1}, r_n) \, dr_{n-1} \, dr_n \qquad (7.158)$$

The joint density function $f_{RR}(r_1, r_2)$ is given by (7.103) and the distribution function $F_R(\eta)$ by (7.100).

When $R((n-1)T)$ and $R(nT)$ are correlated, the integral in (7.158) must be evaluated numerically. Series expansion of the Bessel functions has been used by Yang (1973), and a reformulation in terms of the conditional probability density function was given by Krenk (1979). However, it appears to be more instructive to reformulate the double integral to a single integral in terms of the correlation parameter k. This procedure is similar to that used to evaluate the joint upcrossing intensity $v_{\eta\eta}^Y$ for a normal process. It makes use of the identity

$$\frac{\partial}{\partial k} f_{RR}(r_1, r_2) = \frac{\partial^2}{\partial r_1 \partial r_2} \left| \frac{r_1 r_2}{1 - k^2} I_1 \left| \frac{k r_1 r_2}{1 - k^2} \right| \exp \left| -\frac{1}{2} \frac{r_1^2 + r_2^2}{1 - k^2} \right| \right| \qquad (7.159)$$

which is easily verified when use is made of the fact that $I_1(z) = dI_0(z)/dz$. Note that this relation is similar to (7.145) for the two-dimensional normal distribution, but in the present case $I_1(\ )$ must be used on the right-hand side instead of $I_0(\ )$. The joint density function $f_{RR}(r_1,r_2)$ can then be expressed in the form

$$f_{RR}(r_1,r_2;k) = f_R(r_1)f_R(r_2) + \int_0^k \frac{\partial}{\partial\kappa}f_{RR}(r_1,r_2;\kappa)\,d\kappa \quad (7.160)$$

Substitution of this expression into (7.158) then gives the transition probability

$$p(nT) = 1 - F_R(\eta_n) - \frac{1}{F_R(\eta_{n-1})}\int_0^k \frac{\eta_{n-1}\eta_n}{1-\kappa^2}\,I_1\left|\frac{\kappa\eta_{n-1}\eta_n}{1-\kappa^2}\right| \quad (7.161)$$
$$\times\exp\left|-\frac{1}{2}\frac{\eta_{n-1}^2+\eta_n^2}{1-\kappa^2}\right|d\kappa$$

The integrand is positive and the transition probability $p(nT)$ is therefore a decreasing function of the correlation parameter k. $p(nT)$ represents the conditional probability of a crossing within a period of length $T = 1/v_0^Y$. In Fig. 7.18 the crossing rate pv_0^Y is compared with the equivalent rate v_η^Y from the Poisson assumption. The parameters are as in Fig. 7.17, and the good agreement between these two approximation methods is noted.

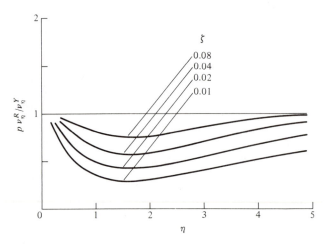

Figure 7.18 Normalized crossing rate of Markov point process.

It should be noted that both of the last two approximations make use of an explicit estimate of the hazard function given entirely by properties of the process at the time of observation, i.e., without consideration of the previous history. In contrast, the integral equation (7.124) accounts for the previous history through the kernel, and the example of Figs. 7.15 and 7.16 indicates considerable memory located at the times $\tau = t - nT$. In fact, this location of the memory for narrow-band processes can be used to obtain explicit step-function approximations to the first-passage probability $f_1(t)$ for narrow-band processes (Krenk and Madsen, 1983; Madsen and Krenk, 1984).

7.5 SUMMARY

This chapter deals with four subjects: statistics of ordered stochastic variables, extreme-value distributions, discrete stochastic processes, and the normal process. The fundamental theory has been presented in order of increasing complexity and a number of useful results derived.

The chapter starts with some simple considerations on a finite number of observations of a stochastic variable with unknown distribution function. It is demonstrated that predictions concerning extremes require a very large number of prior observations if no assumption is made about the distribution function.

On the other hand, rather weak assumptions about the asymptotic behavior of the original distribution may lead to one of three types of distribution for the extreme value of a large number of observations. The properties of these three distributions are discussed in Section 7.2, and the mathematical details are given in Tables 7.1 and 7.2.

Many of the stochastic variables of interest in safety considerations such as load and material strength are functions of time and/or position, and the concept of a stochastic process is therefore needed. It is introduced stepwise by first considering discrete processes of Poisson and renewal type. The extreme-value distributions for Poisson processes of "spike" and step-function type are derived, and their relation to approximate extreme-value distributions of continuous stochastic processes is pointed out. The waiting-time problem is discussed within the more general framework of renewal processes in order to illustrate the limitations of the Poisson assumption of independent arrivals and to provide the necessary background for the traffic load model in Section 10.4.

The main part of the chapter is concerned with the normal process. The theory is kept as simple as possible without excluding a detailed discussion of the crossing problem for nonstationary normal processes. Also, the distribution of local extremes and the role of the bandwidth are discussed in some detail. The importance of the normal process in structural safety is intimately connected with random response of lightly damped structures, and narrow-band processes and their associated envelopes have therefore been given special attention. The chapter concludes with a presentation of several approximate solution methods for the first-passage problem.

REFERENCES

ABRAMOWITZ, M. and I. STEGUN, *Handbook of Mathematical Functions,* Dover, New York, 1965.

BERNARD, M. C. and J. W. SHIPLEY, "The First-Passage Problem for Stationary Random Structural Vibration," *Journal of Sound and Vibration,* Vol. 24, 1972, pp. 121-132.

CARTWRIGHT, D. E. and M. S. LONGUET-HIGGINS, "The Statistical Distribution of the Maxima of a Random Function," *Proceedings of the Royal Society of London,* Vol. A 237, 1956, pp. 212-232.

CHRISTENSEN, J. K. and J. D. SØRENSEN, "Simulering af Gaussiske Processer på Datamaskine med Henblik på Fastlæggelse af Brudsandsynligheder," Report 8007 (in Danish), Institute of Building Technology and Structural Engineering, AUC Aalborg, 1980.

COX, D. R. and H. D. MILLER, *The Theory of Stochastic Processes,* Chapman and Hall, London, 1977.

CRANDALL, S. H., K. L. CHANDIRAMANI and R. G. COOK, "Some First-Passage Problems in Random Vibration," *Journal of Applied Mechanics,* ASME, Vol. 33, 1966, pp. 532-538.

DAVENPORT, A. G., "Note on the Distribution of the Largest Value of a Random Function with Application in Gust Loading," *Proceedings of the Institution of Civil Engineers London,* Vol. 28, 1964, pp. 187-196.

FISHER, R. A. and L. H. C. TIPPETT, "Limiting Forms of the Frequency Distribution of the Largest or Smallest Member of a Sample," *Proceedings of the Cambridge Philosophical Society,* Vol. 24, 1928, pp. 180-190.

GUMBEL, E. J., *Statistics of Extremes,* Columbia University Press, New York, 1958.

HOGG, R. V. and A. T. CRAIG, *Introduction to Mathematical Statistics,* Collier-MacMillan, New York, 1966.

JOHNSON, A. I., *Strength, Safety and Economical Dimensions of Structures,* Statens Kommitte för Byggnadsforskning, Meddelanden No. 22, Stockholm, 1953.

KRENK, S., "A Double Envelope for Stochastic Processes," The Danish Center for Applied Mathematics and Mechanics, Report No. 134, Lyngby, Denmark, 1978.

KRENK, S., "Nonstationary Narrow-Band Response and First-Passage Probability," *Journal of Applied Mechanics,* ASME, Vol. 46, 1979, pp. 919-924.

KRENK, S. and P. H. MADSEN, "Stochastic Response Analysis," in *Reliability Theory and its Applications in Structural and Soil Mechanics,* ed. P. Thoft-Christensen, NATO ASI Series E, Martinus Nijhoff Publishers, The Hague, 1983, pp. 103-172.

LONGUET-HIGGINS, M. S., "The Distribution Intervals Between Zeros of a Stationary Random Function," *Philosophical Transactions of the Royal Society of London,* Vol. A 254, 1962, pp. 557-599.

MADSEN, P. H. and S. KRENK, "An Intergal Method for the First-Passage Problem in Random Vibration," *Journal of Applied Mechanics,* ASME, Vol. 51, 1984, pp. 674-679.

MILLER, K. S., *Complex Stochastic Processes,* Addison-Wesley, Reading, Mass., 1974.

NIELSEN, S. K., "Probability of Failure of Structural Systems under Random Vibration," Report No. 8001, Institute of Building Technology and Structural Engineering, AUC Aalborg, 1980.

PAPOULIS, A., *Probability, Random Variables, and Stochastic Processes,* McGraw-Hill, Tokyo, 1965.

PARZEN, E., *Stochastic Processes,* Holden-Day, San Francisco, 1962.

RICE, J. R. and F. P. BEER, "On the Distribution of Rises and Falls in a Continuous Random Process," *Journal of Basic Engineering,* Vol. 87, 1965, pp. 398-404.

RICE, J. R. and F. P. BEER, "First-Occurrence Time of High-Level Crossings in a Continuous Random Process," *Journal of the Acoustical Society of America,* Vol. 39, 1966, pp. 323-335.

RICE, S. O., "Mathematical Analysis of Random Noise," *Bell System Technical Journal,* Vol. 23., 1944, pp. 282-332 and Vol. 24, 1944, pp. 46-156. Reprinted in *Selected Papers in Noise and Stochastic Processes,* ed. N. Wax, Dover, New York, 1954.

ROBERTS, J. B., "Probability of First-Passage for Nonstationary

Random Vibration," *Journal of Applied Mechanics,* ASME, Vol. 42, 1975, pp. 716-720.

SHIPLEY, J. W. and M. C. BERNARD, "The First Passage Time Problem for Simple Structural Systems," *Journal of Applied Mechanics,* ASME, Vol. 39, 1972, pp. 911-917.

STRATONOVICH, R. L., *Topics in the Theory of Random Noise,* Gordon and Breach, New York, 1963.

VANMARCKE, E. H., "On the Distribution of the First-Passage Time for Normal Stationary Random Processes," *Journal of Applied Mechanics,* ASME, Vol. 42, 1975, pp. 215-220.

WEIBULL, W., "A Statistical Distribution Function of Wide Applicability," *Journal of Applied Mechanics,* ASME, Vol. 18, 1951.

YANG, J.-N., "First-Excursion Probability in Nonstationary Random Vibration," *Journal of Sound and Vibration,* Vol. 27, 1973, pp. 165-182.

YANG, J.-N., "Approximation to First-Passage Probability," *Journal of Engineering Mechanics Division,* ASCE, Vol. 101, 1975, pp. 361-372.

YANG, J.-N. and M. SHINOZUKA, "On the First-Excursion Probability in Stationary Narrow-Band Random Vibration," *Journal of Applied Mechanics,* ASME, Vol. 38, 1971, pp. 1017-1022.

YANG, J.-N. and M. SHINOZUKA, "On the First-Excursion Probability in Stationary Narrow-Band Random Vibration, II," *Journal of Applied Mechanics,* ASME, Vol. 39, 1972, pp. 733-730.

STOCHASTIC RESPONSE
OF STRUCTURES

8.1 STATIONARY RESPONSE

A number of important aspects of stochastic response can be dealt with within the framework of stationary processes. This allows a simple and intuitively clear description in the frequency domain in terms of power spectra and transfer functions. The formulation can then be generalized to excitation processes described in terms of evolutionary power spectra.

8.1.1 Single-Degree-of-Freedom Structure

A linear single-degree-of-freedom structure typically consists of a mass M, the displacement of which is u. The mass is constrained by an elastic force Ku and a viscous damping force $C\dot{u}$, where $\dot{u}$ is the time derivative of u. When the mass is exposed to a time-dependent load $f(t)$, the equation of motion is

$$M\ddot{u} + C\dot{u} + Ku = f(t) \tag{8.1}$$

A normalized form of this equation is

$$\ddot{u} + 2\zeta\omega_0\dot{u} + \omega_0^2 u = q(t) \tag{8.2}$$

in which ω_0 is the undamped natural angular frequency and ζ is the damping ratio.

$$\omega_0^2 = \frac{K}{M}, \quad \zeta = \frac{1}{2}\frac{C}{\sqrt{MK}} \tag{8.3}$$

The solution of (8.2) for $t \geqslant 0$ is

$$u(t) = u(0)g(t) + \dot{u}(0)h(t) + \int_0^t q(\tau)h(t-\tau)d\tau \tag{8.4}$$

in which $g(t)$ and $h(t)$ are the response functions for a unit displacement and a unit impulse at $t = 0$.

$$g(t) = U(t)e^{-\zeta\omega_0 t}\left|\cos(\omega_d t) + \frac{\zeta\omega_0}{\omega_d}\sin(\omega_d t)\right| \tag{8.5}$$

$$h(t) = U(t)e^{-\zeta\omega_0 t}\omega_d^{-1}sin(\omega_d t) \tag{8.6}$$

The damped natural frequency is $\omega_d = \sqrt{1-\zeta^2}\omega_0$, and $U(t)$ is Heaviside's step function.

It follows from the linearity of the differential equation that if the load $q(t)$ is a normal process, the response $u(t)$ will also be a normal process, and therefore characterized by its expectation and covariance function. In the case of stationary excitation from the infinite past, the response is also stationary and given by the integral in (8.1) with the lower limit $-\infty$. Let the expectation and covariance function of the load $q(t)$ be μ^Q and $c^Q(t)$, respectively. The expectation and covariance of the response $u(t)$ then follow from (8.4).

$$\mu = \int_{-\infty}^t E[Q(\tau)]h(t-\tau)d\tau = \mu^Q/\omega_0^2 \tag{8.7}$$

$$c(t_1,t_2) = \int_{-\infty}^{t_1}\int_{-\infty}^{t_2} c^Q(\tau_1-\tau_2)h(t_1-\tau_1)h(t_2-\tau_2)d\tau_1 d\tau_2 \tag{8.8}$$

The double integral is of convolution type, and it may therefore be advantageous to express the relation (8.8) in the frequency domain.

As discussed in Section 7.4.4, the covariance function $c^Q(\tau)$ can be expressed in terms of a real, nonnegative spectral density function $S^Q(\omega)$ in the form (7.92).

$$c^Q(\tau) = \int_{-\infty}^{\infty} S^Q(\omega)e^{i\omega\tau}d\omega \tag{8.9}$$

Substitution of this representation into (8.8) gives

$$c(t_1,t_2) = \int_{-\infty}^{\infty}\left|\int_{-\infty}^{t_1} h(t_1-\tau_1)e^{i\omega\tau_1}d\tau_1\right|$$

$$\times\left|\int_{-\infty}^{t_2} h(t_2-\tau_2)e^{-i\omega\tau_2}d\tau_2\right|S^Q(\omega)d\omega \tag{8.10}$$

Upon introduction of the transfer function

$$H(\omega) = \int_0^{\infty} h(\tau)e^{-i\omega\tau}d\tau = \frac{1}{\omega_0^2-\omega^2+2i\zeta\omega_0\omega} \tag{8.11}$$

a simple variable transform in (8.10) leads to the result

$$c(t_1,t_2) = \int_{-\infty}^{\infty} H(\omega)\,\overline{H(\omega)}\,S^Q(\omega)\,e^{i\omega(t_1-t_2)}\,d\omega \qquad (8.12)$$

An overbar denotes the complex conjugate. By comparison with (8.9) it is seen that the spectral density of the response is

$$S(\omega) = |H(\omega)|^2 S^Q(\omega) \qquad (8.13)$$

i.e., the structure acts as a filter that amplifies the amplitude of a signal associated with the frequency ω by $|H(\omega)|$ shown in Fig. 8.1.

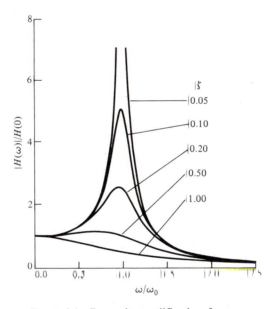

Figure 8.1 Dynamic amplification factor.

Formula (8.12) is often an attractive alternative to (8.8) for calculation of the response covariance function, in particular when $S^Q(\omega)$ is a rational function.

The envelope defined in (7.97) and (7.98) is closely related to the response of a linear oscillator. The equation of motion states that the external force equals the sum of inertial, viscous, and elastic forces generated by the structure. Upon multiplication by $\dot{u}(t)$ it becomes a power balance equation, and by rearrangement of the terms

$$\frac{1}{2}\frac{d}{dt}(\dot{u}^2 + \omega_0^2 u^2) = \frac{1}{2}\omega_0^2 \frac{dR^2}{dt} = q\dot{u} - 2\zeta\omega_0\dot{u}^2 \qquad (8.14)$$

Thus the square of the envelope, $R(t)^2$, represents the mechanical, i.e., kinetic and elastic energy, and the rate of change equals the externally supplied power, $q\dot{u}$, minus the rate of dissipation. For a lightly damped structure with wide-band excitation, it follows from (8.11) and (8.13) that the power spectrum $S(\omega)$ is dominated by resonance near the eigenfrequency ω_0. On the other hand, the fluctuations of the mechanical energy are governed by the correlation function $k(\tau)^2$ from (7.102) and the characteristic frequency is ω_R given by (7.109). For narrow-band response $\omega_0 \gg \omega_R$, with the implication that the mechanical energy of the structure changes slowly, when measured by the period of oscillation.

8.1.2 Multi-Degree-of-Freedom Structure

Let a linear structure have n degrees of freedom. The equation of motion then is

$$\mathbf{M}\ddot{\mathbf{u}} + \mathbf{C}\dot{\mathbf{u}} + \mathbf{K}\mathbf{u} = \mathbf{f}(t) \tag{8.15}$$

in which $\mathbf{M}$, $\mathbf{C}$, and $\mathbf{K}$ are the mass, damping, and stiffness matrices, respectively. $\mathbf{f}(t)$ is the excitation force vector.

In the absence of damping and external forces, the structure can perform vibrations of the form $\mathbf{u}(t) = \mathbf{v}\cos(\omega t)$, where the mode shape vector $\mathbf{v}$ and the corresponding angular frequency ω are a solution to the eigenvalue problem

$$[\mathbf{K} - \omega^2\mathbf{M}]\mathbf{v} = \mathbf{0} \tag{8.16}$$

For positive-definite matrices $\mathbf{K}$ and $\mathbf{M}$, there are n solutions to (8.16), each with real and positive angular frequency ω_j. In the absence of multiple eigenvalues, the mode shape vectors $\mathbf{v}_i$ and $\mathbf{v}_j$ satisfy the orthogonality relations

$$\mathbf{v}_i^T\mathbf{K}\mathbf{v}_j = \omega_j^2\delta_{ij}, \quad \mathbf{v}_i^T\mathbf{M}\mathbf{v}_j = \delta_{ij} \tag{8.17}$$

δ_{ij} is the Kronecker delta, $\delta_{ij} = 1$ for $i = j$ and $\delta_{ij} = 0$ for $i \neq j$. Also in the case of multiple eigenvalues, $\omega_i = \omega_j$, the mode shape vectors $\mathbf{v}_i$ and $\mathbf{v}_j$ can be chosen in accordance with (8.17).

The solution to the full equation (8.15) is now conveniently expressed in terms of the mode shape vectors $\mathbf{v}_j$, $j = 1, 2, \ldots, n$, determined from (8.16).

$$\mathbf{u}(t) = \sum_{j=1}^{n} \mathbf{v}_j T_j(t) \tag{8.18}$$

By substitution into (8.15) and use of the orthogonality relations (8.17) the equation of motion becomes

$$\ddot{T}_i + \sum_{j=1}^{n} \mathbf{v}_i^T \mathbf{C} \mathbf{v}_j \dot{T}_j + \omega_i^2 T_i = \mathbf{v}_i^T \mathbf{f}(t) \tag{8.19}$$

If the mode shape vectors are also orthogonal to the damping matrix $\mathbf{C}$, modal damping ratios ζ_i can be introduced by

$$2\omega_i \zeta_i = \mathbf{v}_i^T \mathbf{C} \mathbf{v}_i, \quad i = 1, 2, \ldots, n \tag{8.20}$$

and the equations of motion uncouple in the form

$$\ddot{T}_i + 2\zeta_i \omega_i \dot{T}_i + \omega_i^2 T_i = q_i(t), \quad i = 1, 2, \ldots, n \tag{8.21}$$

in which the generalized forces are

$$q_i(t) = \mathbf{v}_i^T \mathbf{f}(t), \quad i = 1, 2, \ldots, n \tag{8.22}$$

The solution of (8.21) follows from (8.4) to (8.6) by introduction of appropriate subscripts.

Various forms of the damping matrix $\mathbf{C}$ that lead to uncoupling have been discussed (Clough and Penzien, 1975, Section 13-3). The physical implication of the uncoupling is that each individual mode is in phase. If this assumption is not justified, a more elaborate decomposition in terms of complex eigenvectors can be used (Foss, 1958).

In the multidimensional case the expectation and the covariance must be described by functions of dimension n and $n \times n$, respectively. The expectation μ_j of T_j is analogous to (8.7).

$$\mu_j = \int_{-\infty}^{t} E[Q_j(\tau)] h_j(t - \tau) \, d\tau = \mu_j^Q / \omega_j^2 \tag{8.23}$$

The covariance of $T_i(t_1)$ and $T_j(t_2)$ is

$$\iota_{ij}(t_1, t_2) - \int_{-\infty}^{t_1} \int_{-\infty}^{t_2} c_{ij}^Q(t_1 - t_2) h_i(t_1 - t_1) h_j(t_2 - t_2) \, dt_1 \, dt_2 \tag{8.24}$$

in which the covariance function $c_{ij}^Q(\tau_1 - \tau_2)$ of $Q_i(\tau_1)$ and $Q_j(\tau_2)$ is given in terms of the covariance matrix $\mathbf{c}^F(\tau_1 - \tau_2)$ of $\mathbf{F}(\tau_1)$ and $\mathbf{F}(\tau_2)$ as

$$c_{ij}^Q(\tau_1 - \tau_2) = \mathbf{v}_i^T \text{Cov}[\mathbf{F}(\tau_1), \mathbf{F}(\tau_2)] \mathbf{v}_j = \mathbf{v}_i^T \mathbf{c}^F(\tau_1 - \tau_2) \mathbf{v}_j \tag{8.25}$$

The spectral density is now of dimension $n \times n$, and for the generalized forces Q_i it is defined by

$$c_{ij}^Q(\tau) = \int_{-\infty}^{\infty} S_{ij}^Q(\omega) e^{i\omega\tau} \, d\omega \tag{8.26}$$

Note that the off-diagonal terms $c_{ij}^Q(\tau)$, $i \neq j$, are not necessarily even functions of τ, and that the cross spectra $S_{ij}^Q(\omega)$ can therefore be complex-valued. However, by inversion of the Fourier transform (8.26), the following relations are established for $S_{ij}^Q(\omega)$:

$$S_{ij}^Q(\omega) = \overline{S_{ji}^Q(\omega)} = S_{ji}^Q(-\omega) \tag{8.27}$$

The first equality establishes $S_{ij}^Q(\omega)$ as elements of a Hermitean matrix. The generalized form of (8.12) is

$$c_{ij}(t_1,t_2) = \int_{-\infty}^{\infty} H_i(\omega) \, \overline{H_j(\omega)} \, S_{ij}^Q(\omega) \, e^{i\omega(t_1-t_2)} \, d\omega \tag{8.28}$$

in which

$$H_j(\omega) = \int_{-\infty}^{\infty} h_j(\tau) \, e^{-i\omega\tau} \, d\tau = \frac{1}{\omega_j^2 - \omega^2 + 2i\zeta_j\omega_j\omega} \tag{8.29}$$

This gives the spectral density of the modal responses $T_i(t)$ and $T_j(t)$ as

$$S_{ij}(\omega) = H_i(\omega) \, \overline{H_j(\omega)} \, S_{ij}^Q(\omega) \tag{8.30}$$

Clearly, the elements $S_{ij}(\omega)$ also satisfy the relations (8.27).

8.1.3 Rational Power Spectra

When the spectral density $S_{ij}^Q(\omega)$ of the generalized forces is a rational function, the modal covariance function $c_{ij}(t_1,t_2)$ can be evaluated explicitly from (8.28) by complex contour integration. The modal transfer function $H_j(\omega)$ can be written in the form

$$H_j(\omega) = \frac{1}{(\Omega_j - \omega)(\overline{\Omega}_j + \omega)} \tag{8.31}$$

exhibiting a pole in the first quadrant

$$\Omega_j = \omega_j \, e^{i\theta_j}, \qquad \sin\theta_j = \zeta_j \tag{8.32}$$

and a symmetrically located pole $\omega = -\overline{\Omega}_j$ in the second quadrant. The second factor in (8.28) is the complex conjugate of (8.31), and on the real axis

$$\overline{H_j(\omega)} = H_j(-\omega) \tag{8.33}$$

Thus the two poles of the second factor are located in the half-plane $\mathrm{Im}\,[\omega] < 0$.

As a consequence of the symmetry relations (8.27) and (8.33), the relation (8.28) can be written in the form

$$c_{ij}(t_1,t_2) = \int_0^{\infty} [H_i(\omega) \, H_j(-\omega) \, S_{ij}^Q(\omega) \, e^{i\omega(t_1-t_2)}$$

$$+ \, H_i(-\omega) \, H_j(\omega) \, S_{ij}^Q(-\omega) \, e^{-i\omega(t_1-t_2)}] \, d\omega$$

$$= 2\,\mathrm{Re}\left| \int_0^{\infty} H_i(\omega) \, H_j(-\omega) \, S_{ij}^Q(\omega) \, e^{i\omega(t_1-t_2)} \, d\omega \right| \tag{8.34}$$

Now let $S_{ij}^Q(\omega)$ be a rational function of ω. $S_{ij}^Q(\omega)$ is then defined in the full complex plane, and the last of the symmetry relations (8.27) can be written for complex argument ω as

$$S_{ij}(\omega) = \bar{S}_{ij}(-\omega) \tag{8.35}$$

where only the constants in S_{ij} are conjugated. On the imaginary axis $-\omega = \bar{\omega}$, and consequently $S_{ij}(\omega)$ is real for imaginary ω. Also $H_i(\omega)$, $H_j(-\omega)$, and the exponential factor are real for imaginary ω. On the imaginary axis the differential element $d\omega$ is imaginary, and the integral in (8.34) can therefore be considered as an integral along a closed contour containing the first or the fourth quadrant, depending on the sign of $t_1 - t_2$. For $t_1 \geqslant t_2$ the integral is closed in the first quadrant, and the value is $2\pi i$ times the sum of the residues (see Fig. 8.2).

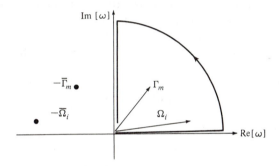

Figure 8.2   Contour and poles in the complex ω plane.

When the poles of $S_{ij}^Q(\omega)$ in the first quadrant are called Γ_m, the result is

$$c_{ij}(t_1,t_2) = 2\pi \, \text{Im} \left| \frac{H_j(-\Omega_i)}{\omega_i \sqrt{1-\zeta_i^2}} S_{ij}^Q(\Omega_i) \, e^{i\Omega_i(t_1-t_2)} \right.$$

$$\left. -2\sum_m H_i(\Gamma_m) H_j(-\Gamma_m) \, \text{Res}\,[S_{ij}^Q(\Gamma_m)] \, e^{i\Gamma_m(t_1-t_2)} \right|, \quad t_1 \geqslant t_2 \tag{8.36}$$

The summation only includes poles in the first quadrant. For $t_1 < t_2$ the subscripts i and j are interchanged.

In addition to the computational convenience of formula (8.36) it gives a qualitative description of the modal response covariance function. $c_{ij}(t_1,t_2)$ is a sum of exponentials with complex angular frequencies Ω_i (Ω_j) and Γ_m. The first term may be considered as a resonance term, and its magnitude is determined by the complex continuation of the spectral density at $\omega = \Omega_i = \omega_i(\sqrt{1-\zeta_i^2}+i\zeta_i)$. For small damping, $\zeta_i \ll 1$, this term will decrease slowly with

$(t_1 - t_2)$, and the magnitude will be approximately determined by the spectral density value $S_{ij}^Q(\omega_i)$. The angular frequencies Γ_m of the remaining terms are determined by the complex continuation of the spectral density, and these terms will typically decrease more rapidly with $(t_1 - t_2)$.

In the extreme case where the spectral density $S_{ij}^Q(\omega)$ takes a constant value except for very large values of ω, i.e., $\omega \gg \omega_i, \omega_j$, only the first term is present. $Q_i(t)$ is then called a white noise process. The covariance function of the response is illustrated in Fig. 8.3, showing the normalized covariance functions $\rho_{11}(t_1 - t_2)$ and $\rho_{12}(t_1 - t_2)$ for $\omega_2 = 1.25\omega_1$ and $\zeta_1 = \zeta_2 = 0.05$, where the excitation is given by a single scalar white noise process $n(t)$ in the form $q_i(t) = q_i n(t)$. The correlation functions are exponentially damped cosines with periods corresponding to the modal frequencies. Note that the cross correlation has the period $2\pi/\omega_1$ for $t_1 > t_2$ and the period $2\pi/\omega_2$ for $t_2 > t_1$; i.e., the period corresponds to the mode that is observed at the latest time. Response to white noise excitation is considered in detail in Section 8.3.

The spectral moments for a stationary scalar process were defined by (7.93). The definition is generalized to matrix form by considering a linear combination of the modal responses.

$$Y(t) = \sum_{j=1}^{n} \eta_j T_j(t) \tag{8.37}$$

By using the fact that the spectral moments corresponding to $Y(t)$ are real, the following expression is obtained:

$$\lambda_k^Y = 2\int_0^\infty \omega^k S^Y(\omega)\,d\omega \tag{8.38}$$

$$= \sum_{i=1}^{n}\sum_{j=1}^{n} \eta_i \eta_j \int_0^\infty \omega^k [S_{ij}(\omega) + \overline{S_{ij}(\omega)}]\,d\omega$$

This leads to the real-valued modal spectral moment matrix

$$\lambda_k^{ij} = 2\int_0^\infty \omega^k \operatorname{Re}[S_{ij}(\omega)]\,d\omega \tag{8.39}$$

The moments of even order follow directly from $c_{ij}(t_1, t_2)$ by differentiation.

$$\lambda_{2k}^{ij} = \left(\frac{\partial^2}{\partial t_1 \partial t_2}\right)^k c_{ij}(t_1, t_2)_{t_1 = t_2} \tag{8.40}$$

In the case of rational spectral density $S_{ij}^Q(\omega)$, the spectral moments of the response therefore follow directly from (8.36), provided that the integral (8.39) converges.

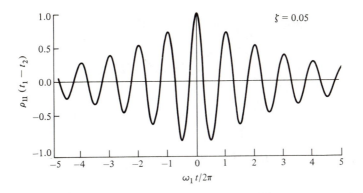

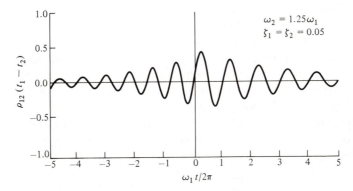

Figure 8.3 Modal correlation functions for white noise excitation, $\omega_2 = 1.25\omega_1$, $\zeta_1 = \zeta_2 = 0.05$.

$$\lambda_{2k}^{ij} = 2\pi \text{Im} \left| \frac{H_j(-\Omega_i)}{\omega_i \sqrt{1-\zeta_i^2}} \Omega_i^{2k} S_{ij}^Q(\Omega_i) \right.$$

$$\left. - 2\sum_m H_i(\Gamma_m) H_j(-\Gamma_m) \Gamma_m^{2k} \text{Res} [S_{ij}^Q(\Gamma_m)] \right| \tag{8.41}$$

Each power is multiplied with the appropriate power of the corresponding complex angular frequency.

The moments of odd order are generally more difficult to evaluate. Usually only the first moment is needed, notably for characterizing the envelope defined in terms of the complex conjugate process as discussed in Section 7.4.4. When $S_{ij}^Q(\omega)$ is a rational function, it follows from (8.30) that $\omega S_{ij}(\omega)$ can be expanded in partial fractions, and λ_1^{ij} can be evaluated by direct integration of (8.39). The general result is

$$\lambda_i^{ij} = 2 \operatorname{Re} \left| \frac{H_j(-\Omega_i)}{\omega_i \sqrt{1-\zeta_i^2}} \Omega_i \, S_{ij}^Q(\Omega_i) \log(\Omega_i/i) \right.$$

$$+ \frac{H_i(-\Omega_j)}{\omega_j \sqrt{1-\zeta_j^2}} \Omega_j \, S_{ij}^Q(-\Omega_j) \log(\Omega_j/i)$$

$$- 2 \sum_m (H_i(\Gamma_m) \, H_j(-\Gamma_m) \operatorname{Res}[S_{ij}^Q(\Gamma_m)]$$

$$\left. - H_i(-\Gamma_m) \, H_j(\Gamma_m) \operatorname{Res}[S_{ij}^Q(-\Gamma_m)]) \Gamma_m \log(\Gamma_m/i) \right| \tag{8.42}$$

The apparent dimensional inconsistency implied by the appearance of logarithms of angular frequencies disappears when it is observed that the asymptotic behavior of $S_{ij}^Q(\omega)$ for large ω implies that all arguments of the logarithms may be divided by a common factor, e.g., $\sqrt{\omega_i \omega_j}$, without changing the result. The envelope concept is most useful when a single frequency is dominating. Then only the simpler diagonal term $i = j$ is needed (Krenk et al., 1983).

The rational power spectra considered in this section are closely related to simple systems excited by white noise. Let the stochastic process $q(t)$ be the response to white noise excitation of a system governed by the differential equation

$$\sum_m a_m \left(\frac{d}{dt}\right)^m q(t) = n(t) \tag{8.43}$$

The transfer function $H_q(\omega)$ is found as the complex amplitude to a solution with angular frequency ω. Substitution of $n(t) = e^{i\omega t}$ and $q(t) = H_q(\omega) e^{i\omega t}$ into (8.43) gives

$$H_q(\omega) = \left| \sum_m a_m (i\omega)^m \right|^{-1} = \sum_m c_m \frac{1}{\omega - \Gamma_m} \tag{8.44}$$

The last equality is valid only when all poles Γ_m of $H_q(\omega)$ are distinct; the result follows from expansion in partial fractions. In analogy with (8.29) the system response function $h_q(t)$ is the Fourier transform of $H_q(\omega)$. When the coefficients a_m and ω are real, it follows from (8.44) that $H_q(-\omega) = \overline{H}_q(\omega)$ and the system response function $h_q(t)$ then is

$$h_q(t) = \frac{1}{2\pi} \int_{-\infty}^{\infty} H_q(\omega) \, e^{i\omega t} \, d\omega = \operatorname{Re} \left| \frac{1}{\pi} \int_0^{\infty} H_q(\omega) \, e^{i\omega t} \, dt \right| \tag{8.45}$$

For $t \geq 0$ the contour is closed around the first quadrant, giving the solution

$$h_q(t) = \operatorname{Re} \left| 2i \sum_m c_m \, e^{i\Gamma_m t} \right|, \quad t > 0 \tag{8.46}$$

where the summation includes all Γ_m in the first quadrant. To get $h_q(t) \equiv 0$ for $t < 0$, there cannot be any poles in the fourth quadrant. Symmetry requires that for each pole Γ_m, $-\overline{\Gamma_m}$ must also be a pole. A stable system (8.43) thus has all poles of the transfer function $H_q(\omega)$ symmetrically located in the upper half-plane. If $N(t)$ is a white noise process of unit intensity, the spectral density of the process $Q(t)$ is

$$S^Q(\omega) \ = \ H_q(\omega)\,\overline{H_q(\omega)} \tag{8.47}$$

This procedure may be extended to vector processes and provides a physical interpretation of processes with rational power spectra. The possibility of reducing problems with rational spectra to white noise excitation of the combined system is related to the theory of Markov vector processes dealt with in Section 8.4.3.

The method of complex contour integration gives the full covariance function depending on the poles and the residues. If only the spectral moments are needed, they can be determined explicitly from the coefficients of the factored form (8.47) of the spectrum as described by Spanos (1983).

Three simple rational spectra are listed in Table 8.1. They are shown in Fig. 8.4. Typically, spectrum a exhibits a maximum near ω_a, and it will be more pronounced for decreasing values of ζ_a. This spectrum, often called the Tajimi spectrum, has found application for earthquake excitation. Spectrum b is a high-pass filter, leaving the high frequencies nearly undisturbed for a suitable choice of the parameter ζ_b, e.g., $\zeta_b = \sqrt{2}$. Spectrum c is a simple low-pass filter. The pole is now on the imaginary axis and the residue should be multiplied by $1/2$ before it is used in the formulas of this section.

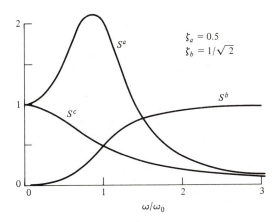

Figure 8.4 The rational spectra of Table 8.1.

TABLE 8.1   Rational Spectra
Spectrum a:
$$S^a(\omega) = \frac{\omega_a^4 + 4\zeta_a^2\omega_a^2\omega^2}{(\omega_a^2-\omega^2)^2 + 4\zeta_a^2\omega_a^2\omega^2}$$
$$\Gamma_a = \omega_a e^{i\theta_a}, \quad \sin\theta_a = \zeta_a$$
$$\text{Res}[S^a(\Gamma_a)] = -\frac{i}{\Gamma_a}\frac{\omega_a^2 + 4\zeta_a^2\Gamma_a^2}{8\zeta_a\sqrt{1-\zeta_a^2}}$$
Spectrum b:
$$S^b(\omega) = \frac{\omega^4}{(\omega_b^2-\omega^2)^2 + 4\zeta_b^2\omega_b^2\omega^2}$$
$$\Gamma_b = \omega_b e^{i\theta_b}, \quad \sin\theta_b = \zeta_b$$
$$\text{Res}[S^b(\Gamma_b)] = \frac{i}{\Gamma_b}\frac{\omega_b^2 - 2(1-2\zeta_b^2)\Gamma_b^2}{8\zeta_b\sqrt{1-\zeta_b^2}}$$
Spectrum c:
$$S^c(\omega) = \frac{\omega_c^2}{\omega_c^2 + \omega^2}$$
$$\Gamma_c = i\omega_c$$
$$\text{Res}[S^c(\Gamma_c)] = \frac{\omega_c^2}{2\Gamma_c}$$

8.2 NONSTATIONARY RESPONSE

The simplicity of describing stationary stochastic response as compared with the general nonstationary case is intimately connected with the covariance function $c_{ij}(t_1,t_2)$, being described completely by one argument, $t_1 - t_2$. This allows the spectral representation (8.12), by which the problem is reduced to algebraic form (8.13). For linear structures the frequency representation also provides a physical interpretation by means of the superposition principle.

Before turning to the problem of nonstationary excitation, the problem of stationary excitation of a structure, which is at rest at $t = 0$, is considered. This problem can be solved by elementary means, and the solution provides a decomposition of the covariance function in terms of simple one-dimensional functions.

A different, and more general type of nonstationarity is obtained by modulating intensity and frequency of stationary processes. This leads to the introduction of the evolutionary power spectrum (Priestley, 1965, 1967). The evolutionary power spectrum depends on one frequency only, but also includes explicit dependence on time in the nonstationary case. While the preceding section demonstrated how filtering through a time-independent system can be used to generate rational spectra from white noise, Section 8.2.2 demonstrates how a nonstationary process can be obtained by filtering a stationary process through a time-dependent system.

8.2.1 Transient Response

The term transient response is used here to describe the response of a system that is excited by stationary processes from time $t = 0$, and satisfies homogeneous initial conditions. For the linear second-order systems treated here, the expectation and covariance function of transient response are closely related to those of the stationary response that are approached as $t \to \infty$.

The simplest way to obtain the transient response may well be to consider the corresponding stationary response $T_i^{st}(t)$ and observing that whereas it satisfies the differential equations

$$\ddot{T}_i(t) + 2\zeta_i\omega_i\dot{T}_i(t) + \omega_i^2 T_i(t) = Q_i(t) \tag{8.48}$$

in the relevant time interval $t > 0$, it fails to satisfy the homogeneous initial conditions at $t = 0$. However, it follows from (8.4) that the function

$$T_i^{st}(0)\,g_i(t) + \dot{T}_i^{st}(0)\,h_i(t) \tag{8.49}$$

gives the same initial values as $T_i^{st}(t)$ while satisfying the homogeneous form of the differential equation (8.48). Thus the stochastic process

$$T_i(t) = T_i^{st}(t) - T_i^{st}(0)\,g_i(t) - \dot{T}_i^{st}(0)\,h_i(t), \quad t \geq 0 \tag{8.50}$$

is the desired transient response process.

The expectation follows directly from (8.50) and (8.23) in the form

$$\mu_i(t) = [1 - g_i(t)]\mu_i^Q/\omega_i^2 \tag{8.51}$$

The ratio of $\mu_i(t)$ to its stationary value $\mu_i^{st} = \mu_i(\infty)$ is shown in Fig. 8.5 for $\zeta_i = 0.05$. It is seen that $\mu_i(t)$ overshoots its final value by nearly a factor of 2 and through regular oscillations eventually reaches μ_i^{st}. This is the well-known deterministic response to an instantaneously applied load μ_i^Q.

The covariance function of the transient response follows from (8.50) in the form

$$C_{ij}(t_1,t_2) = \tag{8.52}$$

$$[1 \; -g_i(t_1) \; -h_i(t_1)] \begin{vmatrix} c_{ij}(t_1,t_2) & c_{ij}(t_1,0) & \dfrac{\partial}{\partial t_2}c_{ij}(t_1,0) \\[2ex] c_{ij}(0,t_2) & c_{ij}(0,0) & \dfrac{\partial}{\partial t_2}c_{ij}(0,0) \\[2ex] \dfrac{\partial}{\partial t_1}c_{ij}(0,t_2) & \dfrac{\partial}{\partial t_1}c_{ij}(0,0) & \dfrac{\partial^2}{\partial t_1\partial t_2}c_{ij}(0,0) \end{vmatrix} \begin{vmatrix} 1 \\[2ex] -g_j(t_2) \\[2ex] -h_j(t_2) \end{vmatrix}$$

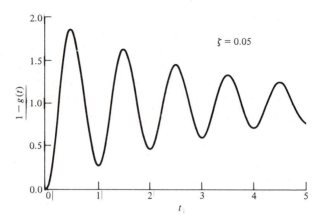

Figure 8.5 Transient expected value $\mu_i(t)$ from (8.51).

where the covariances of $\dot{T}_i(0)$ and $T_j(t)$, etc., were given in (7.61) and (7.62). The result (8.52) can be written in operator form as

$$C_{ij}(t_1,t_2) \;=\; \left\{1 - \left[g_i(t_1)+h_i(t_1)\frac{\partial}{\partial t_1}\right]_{t_1=0}\right\} \tag{8.53}$$

$$\times \left\{1 - \left[g_j(t_2)+h_j(t_2)\frac{\partial}{\partial t_2}\right]_{t_2=0}\right\} c_{ij}(t_1,t_2)$$

The subscripts $t_1=0$ and $t_2=0$ refer to the arguments of the covariance function $c_{ij}(t_1,t_2)$ of the stationary response $T_i^{st}(t)$.

If the excitation terminates at the time $T>0$, the modal response is given in terms of $T_i(T)$ and $\dot{T}_i(T)$ as

$$T_i(t) \;=\; T_i(T)\,g_i(t-T) + \dot{T}_i(T)\,h_i(t-T), \quad t \geqslant T \tag{8.54}$$

The expectation and covariance function can therefore be calculated by use of (8.51) and (8.53). For $t_1,t_2 > T$ the result is

$$\mu_i(t) \;=\; \left|g_i(t-T) + h_i(t-T)\frac{d}{d\tau}\right|_{\tau=T} \mu_i(\tau), \quad t > T \tag{8.55}$$

$$C_{ij}(t_1,t_2) \;=\; \left\{\left|g_i(t_1-T)+h_i(t_1-T)\frac{\partial}{\partial\tau_1}\right|_{\tau_1=T}\right\} \tag{8.56}$$

$$\times \left\{\left|g_j(t_2-T)+h_j(t_2-T)\frac{\partial}{\partial\tau_2}\right|_{\tau_2=T}\right\} C_{ij}(\tau_1,\tau_2), \quad t_1,t_2 > T$$

In (8.53) the two-dimensional function $C_{ij}(t_1,t_2)$ is expressed in terms of one-dimensional functions as $c_{ij}(t_1,t_2)$ depends only on

$\tau = t_1 - t_2$. In spite of the number of different terms needed, this represents a considerable computational advantage. In the general case the derivatives of $c_{ij}(t_1, t_2)$ can be evaluated from (8.28), and when the spectrum is rational the derivatives are available in closed form from (8.36). In the special case of white noise excitation considerable simplifications are possible, as demonstrated in Section 8.3.

Typical features of the covariance functions of transient response are illustrated in Fig. 8.6, showing the transient equivalent of Fig. 8.3. It is seen that the covariances grow gradually over several periods. A closer examination of wide-band excitation of lightly damped structures is presented in Section 8.3.

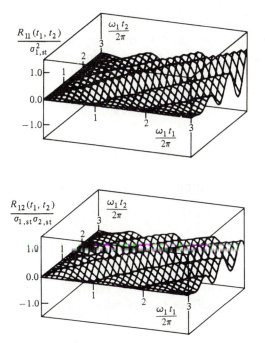

Figure 8.6   Modal covariance functions for transient response to white noise excitation, $\omega_2 = 1.25\omega_1$, $\zeta_1 = \zeta_2 = 0.05$.

8.2.2 Modulated Excitation

A very useful type of nonstationary stochastic process is obtained by modulation of a stationary process. It is important to include time variation of frequency content as well as intensity of the process. Let the stochastic process $Y(t)$ be obtained from the stationary process $X(t)$ through the integral

$$Y(t) = \int_0^\infty a(t,\tau) X(t-\tau) d\tau \qquad (8.57)$$

$Y(t)$ may be considered as the output of a linear time-dependent system with response function $a(t,\tau)$ and input $X(t-\tau)$. At any time t the output, $Y(t)$, is a weighted average of the earlier input. The response function $a(t,\tau)$ creates nonstationarity through the dependence on t, while the dependence on τ represents an averaging effect.

The expectation of $Y(t)$ follows immediately from (8.57)

$$\mu^Y(t) = \mu^X \int_0^\infty a(t,\tau) d\tau \qquad (8.58)$$

Similarly the covariance function of $Y(t)$ can be expressed in terms of the covariance function of $X(t)$ as

$$C^Y(t_1,t_2) = \int_0^\infty \int_0^\infty a(t_1,\tau_1) a(t_2,\tau_2) c^X(t_1-\tau_1,t_2-\tau_2) d\tau_1 d\tau_2 \qquad (8.59)$$

The covariance function of $X(t)$ only depends on the difference of its arguments, and it can be expressed in terms of its spectral density $S^X(\omega)$ as in (8.9). After interchange of the order of integration

$$C^Y(t_1,t_2) = \int_{-\infty}^\infty \left| \int_0^\infty a(t_1,\tau_1) e^{-i\omega\tau_1} \right| \qquad (8.60)$$

$$\times \left| \int_0^\infty a(t_2,\tau_2) e^{i\omega\tau_2} d\tau_2 \right| S^X(\omega) e^{i\omega(t_1-t_2)} d\omega$$

Now introduce the notation

$$A(\omega,t) = \int_0^\infty a(t,\tau) e^{-i\omega\tau} d\tau \qquad (8.61)$$

by which (8.60) takes the form

$$C^Y(t_1,t_2) = \int_{-\infty}^\infty A(\omega,t_1) \overline{A(\omega,t_2)} S^X(\omega) e^{i\omega(t_1-t_2)} d\omega \qquad (8.62)$$

This is a generalization of (8.12) for stationary response of a linear system. For $t_1 = t_2$ it defines the so-called "evolutionary power spectrum"

$$S^Y(\omega,t) = |A(\omega,t)|^2 S^X(\omega) \qquad (8.63)$$

The time-dependent factor $|A(\omega,t)|^2$ can redistribute the spectral density, e.g., effect a gradual change to higher frequencies and increased intensity as shown in Fig. 8.7.

It follows from combination of (8.58) and (8.61) that the expectation of the modulated process $Y(t)$ is

$$\mu^Y(t) = A(0,t)\mu^X \qquad (8.64)$$

In this connection it is observed that intensity modulation alone by

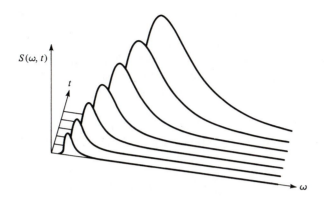

Figure 8.7 Evolutionary spectrum.

(8.61) requires $A(\omega,t)$ independent of ω, i.e., $A(\omega,t) \equiv A(0,t)$. Thus the expectation of Y is independent of the frequency modulation.

Now let the vector-valued excitation process $Q_i(t)$, $i = 1, \ldots, n$, be generated by modulation of a vector-valued stationary process $P_k(t)$. In analogy with (8.57),

$$Q_i(t) = \sum_k \int_0^\infty a_{ik}(t,\tau) P_k(t-\tau)\, d\tau \tag{8.65}$$

In the most general case the summation extends over $k = 1, \ldots, n$, but in practice $P_k(t)$ may have only one nonvanishing component. The covariance function of the generalized forces $Q_i(t)$ is

$$C_{ij}^Q(t_1,t_2) = \sum_k \sum_l \int_0^\infty \int_0^\infty a_{ik}(t_1,\tau_1) a_{jl}(t_2,\tau_2) c_{kl}^P(t_1-\tau_1, t_2-\tau_2)\, d\tau_1 d\tau_2 \tag{8.66}$$

Upon Introduction of the functions

$$A_{ik}(\omega,t) = \int_0^\infty a_{ik}(t,\tau)\, e^{-i\omega\tau}\, d\tau \tag{8.67}$$

the expectation of $Q_i(t)$ becomes

$$\mu_i^Q(t) = \sum_k A_{ik}(0,t)\, \mu_k^P \tag{8.68}$$

The covariance function can be rewritten by use of the power spectrum of $P_k(t)$ as

$$C_{ij}^Q(t_1,t_2) = \sum_k \sum_l \int_{-\infty}^\infty A_{ik}(\omega,t_1)\, \overline{A_{jl}(\omega,t_2)}\, S_{kl}^P(\omega)\, e^{i\omega(t_1-t_2)}\, d\omega \tag{8.69}$$

In practice, stochastic independence of the components $P_k(t)$ can be assumed, whereby the correlation between the components $Q_i(t)$ is produced solely by the filter functions $a_{jl}(t,\tau)$. This reduces the matrix $S_{kl}^P(\omega)$ to diagonal form and results in a single summation in (8.69).

The modal response $T_i(t)$ generated by the generalized forces $Q_i(t)$ satisfies the equation of motion (8.48), and the determination of expectation and covariances of $T_i(t)$ follows from the procedure used previously. If the excitation starts at a definite time, the structure is assumed to satisfy homogeneous initial conditions, and the modal response is then determined by the integral

$$T_i(t) = \int_{-\infty}^{t} Q_i(\tau) h_i(t-\tau) d\tau \qquad (8.70)$$

The expectation of $T_i(t)$ follows immediately from (8.70) and (8.68) in the form

$$\mu_i(t) = \sum_k \mu_k^P \int_{-\infty}^{t} A_{ik}(0,\tau) h_i(t-\tau) d\tau \qquad (8.71)$$

Comparison with the corresponding result (8.23) for stationary response shows that the integral of $h_i(t-\tau)$ has been replaced with a nontrivial convolution integral.

The covariance function of the modal response is given by the convolution integral

$$C_{ij}(t_1,t_2) = \int_{-\infty}^{t_1} \int_{-\infty}^{t_2} C_{ij}^Q(\tau_1,\tau_2) h_i(t_1-\tau_1) h_j(t_2-\tau_2) d\tau_1 d\tau_2 \qquad (8.72)$$

The only difference from the stationary case, (8.24), is that the covariance function of the generalized forces now depends on τ_1 and τ_2 independently and not only through the differences. When the covariance function $C_{ij}^Q(\tau_1,\tau_2)$ is substituted from (8.69), the result contains convolution integrals of $A_{ik}(\omega,\tau)$ and $h_i(\tau)$.

$$C_{ij}(t_1,t_2) = \sum_k \sum_l \int_{-\infty}^{\infty} \left[\int_{-\infty}^{t_1} A_{ik}(\omega,\tau_1) h_i(t_1-\tau_1) e^{i\omega\tau_1} d\tau_1 \right] \qquad (8.73)$$

$$\times \left[\int_{-\infty}^{t_2} \overline{A_{jl}(\omega,\tau_2)} h_j(t_2-\tau_2) e^{-i\omega\tau_2} d\tau_2 \right] S_{kl}^P(\omega) d\omega$$

Clearly, this formula generalizes (8.10), and a complete analogy with the frequency representation of stationary response is obtained by introduction of the time-dependent transfer function

$$H_{ik}(\omega,t) = \int_0^{\infty} A_{ik}(\omega,t-\tau) h_i(\tau) e^{-i\omega\tau} d\tau \qquad (8.74)$$

The expectation (8.71) then takes the form

$$\mu_i(t) = \sum_k \mu_k^P H_{ik}(0,t) \qquad (8.75)$$

and the covariance function (8.73) is rewritten as

$$C_{ij}(t_1,t_2) = \sum_k \sum_l \int_{-\infty}^{\infty} H_{ik}(\omega,t_1) \overline{H}_{jl}(\omega,t_2) S_{kl}^P(\omega) e^{i\omega(t_1-t_2)} d\omega \qquad (8.76)$$

The computational effort necessary to evaluate the covariance function of (8.74) and (8.76) is reduced considerably if $H_{ik}(\omega,t)$ can be expressed in closed form. A special case is considered in the following example.

Example 8.1

Let the excitation processes $Q_i(t)$ be generated by amplitude modulation of the stationary processes $P_k(t)$, and assume the amplitude to be zero for $t<0$. Thus (8.65) gives

$$Q_i(t) = \sum_k U(t)\, b_{ik}(t)\, P_k(t)$$

in which $U(t)$ is Heaviside's step function, and $b_{ik}(t)$ represents the amplitude modulation. The dependence on τ is in the form of a delta function, and it then follows from (8.67) that

$$A_{ik}(\omega,t) = U(t)\, b_{ik}(t)$$

and the time-dependent transfer function $H_{ik}(\omega,t)$ follows from (8.74)

$$H_{ik}(\omega,t) = \int_0^\infty U(t-\tau)\, b_{ik}(t-\tau)\, h_i(\tau)\, e^{-i\omega\tau}\, d\tau$$
$$= \int_0^t b_{ik}(t-\tau)\, h_i(\tau)\, e^{-i\omega\tau}\, d\tau$$

The simplest case is that of constant intensity for $t>0$, i.e., $b_{ik}(t-\tau)=b_{ik}$. The time-dependent transfer function is

$$H_{ik}(\omega,t) = b_{ik} \int_0^t h_i(\tau)\, e^{-i\omega\tau}\, d\tau$$
$$= b_{ik}\{ 1 - [g_i(t)+i\omega h_i(t)]\, e^{-i\omega t}\}\, H_i(\omega)$$

In this case the matrix b_{ik} only serves to define the processes $Q_i(t)$, $i=1,2,\ldots$, as linear combinations of the processes $P_k(t)$, $k=1,2,\ldots$, and by suitable choice of these processes b_{ik} is the unit matrix, $h_{ik}=h_{ik}$. Tin (1975) has made extensive use of the function $H_{ii}(\omega,t)$ for analysis of transient response. The simple relation between $H_{ii}(\omega,t)$ and $H_i(\omega)$ for constant amplitude immediately reduces the general formula (8.76) to the form (8.53), in which the transient covariance function is expressed by the stationary functions and the system functions (Madsen and Krenk, 1982).

This procedure can be generalized to the case in which $b_{ik}(t)$ is expanded in terms of polynomials and exponentials. For simplicity, only the exponential case is considered here. Introduce the expansion

$$b_{ik}(t) = \sum_m B_{ik}^m\, e^{\alpha_m t}$$

in the formula (8.74) for the time-dependent transfer function, whereby

$$H_{ik}(\omega,t) = \sum_m B_{ik}^m \int_0^\infty h_i(\tau)\, e^{\alpha_m(t-\tau)}\, e^{-i\omega\tau} d\tau = \sum_m B_{ik}^m\, e^{\alpha_m t} \int_0^t h_i(\tau)\, e^{-(\alpha_m+i\omega)\tau} d\tau$$
$$= \sum_m B_{ik}^m\, e^{\alpha_m t}\{ 1 - [g_i(t)+(\alpha_m+i\omega)h_i(t)]\, e^{-(\alpha_m+i\omega)t}\}\, H_i(\omega-i\alpha_m)$$

The integration follows directly from the result for constant intensity, when ω is replaced by $\omega - i\alpha_m$. This form of the time-dependent transfer function enables a straightforward, although perhaps somewhat laborious extension of the closed-form result (8.36) for the covariance function in the case of rational power spectra $S_{kl}^P(\omega)$.

8.3 WHITE NOISE APPROXIMATION

It may be argued intuitively that if the excitation process is correlated only over very short time intervals compared with the natural period of the structure, the effect would be equivalent to that produced by a large number of uncorrelated pulses. This leads to the useful concept of a white noise process already mentioned briefly in Section 8.1.3. In the following some useful results are derived for excitation of linear structures by ideal white noise processes, and the application of these results to excitation by general wideband processes is treated. Representation of the generalized forces by white noise processes not only simplifies response calculation of linear structures considerably, but also enables some nonlinear effects to be included as discussed in Section 8.4.3.

8.3.1 Ideal White Noise

The white noise process is an idealization in which the value of the covariance function of the process is zero at different times t_1 and t_2. In the stationary case this implies a covariance function of the form

$$c_{ij}^Q(t_1,t_2) = 2\pi S_{ij}^Q \delta(t_1 - t_2) \qquad (8.77)$$

in which $\delta(\ )$ is Dirac's delta function. By inversion of (8.26) the constants S_{ij}^Q are seen to constitute the spectral matrix, i.e., $S_{ij}^Q(\omega) = S_{ij}^Q$. The fact that $S_{ij}^Q(\omega)$ is constant implies that the spectral density is not integrable, and as a consequence the covariance function takes an infinite value at $t_1 = t_2$. This seeming inconsistency is a result of replacing a covariance function with nonvanishing values only for small arguments $(t_1 - t_2)$ with the expression (8.77). The difference is likely to be small for integrals such as (8.24) for the covariance of the response, and it is also seen from (8.30) that the spectral density of the response is integrable.

In the case of stationary ideal white noise excitation, the covariance matrix of the modal response is given by the first term in (8.36). The constants S_{ij}^Q are real, and the covariance function can then be expanded in terms of the modal response functions $g_i(t)$ and $h_i(t)$ as

$$c_{ij}(t_1,t_2) = \pi S_{ij}^Q[\alpha_{ij}g_i(t_1-t_2) + \beta_{ij}h_i(t_1-t_2)], \qquad t_1 \geqslant t_2 \quad (8.78)$$

The coefficients α_{ij} and β_{ij} are independent of t_1 and t_2 and given by

$$\alpha_{ij} = \frac{4(\omega_i\zeta_i + \omega_j\zeta_j)}{(\omega_i^2 - \omega_j^2)^2 + 4\omega_i\omega_j(\omega_i\zeta_j + \omega_j\zeta_i)(\omega_i\zeta_i + \omega_j\zeta_j)} \quad (8.79)$$

and

$$\beta_{ij} = \frac{2(\omega_j^2 - \omega_i^2)}{(\omega_i^2 - \omega_j^2)^2 + 4\omega_i\omega_j(\omega_i\zeta_j + \omega_j\zeta_i)(\omega_i\zeta_i + \omega_j\zeta_j)} \quad (8.80)$$

For $t_1 < t_2$ the indices i and j are interchanged in (8.78). The symmetry relations $\alpha_{ij} = \alpha_{ji}$ and $\beta_{ij} = -\beta_{ji}$ are noted. The coefficients α_{ij} and β_{ij} are shown in Figs. 8.8 and 8.9 as functions of the frequency ratio ω_j/ω_i for selected values of the damping ratios ζ_i and ζ_j. Generally, the correlation between modes decreases with increasing relative difference in modal frequencies.

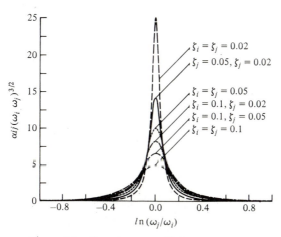

Figure 8.8 Covariance coefficient α_{ij}, (8.79).

The first three spectral moments of the modal response follow from (8.41) and (8.42). They are given in Table 8.2, where it has been convenient to introduce the symmetric coefficients

$$\gamma_{ij} = \omega_i^2\alpha_{ij} + 2\zeta_i\omega_i\beta_{ij} \quad (8.81)$$

$$= \frac{4\omega_i\omega_j(\omega_i\zeta_j + \omega_j\zeta_i)}{(\omega_i^2 - \omega_j^2)^2 + 4\omega_i\omega_j(\omega_i\zeta_j + \omega_j\zeta_i)(\omega_i\zeta_i + \omega_j\zeta_j)}$$

The coefficient γ_{ij} is shown in Fig. 8.10. These results are from Madsen and Krenk (1982). Similar results have been obtained by

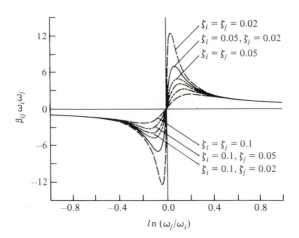

Figure 8.9 Covariance coefficient β_{ij}, (8.80).

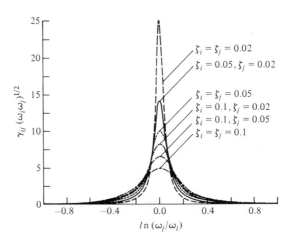

Figure 8.10 Covariance coefficient γ_{ij}, (8.81).

Der Kiureghian (1980). The variance of mode i is given by

$$\lambda_0^{ii} = \frac{\pi}{2\omega_i^3 \zeta_i} \tag{8.82}$$

and the full autocovariance function follows from (8.78):

$$c_{ii}(t_1, t_2) = \lambda_0^{ii} g_i(\,|\,t_1 - t_2\,|\,) \tag{8.83}$$

Thus, for white noise excitation the autocorrelation function for any mode is simply the modal response function $g_i(\,|\,t_1 - t_2\,|\,)$. Each individual mode exhibits narrow-band properties for low damping,

$\zeta_i \ll 1$. Only the first three spectral moments exist and the appropriate bandwidth parameter therefore is δ, from (7.117). A combination of the spectral moments in Table 8.2 yields:

$$\delta = \frac{2}{\pi} \frac{\arccos \zeta_i}{\sqrt{1-\zeta_i^2}} u \approx 1 - \frac{2\zeta_i}{\pi} \tag{8.84}$$

Thus for decreasing damping ratio, $\zeta_i \to 0$, the variance increases without bound by (8.82), and the narrow-band properties of each modal response are enhanced by (8.84).

	TABLE 8.2 Spectral Moments, λ_k^{ii}/S_{ij}^Q	
k	$i = j$	$i \neq j$
0	$\dfrac{\pi}{2\omega_i^3 \zeta_i}$	$\pi \alpha_{ij}$
1	$\dfrac{\arccos \zeta_i}{\omega_i^2 \zeta_i \sqrt{1-\zeta_i^2}}$	$\dfrac{(\omega_i^2 \alpha_{ij} + \gamma_{ij})\arccos \zeta_i}{2\omega_i \sqrt{1-\zeta_i^2}}$ $+ \dfrac{(\omega_j^2 \alpha_{ij} + \gamma_{ij})\arccos \zeta_j}{2\omega_j \sqrt{1-\zeta_j^2}} + \beta_{ij}\log\left(\dfrac{\omega_i}{\omega_j}\right)$
2	$\dfrac{\pi}{2\omega_i \zeta_i}$	$\pi \gamma_{ij}$

The transient response to ideal white noise excitation constitutes a convenient starting point for the more general nonstationary problem. When the covariance function (8.78) for stationary response is substituted into the expression (8.53) for the transient covariance function, the following formula is obtained:

$$C_{ij}(t_1,t_2) = c_{ij}(t_1,t_2) - \pi S_{ij}^Q \{\alpha_{ij} g_i(t_1)g_j(t_2) + \gamma_{ij} h_i(t_1)h_j(t_2) \tag{8.85}$$
$$- \beta_{ij}[g_i(t_1)h_j(t_2) - h_i(t_1)g_j(t_2)]\}, \quad t_1, t_2 \geqslant 0$$

This is the function illustrated in Fig. 8.6. The transient autocovariance function is obtained for $j = i$.

$$C_{ii}(t_1,t_2) = \lambda_0^{ii}[g_i(|t_1 - t_2|) - g_i(t_1)g(t_2) - \omega_i^2 h_i(t_1)h_i(t_2)] \tag{8.86}$$

The transient variance takes a particularly simple form when it is observed that $h_i(t) = -\dot{g}_i(t)/\omega_i^2$, whereby

$$\sigma_i(t)^2 = \lambda_0^{ii}\{1 - [g_i(t)^2 + (\dot{g}_i(t)/\omega_i)^2]\} \tag{8.87}$$

The term in the brackets represents the transient. It is the squared value of the amplitude of $g_i(t)$, and therefore the stationary limit is approached more rapidly by highly damped structures. However, this should not be confused with a more rapid growth of response intensity in highly damped structures, because by (8.82) the stationary limit is simply lower for large damping ratios.

Now let the generalized forces consist of amplitude-modulated white noise in the form

$$Q_i(t) = U(t) \sum_k b_{ik}(t) P_k(t) \qquad (8.88)$$

The covariance functions of the white noise processes $P_k(t)$ are

$$c_{kl}^P(\tau_1, \tau_2) = 2\pi S_{kl}^P \delta(\tau_1 - \tau_2) \qquad (8.89)$$

The covariance function for $Q_i(t)$ follows by combination of (8.88) and (8.89):

$$C_{ij}^Q(\tau_1, \tau_2) = U(\tau_1) U(\tau_2) 2\pi \sum_k \sum_l S_{kl}^P b_{ik}(\tau_1) b_{jl}(\tau_2) \delta(\tau_1 - \tau_2) \qquad (8.90)$$

The simple form of this expression favors direct calculation of the response covariance function in the time domain by use of (8.72). Integration of the delta function reduces the expression to a single integral with $\tau_1 = \tau_2 = \tau$ and the upper limit $t_m = \min(t_1, t_2)$.

$$C_{ij}(t_1, t_2) = 2\pi \sum_k \sum_l S_{kl}^P \int_0^{t_m} b_{ik}(\tau) b_{jl}(\tau) h_i(t_1 - \tau) h_j(t_2 - \tau) \, d\tau \qquad (8.91)$$

By suitable choice of the representation (8.88) diagonal form of S_{kl}^P can be achieved, whereby only a single summation in (8.91) is needed. The resultant expression,

$$C_{ij}(t_1, t_2) = 2\pi \sum_k S_{kk}^P \int_0^{t_m} b_{ik}(\tau) h_i(t_1 - \tau) b_{jk}(\tau) h_j(t_1 - \tau) \, d\tau \qquad (8.92)$$

clearly shows the response covariance as the result of pulses generated by the processes $P_k(t)$, $k = 1, \ldots,$ in the time interval $0 \leqslant \tau \leqslant \min(t_1, t_2)$.

8.3.2 Approximation of Wide-Band Excitation

The response analysis is considerably simplified if the excitation is by amplitude-modulated white noise, and it is therefore of interest to consider the possibility of approximating the response to general wide-band excitation by the response to a suitably calibrated white noise process. First, the excitation must be of wide-band character, i.e., the spectral density must be distributed evenly over a wide range of frequencies, including the eigenfrequencies of the structure. This implies that the excitation process is only lightly correlated over time intervals corresponding to natural periods of the structure, and an approximation based on stationary response may therefore also prove adequate when extended to nonstationary response.

Second, it is observed that stationary white noise excitation is described in terms of the real-valued matrix S_{ij}^Q, whereas the actual spectral densities $S_{ij}^Q(\omega)$ may vary with ω and take complex values in accordance with the symmetry relations (8.27). A nonvanishing imaginary part of the cross spectral density $S_{ij}^Q(\omega)$ is equivalent to an odd component of the covariance function $c_{ij}^Q(t_1 - t_2)$ and thereby represents an influence of the sign of the time difference $(t_1 - t_2)$ (see, e.g., Bendat and Piersol, 1966). This influence is not present in white noise, and the following considerations are based on the assumption that the spectral densities $S_{ij}^Q(\omega)$ are real-valued. This assumption is satisfied if the processes $Q_i(t)$, $i = 1, \ldots,$ are linear combinations of stochastically independent processes.

The spectral matrix S_{ij}^Q of the equivalent white noise excitation process is determined by matching the response covariance matrix $c_{ij}(t,t)$ (Madsen and Krenk, 1982). The response covariance matrix for ideal white noise is given by (8.78).

$$c_{ij}(t,t) = \pi \alpha_{ij} S_{ij}^Q \tag{8.93}$$

The similar result for stationary excitation with real valued spectral densities follow from (8.28).

$$c_{ij}(t,t) = \int_{-\infty}^{\infty} \text{Re}[H_i(\omega)H_j(-\omega)] S_{ij}^Q d\omega \tag{8.94}$$

It is clear from Fig. 8.1 showing $|H_i(\omega)|$ that for lightly damped structures the integral in (8.94) is dominated by the behavior of the spectral density $S_{ij}^Q(\omega)$ around the natural frequencies ω_i and ω_j. This suggests a decomposition of the first factor in (8.94) to separate the influence of the two resonance frequencies. By expansion in partial fractions and recombination of terms belonging to the same mode, the following decomposition is obtained

$$\text{Re}[H_i(\omega)H_j(-\omega)] = |H_i(\omega)|^2 \left| \zeta_i \omega_i^3 \alpha_{ij} + \frac{1}{2}\beta_{ij}(\omega_i^2 - \omega^2) \right| \tag{8.95}$$

$$+ |H_j(\omega)|^2 \left| \zeta_j \omega_j^3 \alpha_{ji} + \frac{1}{2}\beta_{ji}(\omega_j^2 - \omega^2) \right|$$

The last terms are seen to vanish at the appropriate natural frequencies, and for nearly constant spectral densities these terms only give small contributions. Therefore, the integral (8.94) may be given approximately as

$$c_{ij}(t,t) \approx \zeta_i \omega_i^3 \alpha_{ij} \int_{-\infty}^{\infty} |H_i(\omega)|^2 S_{ij}^Q(\omega) d\omega \tag{8.96}$$

$$+ \zeta_j \omega_j^3 \alpha_{ji} \int_{-\infty}^{\infty} |H_j(\omega)|^2 S_{ij}^Q(\omega) d\omega$$

A further step is to replace the spectral density function with its value at the appropriate resonance frequencies, whereby (8.96) reduces to

$$c_{ij}(t,t) \approx \frac{\pi}{2} \alpha_{ij} [S_{ij}^{Q}(\omega_i) + S_{ij}^{Q}(\omega_j)] \qquad (8.97)$$

Comparison with (8.93) gives the equivalent white noise spectral density S_{ij}^{Q} as the average of the actual values at the two natural frequencies.

$$S_{ij}^{Q} = \frac{1}{2}[S_{ij}^{Q}(\omega_i) + S_{ij}^{Q}(\omega_j)] \qquad (8.98)$$

Direct use of (8.98) in the nonstationary case replaces (8.91) with

$$C_{ij}(t_1,t_2) = \pi \int_0^{t_m} [S_{ij}^{Q}(\omega_i,\tau) + S_{ij}^{Q}(\omega_j,\tau)] \, h_i(t_1-\tau) \, h_j(t_2-\tau) \, d\tau \quad (8.99)$$

in which $S_{ij}^{Q}(\omega,\tau)$ is the evolutionary spectrum from (8.69).

$$S_{ij}^{Q}(\omega,\tau) = \sum_k \sum_l A_{ik}(\omega,\tau) \, \overline{A_{jl}(\omega,\tau)} \, S_{kl}^{P}(\omega) \qquad (8.100)$$

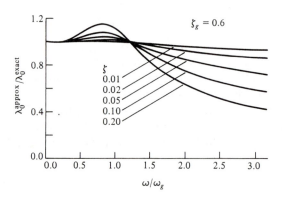

Figure 8.11   Accuracy of variance approximation by (8.97).

The white noise calibration described here is subject to two essentially different types of error. The first type is the error involved in the approximate evaluation of the covariances $c_{ij}(t,t)$ from the spectral densities near the natural frequencies by (8.97). The approximate and exact values of the modal variance, $c_{ii}(t,t)$, are compared in Fig. 8.11 for excitation with a spectral density of Tajimi type — type a in Table 8.1. The value $\zeta_a = 0.6$ often associated with earthquake excitation is used. It is seen that for a lightly damped structure the approximation is acceptable if the natural frequency ω_i is not very much larger than the typical excitation frequency ω_a. If this is the case, the response intensity is underestimated because the spectral density $S^a(\omega_i)$ is not representative. The approximation of the covariances $c_{ij}(t,t)$, $i \neq j$, follows the same trend but includes an additional uncertainty due to the omission of the last term in (8.95).

The other type of error concerns the time dependence of the covariances $c_{ij}(t,t)$ in the nonstationary case. It is illustrated in Fig. 8.12, showing the normalized transient response of two modes to excitation by a single process with spectral density of type a in Table 8.1 and with parameters $\omega_2 = 1.25\omega_1$, $\omega_a = 2\omega_1$, and $\zeta_a = 0.6$. The figure also shows the white noise approximation, and the agreement is very satisfactory.

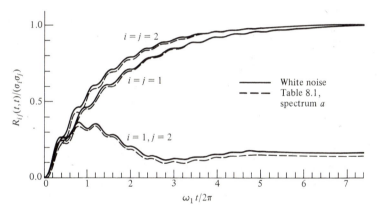

Figure 8.12 Covariances of transient response to excitation with Tajimi spectrum and white noise, $\omega_2 = 1.25\omega_1$, $\zeta_{s1} = \zeta_{s2} = 0.05$.

8.4 NONLINEAR EFFECTS

The response of linear structures to random excitation was treated entirely within the framework of normal processes. This is possible, because a normal excitation of a linear structure produces a normal response, and the problem is therefore reduced to that of relating the respective mean and covariance functions.

This section gives a brief discussion of some methods that can be used to characterize nonlinear and thereby nonnormal response. It is common for these methods that they give only partial answers and most of the results are only approximations.

The simplest way to treat nonlinear behavior is to replace the nonlinear governing equation with an equivalent set of linear equations, and in this case the problem consists in devising a suitable procedure for selecting the equivalent linear system. The equivalent linearization procedure is discussed in Section 8.4.1. If the load process is nonnormal but can be expressed as a function of a normal process, it is possible to retain the characteristics of the load spectrum, if the response is assumed to be normal. Problems of this type are conveniently handled by use of Hermite polynomial expansions as described in Section 8.4.2. Application to wave loads is dealt with in Section 10.2.

Both of the mentioned techniques depend on special properties of the normal distribution, and the response is approximated by normal processes. While this may be adequate for estimating intensities and spectra, it does not provide accurate information about extreme events. For this purpose use can be made of the theory of Markov processes for which some exact results are known concerning the distribution of the response. The central point is the limited memory that leads to an equation of diffusion type for the probability density. Section 8.4.3 gives a brief derivation of the Fokker-Planck equation and discusses its application to response of structures with nonlinear stiffness.

8.4.1 Equivalent Linearization

The method of equivalent linearization is closely linked to properties of the normal distribution. Let $X_1, \ldots, X_n$ be joint normal variables with expectations μ_j and covariance matrix c_{jk}. As described in Section 7.4.1, the probability density function then is

$$f_\mathbf{X}(\mathbf{x}) = \frac{1}{(2\pi)^{n/2} |c_{jk}|^{1/2}} \exp\left|-\frac{1}{2}\sum_j \sum_k (x_j - \mu_j)c_{jk}^{-1}(x_k - \mu_k)\right| \quad (8.101)$$

Differentiation gives

$$\frac{\partial}{\partial x_j} f_\mathbf{X}(\mathbf{x}) = -\sum_k c_{jk}^{-1}(x_k - \mu_k) f_\mathbf{X}(\mathbf{x}) \quad (8.102)$$

Upon multiplication with the covariance matrix c_{jk},

$$\sum_j c_{jk} \frac{\partial}{\partial x_j} f_\mathbf{X}(\mathbf{x}) = -(x_k - \mu_k) f_\mathbf{X}(\mathbf{x}) \quad (8.103)$$

This relation enables integration by parts of expectations of functions of joint normal variables. The basic step consists in the evaluation of the expectation

$$E[(X_k - \mu_k)g(\mathbf{X})] \tag{8.104}$$

$$= \int_{-\infty}^{\infty} \cdots \int_{-\infty}^{\infty} (x_k - \mu_k) g(\mathbf{x}) f_{\mathbf{X}}(\mathbf{x}) \, dx_1 \cdots dx_n$$

where $g(\mathbf{X})$ is an arbitrary function of $X_1, \ldots, X_n$. By use of (8.103) and integration by parts

$$E[(X_k - \mu_k)g(\mathbf{X})] \tag{8.105}$$

$$= -\sum_j c_{kj} \int_{-\infty}^{\infty} \cdots \int_{-\infty}^{\infty} g(\mathbf{x}) \frac{\partial}{\partial x_j} f_{\mathbf{X}}(\mathbf{x}) \, dx_1 \cdots dx_n$$

$$= \sum_j c_{kj} \int_{-\infty}^{\infty} \cdots \int_{-\infty}^{\infty} f_{\mathbf{X}}(\mathbf{x}) \frac{\partial}{\partial x_j} g(\mathbf{x}) \, dx_1 \cdots dx_n$$

where contributions from $\pm\infty$ are assumed to vanish. This establishes the important result

$$E[(X_k - \mu_k)g(\mathbf{X})] = \sum_j c_{kj} E\left|\frac{\partial}{\partial x_j} g(\mathbf{X})\right| \tag{8.106}$$

for joint normal variables.

Example 8.2

In particular, the formula (8.106) enables the evaluation of higher moments of joint normal variables in terms of the covariance matrix c_{ij}. The odd central moments vanish, while for a single variable with zero mean

$$E[X^{2n}] = E[X^2](2n-1)E[X^{2(n-1)}] = E[X^2]^2(2n-1)(2n-3)E[X^{2(n-2)}]$$

$$= \cdots = (2n-1)(2n-3)\times \cdots \times 3 \times 1 \, E[X^2]^n$$

The formula (7.101) for the correlation function of the squared envelope process follows from the result

$$E[X_1^2 X_2^2] = c_{11}E[X_2^2] + 2c_{12}E[X_1 X_2] = c_{11}c_{22} + 2c_{12}^2$$

for joint normal variables with zero mean.

Let a nonlinear equation of motion be of the form

$$g(\ddot{\mathbf{X}}, \dot{\mathbf{X}}, \mathbf{X}) = \mathbf{F}(t) \tag{8.107}$$

where $\mathbf{X}(t)$ is the unknown generalized displacement vector. For simplicity it is assumed that $\mathbf{F}(t)$ is a stationary, zero-mean normal process and that $g(\ ,\ ,\ )$ is an odd function of its arguments. The equivalent linearization technique consists in the construction of an equivalent linear system of equations. The solution $\mathbf{Y}(t)$ to the equivalent linear system of equations is then taken as an approximation to $\mathbf{X}(t)$.

The problem is to determine the coefficients of the equation

$$\mathbf{M}\ddot{\mathbf{Y}} + \mathbf{C}\dot{\mathbf{Y}} + \mathbf{K}\mathbf{Y} = \mathbf{F}(t) \tag{8.108}$$

In the following analysis, developed by Atalik and Utku (1976), it is convenient to consider the vector

$$\mathbf{Z}^T = [\ddot{\mathbf{Y}}^T, \dot{\mathbf{Y}}^T, \mathbf{Y}^T] \tag{8.109}$$

which by the linearity of (8.108) has joint normal components with zero mean. It is convenient to match (8.108) to (8.107) by considering the error that would result from substituting $\mathbf{Y}(t)$ into (8.107).

$$\mathbf{\Delta} = \mathbf{g}(\mathbf{Z}) - (\mathbf{M}\ddot{\mathbf{Y}} + \mathbf{C}\dot{\mathbf{Y}} + \mathbf{K}\mathbf{Y}) \tag{8.110}$$

The prime reason for defining this error measure is that the process $\mathbf{Y}(t)$ will be available and normal for any particular choice of the matrices in (8.108), while no general results exist for $\mathbf{X}(t)$.

A matching procedure can now be defined consisting in minimizing the mean square error $E[\mathbf{\Delta}^T\mathbf{\Delta}]$. This requires

$$\frac{\partial}{\partial M_{ij}} E[\mathbf{\Delta}^T\mathbf{\Delta}] = 0$$

$$\frac{\partial}{\partial C_{ij}} E[\mathbf{\Delta}^T\mathbf{\Delta}] = 0, \qquad i,j = 1, \ldots, n \tag{8.111}$$

$$\frac{\partial}{\partial K_{ij}} E[\mathbf{\Delta}^T\mathbf{\Delta}] = 0$$

and upon substitution of $\mathbf{\Delta}$ from (8.110), these equations can be written in the form

$$E[\mathbf{Z}\mathbf{Z}^T][\mathbf{M}\ \mathbf{C}\ \mathbf{K}]^T = E[\mathbf{Z}\mathbf{g}^T(\mathbf{Z})] \tag{8.112}$$

When use is made of the formula (8.106), the factor $E[\mathbf{Z}\mathbf{Z}^T]$ appears on both sides of (8.112) and it can be removed. The resultant defining equations for $\mathbf{M}$, $\mathbf{C}$, and $\mathbf{K}$ are

$$M_{ij} = E\left|\frac{\partial}{\partial \ddot{Y}_j} g_i(\ddot{\mathbf{Y}},\dot{\mathbf{Y}},\mathbf{Y})\right| \tag{8.113}$$

$$C_{ij} = E\left|\frac{\partial}{\partial \dot{Y}_j} g_i(\ddot{\mathbf{Y}},\dot{\mathbf{Y}},\mathbf{Y})\right| \tag{8.114}$$

$$K_{ij} = E\left|\frac{\partial}{\partial Y_j} g_i(\ddot{\mathbf{Y}},\dot{\mathbf{Y}},\mathbf{Y})\right| \tag{8.115}$$

A direct interpretation of these relations is that M_{ij}, C_{ij}, and K_{ij} are *typical* values of the slopes of $\mathbf{g}(\ ,\ ,\ )$ defined through mean values. In general, these definitions must be used iteratively as the properties of $\mathbf{Y}(t)$ depend on the initially unknown parameters.

The advantages of the equivalent linearization procedure are its simplicity and the fact that the implicit form of the equations defining the equivalent parameters are often found to give good estimates of the response variance even for quite large nonlinearities. It is interesting to note that taking moments of the nonlinear equation (8.107) with $X(t_1)$ and $F(t_1)$ leads to the same response covariance function if the expectations involving $g(X,\dot{X},\ddot{X})$ are evaluated by use of the normal property (8.106). This so-called Gaussian closure method was used by Iyengar (1975). The coincidence can be explained by use of orthogonal expansions as outlined in the next section. Both methods are capable of several extensions, such as nonstationary excitation, excitation by a function of a normal process, and more general forms of the nonlinearity. As an example, response of hysteretic structures is dealt with in Section 10.3.

Example 8.3

Nonlinearities associated with large deflections can often be described by including an additional cubic term in the stiffness. For a single-degree-of-freedom this results in Duffing's equation,

$$m\ddot{X} + c\dot{X} + k(X+\varepsilon X^3) = F(t)$$

$F(t)$ is assumed to be a normal process that may be nonstationary with nonzero mean. The general linearization problem becomes rather complicated even in this case, if three unknown parameters m_e, c_e, and k_e are introduced. If only the equivalent stiffness is optimized while $m_e = m$ and $c_e = c$, the linear equation is

$$m\ddot{Y} + c\dot{Y} + k_e Y = F(t)$$

The error then is

$$\Delta = k(Y+\varepsilon Y^3) - k_e Y$$

The mean-square error $E[\Delta^2]$ is minimized by

$$E[Y\Delta] = E[k(Y^2+\varepsilon Y^4) - k_e Y^2] = 0$$

i.e.,

$$k_e = k\left(1+\varepsilon\frac{E[Y^4]}{E[Y^2]}\right)$$

Evaluation of the expectations gives the equivalent stiffness:

$$k_e = k\left[1+\varepsilon\frac{\mu_Y^4+6\mu_Y^2\sigma_Y^2+3\sigma_Y^4}{\mu_Y^2+\sigma_Y^2}\right]$$

It is seen that k_e is an increasing function of μ_Y^2 and σ_Y^2. In the case of nonstationary excitation k_e is a function of t and the problem requires iterative numerical solution.

For stationary, zero-mean excitation,

$$k_e = (1+3\varepsilon\sigma_Y^2)k$$

In the particular case of white noise excitation, the variance of the equivalent response, σ_Y^2, is simply related to that of the linear system $\varepsilon = 0$. From the formula for white noise excitation given in Table 8.2, it appears that the variances are

$$\sigma_0^2 = \frac{\pi m}{kc} S^F, \qquad \sigma_Y^2 = \frac{\pi m}{k_e c} S^F$$

The relation between k and k_e leads to

$$\left(\frac{\sigma_Y}{\sigma_0}\right)^2 = \frac{k}{k_e} = \frac{1}{1 + 3\varepsilon\sigma_Y^2}$$

or in explicit form

$$\left(\frac{\sigma_Y}{\sigma_0}\right)^2 = \frac{1}{6\varepsilon\sigma_0^2}(\sqrt{1 + 12\varepsilon\sigma_0^2} - 1) \sim 1 - 3\varepsilon\sigma_0^2 + 18(\varepsilon\sigma_0^2)^2 - 27(\varepsilon\sigma_0^2)^3 + \cdots$$

The theoretical value of $(\sigma_Y/\sigma_0)^2$ can be found by use of the Markov process theory as described in Section 8.4.3. The exact asymptotic expansion has been given by Crandall (1980) as

$$\left(\frac{\sigma_X}{\sigma_0}\right)^2 \sim 1 - 3\varepsilon\sigma_0^2 + 24(\varepsilon\sigma_0^2)^2 - 297(\varepsilon\sigma_0^2)^3 + \cdots$$

Thus the equivalent linearization technique yields the first two terms of the asymptotic expansion of σ_Y^2. However, it is important to note that the expansion of the square root is convergent only for $|12\varepsilon\sigma_0^2| < 1$, and in fact the full expression provides a quite good estimate of σ_Y^2 even for $\varepsilon\sigma_0^2 > 1$ (Smith, 1978).

8.4.2 Hermite Polynomial Expansion

The formula (8.106), which in essence contains the normal property of the variables X, turned out to be very useful in connection with the equivalent linearization technique. Basically, it is a recurrence relation between the central moments of normal variables, and this aspect can be handled efficiently by use of Hermite polynomials.

The Hermite polynomial of degree n is defined by

$$\mathrm{He}_n(x) = \exp\left(\frac{1}{2}x^2\right)\left(-\frac{\partial}{\partial x}\right)^n \exp\left(-\frac{1}{2}x^2\right) \qquad (8.116)$$

and the first few polynomials are

$$\mathrm{He}_0(x) = 1, \qquad\qquad \mathrm{He}_1(x) = x$$
$$\mathrm{He}_2(x) = x^2 - 1, \qquad \mathrm{He}_3(x) = x^3 - 3x \qquad (8.117)$$

The Hermite polynomials satisfy the differentiation rule

$$\frac{d}{dx}\mathrm{He}_n(x) = n\,\mathrm{He}_{n-1}(x) \qquad (8.118)$$

and the orthogonality relation

$$\int_{-\infty}^{\infty} \frac{1}{\sqrt{2\pi}} \exp\left(-\frac{1}{2}x^2\right) \mathrm{He}_n(x)\,\mathrm{He}_m(x)\,dx = n!\,\delta_{mn} \qquad (8.119)$$

where it is noted that the weight function is the normal density function $\varphi(x)$.

The usefulness of Hermite polynomials in the present context is related to the following result. Let $a(x)$ and $b(y)$ be two functions with the polynomial expansions

$$a(x) = \sum_{j=0}^{\infty} a_j \operatorname{He}_j(x) \tag{8.120}$$

$$b(y) = \sum_{k=0}^{\infty} b_k \operatorname{He}_k(y)$$

If X and Y are zero-mean joint normal variables of unit variance, the expectation of the product $a(X)b(Y)$ is

$$\operatorname{E}[a(X)b(Y)] = \sum_j \sum_k a_j b_k \int_{-\infty}^{\infty}\int_{-\infty}^{\infty} \operatorname{He}_j(x) \operatorname{He}_k(y)\, \varphi(x,y;\rho)\, dxdy \tag{8.121}$$

where $\varphi(x,y;\rho)$ is the two-dimensional normal density function

$$\varphi(x,y;\rho) = \frac{1}{2\pi\sqrt{1-\rho^2}}\exp\left|-\frac{x^2-2\rho xy+y^2}{2(1-\rho^2)}\right| \tag{8.122}$$

ρ is the correlation coefficient. $\varphi(x,y;\rho)$ satisfies the identity

$$\frac{\partial\varphi}{\partial\rho} = \frac{\partial^2\varphi}{\partial x\partial y} \tag{8.123}$$

and a power series in ρ for $\operatorname{E}[a(X)b(Y)]$ can now be obtained by continued differentiation

$$\operatorname{E}[a(X)b(Y)] = \sum_{n=0}^{\infty}\frac{1}{n!}\rho^n\left(\frac{d}{d\rho}\right)^n \operatorname{E}[a(X)b(Y)]_{\rho=0} \tag{8.124}$$

Differentiation of a typical term gives, after use of (8.123) and (8.118),

$$\frac{d}{d\rho}\int_{-\infty}^{\infty}\int_{-\infty}^{\infty} \operatorname{He}_j(x)\operatorname{He}_k(y)\,\varphi(x,y;\rho)\,dxdy \tag{8.125}$$

$$= jk\int_{-\infty}^{\infty}\int_{-\infty}^{\infty} \operatorname{He}_{j-1}(x)\operatorname{He}_{k-1}(y)\,\varphi(x,y;\rho)\,dxdy$$

It follows from (8.119) that for $\rho=0$ the right-hand integral vanishes except for $j=k=1$. By continued application of this result, the general series (8.124) takes the form

$$\operatorname{E}[a(X)b(Y)] = \sum_{n=0}^{\infty} n!\, a_n\, b_n\, \rho^n \tag{8.126}$$

This is the desired result that gives the expectation of a product of functions of correlated normal variables as a power series in the correlation coefficient. It is formally equivalent to the following expansion of the normal density function:

$$\varphi(x,y;\rho) = \sum_{n=0}^{\infty} \frac{1}{n!} \rho^n \left(\frac{d}{dx}\right)^n \varphi(x) \left(\frac{d}{dy}\right)^n \varphi(y) \qquad (8.127)$$

$$= \varphi(x)\varphi(y) \sum_{n=0}^{\infty} \frac{1}{n!} \rho^n \text{He}_n(x) \text{He}_n(y)$$

Similar results can be derived for expectations of higher-order products or functions of several variables, the latter by use of generalized Hermite polynomials (Stratonovich, 1963).

In particular, formula (8.126) provides information on mean and covariance functions of a stochastic process $a(X(t))$, where $X(t)$ is a normalized normal process. With the expansion coefficients a_j from (8.120) the result is

$$E[a(X(t))] = a_0 \qquad (8.128)$$

$$\text{Cov}[a(X(t_1)),a(X(t_2))] = \sum_{n=1}^{\infty} n! \, a_n^2 \rho(t_1 - t_2)^n \qquad (8.129)$$

The relation of these results to the equivalent linearization and Gaussian closure methods can be illustrated by considering a single-degree-of-freedom system with nonlinear stiffness, i.e., the equation of motion

$$m\ddot{X} + c\dot{X} + a(X) = F(t) \qquad (8.130)$$

The nonlinear term is expanded in terms of Hermite polynomials

$$a(X) = \sum_{j=1}^{\infty} a_j \text{He}_j(X/\sigma) \qquad (8.131)$$

where it is necessary to normalize the argument of the Hermite polynomials with the as yet unknown standard deviation σ. Thus the coefficients a_j also depend on σ. According to the equivalent linearization procedure

$$k_e = E\left[\frac{\partial}{\partial Y} a(Y)\right] = \sum_{j=1}^{\infty} ja_j E[\text{He}_{j-1}(Y/\sigma)] = a_1 \qquad (8.132)$$

Thus equivalent linearization amounts to truncating the expansion (8.131) after the first term. The Gaussian closure method forms moments by multiplication of (8.130) with $X(t)$ and $F(t)$ combined with the assumption of normal variables. By the result (8.126), this also leads to including only the first term of the expansion (8.131).

The use of Hermite polynomials is illustrated by the Duffing oscillator, for which

$$a(X) = k(X + \varepsilon X^3) \tag{8.133}$$
$$= k(1 + 3\varepsilon\sigma^2)\,\mathrm{He}_1(X/\sigma) + k\varepsilon\sigma^2\,\mathrm{He}_3(X/\sigma)$$

This procedure confirms the results of Example 8.3 but also identifies an additional term $k\varepsilon\sigma^2\mathrm{He}_3(X/\sigma)$, which is omitted by the linearization process. By (8.129) the covariance function for this term is $6(k\varepsilon\sigma^2)^2\,\rho(\tau)^3$. The linearization procedure disregards this term because it is orthogonal to $X(t)$ according to (8.119). However, for narrow-band response this term has its main spectral density near the eigenfrequency of the system, as is easily seen from the relations

$$\rho(t) \approx e^{-\zeta\omega_0 t}\cos(\omega_0 t) \tag{8.134}$$

$$\rho(t)^3 \approx e^{-3\zeta\omega_0 t}\left[\frac{3}{4}\cos(\omega_0 t) + \frac{1}{4}\cos(3\omega_0 t)\right] \tag{8.135}$$

Although the neglected term is formally of order ε^2, its spectral density near the eigenfrequency dominates that of the original input if $\varepsilon\sigma^2 > \zeta$, and a direct pertubation procedure therefore requires extremely small nonlinearity in order to provide additional information on the spectrum, such as the excitation of higher harmonics (Crandall, 1963; Smith, 1978). An improved pertubation procedure accounting for the contribution with frequency around $3\omega_0$ has been presented by Krenk (1985).

Section 10.2 on wave loads provides an additional example of the use of Hermite polynomial expansions.

8.4.3 Markov Vector Methods

The methods of equivalent linearization and Hermite polynomial expansion concentrate on the covariance function of the response, while the problem of the distribution of the response is avoided by introducing an *assumption* of normality. These methods may therefore provide rather accurate estimates of the response variance, without giving any information about the influence of the nonlinearity on the "tails" of the distributions. An alternative approach consists in the introduction of simplifications in the covariance structure in order to obtain accurate information about the distributions. The simplification consists in the assumption of Markov properties, and the probability density function then satisfies the Fokker-Planck equation.

A stochastic process $X(t)$ is called a *Markov process* if the conditional probability density of $X(t_1)$, conditional on $X(t_j) = x_j$, $j = 2,3, \ldots, n$, of the $n-1$ previous times $t_2 > t_3 > \cdots > t_n$, depends only on the latest of these. To be specific,

$$f(x_1 \mid x_2, \ldots, x_n) = p(x_1 \mid x_2) \tag{8.136}$$

where $x_j = x(t_j)$ and $t_1 > t_2 > \cdots > t_n$. The function $p(\ \mid\ )$ is called the *transition probability*. In general, $p(x_1 \mid x_2)$ depends explicitly on both t_1 and t_2, but for a stationary Markov process $p(x_1 \mid x_2)$ depends only on the times through the difference $\tau = t_1 - t_2$. The definition of conditional probability density implies that the joint probability density of a Markov process is given by an initial probability density and the transition probabilities.

$$f(x_1, x_2, \ldots, x_n) = f(x_1 \mid x_2, \ldots, x_n) f(x_2 \mid x_3, \ldots, x_n) \cdots f(x_n) \tag{8.137}$$

$$= p(x_1 \mid x_2) p(x_2 \mid x_3) \cdots p(x_{n-1} \mid x_n) f(x_n)$$

The factored form (8.137) of the joint probability density function $f(x_1, x_2, \ldots, x_n)$ can be used to obtain the Smoluchowski equation by integration with respect to x_2.

$$\int_{-\infty}^{\infty} p(x_1 \mid x_2) p(x_2 \mid x_3) \, dx_2 = p(x_1 \mid x_3), \qquad t_1 > t_2 > t_3 \tag{8.138}$$

The integral is the summation of all possible values at the intermediate time t_2, each weighted by their conditional probability density $p(x_2 \mid x_3)$. This identity can be reformulated as a partial differential equation. The following technique is due to Wang and Uhlenbeck (1945).

First it is convenient to introduce the time difference explicitly in the transition probability. Let $p(x, t \mid y)$ be the probability density of $X(t)$ conditional on $X(0) = y$. A weighted integral of the time derivative $\partial p(x, t \mid y) / \partial t$ is now considered.

$$\int_{-\infty}^{\infty} R(x) \frac{\partial}{\partial t} p(x, t \mid y) \, dx \tag{8.139}$$

$$= \lim_{\tau \to 0} \frac{1}{\tau} \int_{-\infty}^{\infty} R(x) [p(x, t+\tau \mid y) - p(x, t \mid y)] \, dx$$

where $R(x)$ is an arbitrary, well-behaved test function. $p(x, t+\tau \mid y)$ is then expressed by the Smoluchowski identity (8.138) with z as the value at the intermediate time t.

$$\int_{-\infty}^{\infty} R(x) \frac{\partial}{\partial t} p(x, t \mid y) \, dx \tag{8.140}$$

$$= \lim_{\tau \to 0} \frac{1}{\tau} \int_{-\infty}^{\infty} R(x) \left[\int_{-\infty}^{\infty} p(x, \tau \mid z) p(z, t \mid y) \, dz - p(x, t \mid y) \right] dx$$

Now the order of integration is changed in the double integral, and it is used that for small τ the integral containing $p(x, \tau \mid z)$ receives contributions for only small values of $\mid x - z \mid$. In this integral the test function is therefore replaced by its Taylor expansion at $x = z$, whereby the double integral becomes

$$\int_{-\infty}^{\infty} p(z,t \mid y) \left| \sum_{j=0}^{\infty} \frac{1}{j!} R^{(j)}(z) \int_{-\infty}^{\infty} (x-z)^j p(x,\tau \mid z) dx \right| dz \qquad (8.141)$$

In the following only processes for which the last integral is of order $O(\tau)$ for $j=0,1,2$ and $o(\tau)$ for $j \geqslant 3$ will be considered. The limit $\tau \to 0$ is then described in terms of

$$A(z,t) = \lim_{\tau \to 0} \frac{1}{\tau} \int_{-\infty}^{\infty} (x-z) p(x,\tau \mid z) \, dx \qquad (8.142)$$

$$B(z,t) = \lim_{\tau \to 0} \frac{1}{\tau} \int_{-\infty}^{\infty} (x-z)^2 p(x,\tau \mid z) \, dx \qquad (8.143)$$

Upon introduction into (8.140) the following identity is obtained:

$$\int_{-\infty}^{\infty} R(x) \frac{\partial}{\partial t} p(x,t \mid y) \, dx \qquad (8.144)$$

$$= \int_{-\infty}^{\infty} \left[R'(x)A(x,t) + \frac{1}{2} R''(x)B(x,t) \right] p(x,t \mid y) \, dx$$

Integration by parts of the right-hand side in connection with the arbitrariness of the test function $R(x)$ establishes the differential equation

$$\left[\frac{\partial}{\partial t} + \frac{\partial}{\partial x} A(x,t) - \frac{1}{2} \frac{\partial^2}{\partial x^2} B(x,t) \right] p(x,t \mid y) = 0 \qquad (8.145)$$

This is the *Fokker-Planck equation* for the transition probability of a scalar Markov process. It must be solved with appropriate boundary conditions and the initial condition

$$\lim_{t \to 0} p(x,t \mid y) = \delta(x-y) \qquad (8.146)$$

In the particular case of a zero-mean stationary normal process, the transition probability is

$$p(x,t \mid y) = \frac{f(x,y)}{f(y)} = \frac{1}{\sqrt{2\pi(1-\rho^2)}\sigma} \exp \left| -\frac{(x-\rho y)^2}{2\sigma^2(1-\rho^2)} \right| \qquad (8.147)$$

The parameters $A(y)$ and $B(y)$ can then be calculated by use of (8.142) and (8.143).

$$A(y) = \lim_{t \to 0} \frac{1}{t} (\rho y - y) = -\beta y \qquad (8.148)$$

$$B(y) = \lim_{t \to 0} \frac{1}{t} [(1-\rho^2)\sigma^2 + (1-\rho)^2 y^2] = 2\beta\sigma^2 \qquad (8.149)$$

where the constant β is

$$\beta = \lim_{t \to 0} \frac{1-\rho(t)}{t} \qquad (8.150)$$

Substitution of $p(x,t \mid y)$ from (8.147) into the Fokker-Planck equation (8.145) gives an identity if the correlation function $\rho(t)$ satisfies the equation

$$\left(\frac{d}{dt} + \beta\right)\rho(t) = 0, \quad t > 0 \tag{8.151}$$

Thus $\rho(t)$ must be of exponential form,

$$\rho(t) = e^{-\beta \mid t \mid} \tag{8.152}$$

From this result it is clear that the typical structural response is not in itself a Markov process. However, it may be possible to describe a stochastic process $X(t)$ by a set of properties, e.g., $(X(t), \dot{X}(t))$, such that this satisfies the Markov condition (8.136) interpreted in vector form. In principle, this corresponds to a description in the phase plane shown in Fig. 7.14 and already used in connection with the introduction of the envelope concept.

The extension of the previous results to the case of a Markov vector process is straightforward if the argument x is interpreted as a vector. The Taylor expansion in (8.142) then contains mixed derivatives, and (8.142) and (8.143) must be generalized to

$$A_j(\mathbf{z},t) = \lim_{\tau \to 0} \frac{1}{\tau} \int_{-\infty}^{\infty} \cdots \int_{-\infty}^{\infty} (x_j - z_j)\, p(\mathbf{x},\tau \mid \mathbf{z})\, dx_1 \cdots dx_n \tag{8.153}$$

$$B_{jk}(\mathbf{z},t) = \lim_{\tau \to 0} \frac{1}{\tau} \int_{-\infty}^{\infty} \cdots \int_{-\infty}^{\infty} (x_j - z_j)(x_k - z_k) \tag{8.154}$$

$$\times p(\mathbf{x},\tau \mid \mathbf{z})\, dx_1 \cdots dx_n$$

The Fokker-Planck equation then takes the form

$$\left| \frac{\partial}{\partial t} + \sum_j \frac{\partial}{\partial x_j} A_j(\mathbf{x},t) - \frac{1}{2}\sum_j \sum_k \frac{\partial^2}{\partial x_j \partial x_k} B_{jk}(\mathbf{x},t) \right| p(\mathbf{x},t \mid \mathbf{y}) = 0 \tag{8.155}$$

Although simple in principle, the extension severely complicates the solution of the Fokker-Planck equation (see, e.g., Stratonovich, 1963).

The Markov vector method is illustrated by considering the stationary response of a single-degree-of-freedom structure to white noise excitation. The problem is governed by the differential equation

$$m\ddot{X} + c\dot{X} + g(X) = F(t) \tag{8.156}$$

The equation is reformulated as two first-order differential equations by introducing the vector $(Y_1, Y_2) = (X, \dot{X})$. It is convenient to write the first-order equations as expressions for the derivatives $\dot{Y}_1$ and $\dot{Y}_2$.

$$\dot{Y}_1 = Y_2 \tag{8.157}$$

$$\dot{Y}_2 = -\frac{1}{m}g(Y_1) - \frac{1}{m}cY_2 + \frac{1}{m}F(t) \tag{8.158}$$

In this phase-plane description the increments only depend on the current position and the white noise process $F(t)$, and $(Y_1(t),Y_2(t))$ is therefore a Markov vector process.

The parameters A_1 and A_2 follow from (8.153) by use of (8.157) and (8.158).

$$A_1 = \mathrm{E}[\dot{Y}_1 \mid \mathbf{Y}=\mathbf{y}] = y_2 \tag{8.159}$$

$$A_2 = \mathrm{E}[\dot{Y}_2 \mid \mathbf{Y}=\mathbf{y}] = -\frac{1}{m}(g(y_1)+cy_2) \tag{8.160}$$

The determination of the parameters B_{jk} requires more careful consideration of the limit $\tau \to 0$. The integral (8.154) contains a quadratic factor in the increments, $\Delta Y_j \Delta Y_k$, and in order to give finite contributions to B_{jk} the expectation of this factor must be of the order $O(\tau)$. It turns out that the only finite contribution comes from the double integral

$$\mathrm{E}\left[\int_0^\tau \int_0^\tau F(t_1)F(t_2)\,dt_1\,dt_2\right] = \int_0^\tau \int_0^\tau 2\pi\,S^F\,\delta(t_1-t_2)\,dt_1\,dt_2 \tag{8.161}$$

$$= 2\pi\,S^F\,\tau$$

and thus

$$B_{11} = B_{12} = B_{21} = 0 \tag{8.162}$$

$$B_{22} = \frac{2\pi}{m^2}\,S^F \tag{8.163}$$

The density function $f(\mathbf{y})$ is found as the limit of $p(\mathbf{y} \mid \mathbf{z})$ for infinite time separation, i.e., by omitting the time derivative in the Fokker-Planck equation (8.155).

$$\left| \frac{\partial}{\partial y_1}y_2 - \frac{1}{m}\frac{\partial}{\partial y_2}(g(y_1)+cy_2) - \frac{\pi}{m^2}S^F\frac{\partial^2}{\partial y_2^2} \right| f(\mathbf{y}) = 0 \tag{8.164}$$

The solution is found by separation of variables with the result

$$f(\mathbf{y}) = f_{X\dot{X}}(x,\dot{x}) = C_g \exp\left| -\frac{c}{\pi S^F}\left(\frac{1}{2}m\dot{x}^2 + \int_0^x g(y)\,dy\right) \right| \tag{8.165}$$

The constant C_g is determined by the unit value of the double integral of $f_{X\dot{X}}(x,\dot{x})$.

It follows from the result (8.165) that $X(t)$ and $\dot{X}(t)$ are statistically independent and that $\dot{X}(t)$ is normally distributed with variance

$$\sigma^2_{\dot{X}} = \frac{\pi S^F}{mc} \tag{8.166}$$

independent of the particular form of the stiffness $g(x)$. However, when $g(x)$ is a nonlinear function, X is not normally distributed, and as a consequence $\dot{X}(t)$ is not in general a normal process. It is observed that the term in the parentheses in the exponent of (8.165) is the mechanical energy of the structure in the state described by $(x,\dot{x})$.

The zero-upcrossing rate follows from Rice's formula (7.59).

$$v_0^X = C_g \sigma^2_{\dot{X}} \tag{8.167}$$

In the linear case $C_g = (2\pi\sigma_X\sigma_{\dot{X}})^{-1}$, but in contrast to $\sigma_{\dot{X}}$, the upcrossing rate depends on the degree of nonlinearity. This effect is well known for deterministic nonlinear oscillations (Stoker, 1950). As a consequence of the independence of $X(t)$ and $\dot{X}(t)$, the upcrossing rate of the curve $\xi(t)$ is

$$v_\xi^X(t) = v_0^X (2\pi)^{1/2} \Psi(\dot{\xi}/\sigma_{\dot{X}}) f_X(\xi)/f_X(0) \tag{8.168}$$

$$= v_0^X (2\pi)^{1/2} \Psi(\dot{\xi}/\sigma_{\dot{X}}) \exp\left| -\frac{c}{\pi S^F} \int_0^\xi g(y)dy \right|$$

For a constant level the normalized derivative $d(-v_\xi^X/v_0^X)/d\xi$, $\xi \geq 0$, plays an important role as a weight function for evaluation of fatigue damage as explained in Section 9.2. This derivative is

$$\frac{d}{d\xi}(-v_\xi^X/v_0^X) = -\frac{d}{d\xi}\exp\left| -\frac{c}{\pi S^F} \int_0^\xi g(y)dy \right| \tag{8.169}$$

$$= \frac{c}{\pi S^F} g(y) \exp\left| \frac{c}{\pi S^F} \int_0^\xi g(y)dy \right|$$

In the particular case of a linear structure with $g(x) = kx$,

$$\frac{d}{d\xi}(-v_\xi^X/v_0^X) = \frac{kc}{\pi S^F} \xi \exp\left| -\frac{1}{2}\frac{kc}{\pi S^F}\xi^2 \right| \tag{8.170}$$

corresponding to a Rayleigh distribution.

Example 8.4

The exact result of the Markov vector method concerning the distribution and crossing frequency provides a convenient basis for comparison with the approximate methods. In the particular case of the Duffing oscillator, for which linearization was discussed in Example 8.3, the restoring force is

$$g(X) = k(X + \varepsilon X^3)$$

As in Example 8.3 it is convenient to use the standard deviation σ_0 of the response for $\varepsilon=0$ as reference unit. The joint probability density of $X(t)$ and $\dot{X}(t)$ given by (8.165) then takes the form

$$f_{X\dot{X}}(x,\dot{x}) = C_g \exp\left|-\frac{1}{2}\left(\frac{\dot{x}}{\sigma_{\dot{X}}}\right)^2\right|\exp\left|-\frac{1}{2}\left(\frac{x}{\sigma_0}\right)^2 - \frac{\varepsilon\sigma_0^2}{4}\left(\frac{x}{\sigma_0}\right)^4\right|$$

It is seen that also in this exact result a convenient measure of the non-linearity is

$$\varepsilon\sigma_0^2 = \varepsilon\,\frac{\pi m}{kc}\,S^F$$

The nonlinear term appears only explicitly in the exponent, where it is clear that it is of primary importance for extreme values of x, i.e., large values of $|x/\sigma_0|$. However, by means of the normalization C_g also becomes a function of ε. In fact, C_g must be of the form

$$C_g = \frac{C_\varepsilon(\varepsilon\sigma_0^2)}{2\pi\sigma_0\sigma_{\dot{X}}}$$

and the zero-crossing rate from (8.167) then is

$$v_0^X = \frac{1}{2\pi}\frac{\sigma_{\dot{X}}}{\sigma_0}\,C_\varepsilon(\varepsilon\sigma_0^2) = (v_0^X)_0\,C_\varepsilon(\varepsilon\sigma_0^2)$$

$(v_0^X)_0$ is the zero-upcrossing rate for the response for $\varepsilon=0$, and the parameter C_ε is identified as the relative magnitude of the zero-upcrossing frequency of the nonlinear response.

These results enable an evaluation of the linearization procedure in this specific case. $C_\varepsilon(\varepsilon\sigma_0^2)$ is calculated by numerical integration of the density function, and then also $(\sigma_X/\sigma_0)^2$ follows from numerical integration. In the linearization procedure the normalized variance was given as

$$\left(\frac{\sigma_Y}{\sigma_{||}}\right)^2 = \frac{k}{k_e} = \frac{1}{6\varepsilon\sigma_0^2}(\sqrt{1+12\varepsilon\sigma_0^2}-1)$$

The zero-upcrossing frequency of the linearized response is

$$v_0^Y/(v_0^X)_0 = \sqrt{k_e/k} = \sigma_0/\sigma_Y$$

Thus the linearization procedure introduces a simple relation between the response magnitude and the crossing rate.

A comparison between exact and linearized estimates of σ_X and v_0^X is given in Table 8.3 based on the calculations of Smith (1978). It is seen that in this case equivalent linearization leads to a quite good estimate of the variance for the nonlinearity parameter $\varepsilon\sigma_0^2$ in the interval 0 to 5, while the zero-upcrossing rate is overestimated, starting at small values of $\varepsilon\sigma_0^2$. This trend continues also for $\varepsilon\sigma_0^2>5$. Thus Crandall (1980) has given the asymptotic result

$$(\sigma_X/\sigma_0)^2 \sim 0.676\,(\varepsilon\sigma_0^2)^{-1/2}$$

for large values of $\varepsilon\sigma_0^2$, while equivalent linearization leads to

$$(\sigma_Y/\sigma_0)^2 \sim 0.577\,(\varepsilon\sigma_0^2)^{-1/2}$$

The similar results for the zero-upcrossing frequency are

$$v_0^X/(v_0^X)_0 \sim 0.978\,(\varepsilon\sigma_0^2)^{1/4}$$

and

$$v_0^Y/(v_0^X)_0 \sim 1.316\,(\varepsilon\sigma_0^2)^{1/4}$$

Thus the relative asymptotic errors are around 0.14 and -0.35, respectively. It is notable that relative errors are bounded, thereby lending some measure of uniformity to the equivalent linearization procedure.

TABLE 8.3 Variance and Zero-Upcrossing Rates of the Duffing Oscillator									
$\varepsilon\sigma_0^2$	0	0.01	0.05	0.1	0.2	0.3	1.0	3.0	5.0
$(\sigma_X/\sigma_0)^2$	1.00	0.97	0.89	0.82	0.72	0.66	0.47	0.31	0.25
$(\sigma_Y/\sigma_0)^2$	1.00	0.97	0.88	0.81	0.70	0.64	0.43	0.28	0.23
$v_0^X/(v_0^X)_0$	1.00	1.01	1.03	1.06	1.10	1.14	1.30	1.53	1.68
$v_0^Y/(v_0^X)_0$	1.00	1.01	1.06	1.11	1.19	1.25	1.52	1.88	2.10

The present description of the Markov vector method is very brief and touches on only some of the possible applications to structural response analysis. The results for rational power spectra derived in Section 8.1.3 can also be treated within the framework of Markov processes (see Section 10.3.1 and Gasparini, 1979). Some results have also been obtained for nonlinear multi-degree-of-freedom structures (Caughey, 1963; Lin, 1976). Finally, it should be mentioned that the theory of Markov processes also serves as a framework for approximations, for example, for the envelope. Details about this type of approximation have been given by Stratonovich (1963).

8.5 SUMMARY

The chapter deals with four aspects of stochastic response of structures: stationary response, nonstationary response produced by modulated stationary excitation, the white noise approximation, and nonlinear effects. Special properties of the normal process are used throughout the chapter.

In the case of stationary random excitation of a structure for extended periods of time the response becomes stationary, and the alternatives of covariance analysis in the time or frequency domain are presented. Considerable attention is given to the case where the excitation process has a rational power spectrum, and it is demonstrated that the excitation may be considered as output of a linear system with white noise input.

Nonstationarity is introduced via its simplest form, transient response, where an otherwise stationary excitation is applied from a certain instant of time. A simple relation between the stationary and the transient covariance functions is established. Then a more general form of nonstationarity is introduced, in which the excitation is represented as a time-dependent convolution of a stationary process. It is demonstrated how a Fourier transform of the convolution kernel with respect to the delay time leads to the notion of an evolutionary power spectrum, i.e., a spectrum that changes with time.

A full nonstationary analysis in terms of evolutionary spectra is quite complicated and time consuming, and a section is therefore devoted to the approximation of wide-band excitation by white noise processes. A system of simple formulas for the covariance function is given, and a calibration procedure for the white noise processes is proposed. The main result is that a simple fit should be based on the mean value of the spectral densities at the natural frequencies of the two modes in question.

The final section on nonlinear effects describes the method of equivalent linearization and then proceeds to discuss the connection between this method and use of a truncated expansion in terms of Hermite polynomials. The chapter concludes with a brief presentation of Markov vector methods and the Fokker-Planck equation.

REFERENCES

ATALIK, T. S. and S. UTKU, "Stochastic Linearization of Multi-Degree-of-Freedom Nonlinear Systems," *Earthquake Engineering and Structural Dynamics,* Vol. 4, 1976, pp. 411-420.

BENDAT, J. S. and A. G. PIERSOL, *Measurement and Analysis of Random Data,* John Wiley, New York, 1966.

CAUGHEY, T. K., "Equivalent Linearization Techniques," *Journal of the Acoustical Society of America,* Vol. 35, 1963, pp. 1706-1711.

CLOUGH, R. W. and J. PENZIEN, *Dynamics of Structures,* McGraw-Hill, New York, 1975.

CRANDALL, S. H., "Pertubation Techniques for Random Vibration of Nonlinear Systems," *Journal of the Acoustical Society of America,* Vol. 35, 1963, pp. 1700-1705.

CRANDALL, S. H., "Non-Gaussian Closure for Random Vibration of Nonlinear Oscillators," *International Journal of Nonlinear Mechanics,* Vol. 15, 1980, pp. 303-313.

DER KIUREGHIAN, A., "Structural Response to Stationary Excitation," *Journal of the Engineering Mechanics Division,* ASCE, Vol. 106, 1980, pp. 1195-1213.

FOSS, K. A., "Co-Ordinates which Uncouple the Equations of Motion of Damped Linear Dynamic Systems," *Journal of Applied Mechanics,* ASME, Vol. 25, 1958, pp. 361-364.

GASPARINI, D. A., "Response of MDOF Systems to Nonstationary Random Excitation," *Journal of the Engineering Mechanics Division,* ASCE, Vol. 105, 1979, pp. 13-27.

IYENGAR, R. N., "Random Vibration of a Second-Order Nonlinear Elastic System," *Journal of Sound and Vibration,* Vol. 40, 1975, pp. 155-165.

KRENK, S., "Generalized Hermite Polynomials and the Spectrum of Nonlinear Random Vibration," in *Probabilistic Methods in the Mechanics of Solids and Structures,* Julius Springer Verlag, 1985.

KRENK, S., H. O. MADSEN and P. H. MADSEN, "Stationary and Transient Response Envelopes," *Journal of Engineering Mechanics,* ASCE, Vol. 109, 1983, pp. 263-278.

LIN, Y. K., *Probabilistic Theory of Structural Dynamics,* Krieger Publishing, Huntington, N. Y., 1976.

MADSEN, P. H. and S. KRENK, "Stationary and Transient Response Statistics," *Journal of the Engineering Mechanics Division,* ASCE, Vol. 108, 1982, pp. 622-635.

PRIESTLY, M. B., "Evolutionary Spectra and Nonstationary Processes," *Journal of the Royal Statistical Society,* Vol. B27, 1965, pp. 204-237.

PRIESTLY, M. B., "Power Spectral Analysis of Nonstationary Random Processes," *Journal of Sound and Vibration,* Vol. 6, 1967, pp. 86-97.

SMITH, E., "On Nonlinear Random Vibrations," Report No. 78-3, Division of Structural Mechanics, The Norwegian Institute of Technology, Trondheim, Norway, 1978.

SPANOS, P.-T. D., "Spectral Moments Calculation of Linear System Output," *Journal of Applied Mechanics,* ASME, Vol. 50, 1983, pp. 901-903.

STOKER, J. J., *Nonlinear Vibration,* Interscience, New York, 1950.

STRATONOVICH, R. L., *Topics in the Theory of Random Noise,* Gordon and Breach, New York, 1963.

WANG, M. C. and G. E. UHLENBECK, "On the Theory of the Brownian Motion II," *Review of Modern Physics,* Vol. 17, 1945, pp. 323-342. Reprinted in *Selected Papers on Noise and Stochastic Processes,* ed. N. Wax, Dover, New York, 1954.

9

STOCHASTIC MODELS
FOR MATERIAL STRENGTH

9.1 CLASSICAL STRENGTH MODELS

This section describes three classical stochastic strength models: the ideal brittle material model, the ideal plastic model, and the fiber bundle model. These models are well suited to illustrate the principal difference between different material behavior.

An *ideal brittle material* is defined as a material that fails if a single particle fails. The strength of the material is thus governed by the smallest strength of a particle. Ideal brittle materials are considered in Section 9.1.1.

When the load effect on a particle in an *ideal plastic material* reaches the yield capacity of the particle, yielding takes place. The particle is still capable of carrying a yield load and a load increase is transferred to other particles in the cross section. The cross section carries its maximum load when yielding takes place for all particles. The strength of the material is thus the sum of the particle strengths. Ideal plastic materials are considered in Section 9.1.2.

The third strength model is a *fiber bundle* where the failure of a particle leads to a stress redistribution but not necessarily to a global failure of the body. The fiber bundle model represents a model between the ideal brittle model and the ideal plastic model. Fiber bundles are treated in Section 9.1.3.

9.1.1 Ideal Brittle Material

To introduce the ideal brittle material model a chain is first considered. A chain represents a series system or weakest-link system with the structure function (5.51). The strengths of the individual elements are assumed to be mutually independent and identically distributed with distribution function $F(x)$. The distribution function $F_n(x)$ of the chain strength is [(7.9)]

$$F_n(x) = 1 - (1 - F(x))^n \tag{9.1}$$

Here n is the number of elements and it has been used that the load is the same for each element. The distribution function can be rewritten as

$$F_n(x) = 1 - \exp[n \log(1 - F(x))] \tag{9.2}$$

which shows that the strength distribution is governed by the behavior of $\log(1 - F(x))$ for small arguments x of interest. Furthermore, the dependence of size, here the number of elements n, appears as a factor in the exponential function. As $n \to \infty$ the distribution function $F_n(x)$ tends to one of the three asymptotic extreme-value distribution described in Section 7.2. Since the strength is bounded downward, only the extreme-value distribution type 3 is a possibility.

Next, a one-dimensional structure of ideal brittle material under uniaxial loading is considered. The structure is divided into a number k of parts of equal length l_0. The strengths of these k parts are assumed mutually independent and identically distributed with distribution function $F_0(x)$. For the strength R of the structure the distribution function becomes

$$F_R(x) = 1 - (1 - F_0(x))^k = 1 - \exp\left|\frac{l}{l_0} \log(1 - F_0(x))\right| \tag{9.3}$$

where l is the total length of the structure. Again, the dependence of the size l/l_0 appears as a factor in the exponential function.

The validity of (9.3) assumes an integer ratio l/l_0, and to extend the result it is convenient to view $F_0(x)$ as the minimum distribution for the strengths of a large number of particles comprising the length l_0. Assuming independent particle strengths with common distribution function $F(x)$ and ν particles per unit length, the distribution functions $F_0(x)$ and $F_R(x)$ becomes

$$F_0(x) = 1 - \exp[\nu l_0 \log(1 - F(x))] \tag{9.4}$$

$$F_R(x) = 1 - \exp\left|\nu l_0\left(\frac{l}{l_0}\right) \log(1 - F(x))\right| \tag{9.5}$$

which leads to the relation (9.3) for all l/l_0 independent of ν. The notion of a particle is thus somewhat fictitious and simply helps to extend the validity of (9.3). A fundamental criticism of the particle description is the independence assumption of particle strengths. This can be resolved by considering v as an effective number of particles with independent strengths. The dependence between particle strengths generally decreases very rapidly with the distance between the particles. The effective number of particles with independent strengths is thus still very large.

The result (9.3) is extended to an isotropic brittle solid under a *homogeneous* stress distribution. This stress is a scalar measure of the stress level. A reference volume V_0 is introduced, e.g., as the volume of a standard test specimen, and the distribution function of the reference volume is $F_0(x)$. The distribution function of the strength R of a solid of volume V is then in analogy with (9.3)

$$F_R(x) = 1 - \exp\left|\frac{V}{V_0}\log(1-F_0(x))\right| \qquad (9.6)$$

Often, the behavior of $\log(1-F_0(x))$ for small arguments can be represented as

$$\log(1-F_0(x)) \approx -\left(\frac{x-x_0}{x_c}\right)^k, \quad x \geqslant x_0 \qquad (9.7)$$

where x_0 is the lower limit on the strength and x_c and k are positive constants. The distribution function is thus written as

$$F_R(x) = 1 - \exp\left|-\frac{V}{V_0}\left(\frac{x-x_0}{x_c}\right)^k\right|, \quad x \geqslant x_0 \qquad (9.8)$$

This is a type 3 extreme-value distribution shifted an amount x_0. The expected value and variance of R follow from Table 7.2 as

$$E[R] = x_0 + x_c\Gamma\left(1+\frac{1}{k}\right)\left(\frac{V}{V_0}\right)^{-1/k} \qquad (9.9)$$

$$\mathrm{Var}[R] = x_c^2\left|\Gamma\left(1+\frac{2}{k}\right)-\Gamma^2\left(1+\frac{1}{k}\right)\right|\left(\frac{V}{V_0}\right)^{-2/k} \qquad (9.10)$$

These results were given by Weibull (1939) for $x_0=0$. The mean-value function decreases with the volume V and for $x_0=0$ the coefficient of variation is independent of V. This *size effect* is known for many brittle materials. The coefficient of variation for a reference volume strength determines the importance of the size effect. The larger the coefficient of variation, i.e., the smaller value of k, the larger is the size effect. Typical values for k have been reported as: $k=58$ for mild steel (Richards, 1954), $k=12$ for concrete (Zech and Wittman, 1977), and $k=6$ for stiff clay (George and Basma, 1984). Figure 9.1 summarizes results for glass threads.

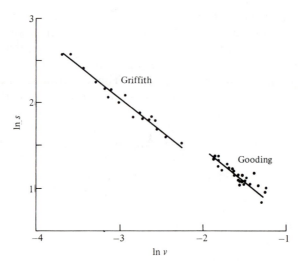

Figure 9.1   Variation of tensile failure stress with volume for glass threads, according to tests by Griffith (1920) on threads with length approximately 5 mm and by Gooding (1932) on threads with length approximately 180 mm, from Johnson (1953).

A fundamental assumption so far has been that the material is homogeneous and isotropic. Often the properties of the surface material deviate significantly from the interior material properties due to fabrication procedures and use. Consider, therefore, a solid with volume V and surface area A under a homogeneous stress. With $F_1(x)$ being the strength distribution function for a reference surface area A_0 and $F_2(x)$ being the strength distribution for a reference volume V_0, the distribution function of the strength becomes

$$F_R(x) = 1 - \exp\left[-\frac{A}{A_0}\log(1 - F_1(x)) - \frac{V}{V_0}\log(1 - F_2(x)) \right] \quad (9.11)$$

The size effect is now a mixture of a volume effect and a surface area effect. For a behavior of $\log(1 - F_1(x))$ and $\log(1 - F_2(x))$, in analogy with (9.7) the strength distribution function becomes

$$F_R(x) = \begin{cases} 1 - \exp\left[-\frac{A}{A_0}\left(\frac{x - x_{01}}{x_{c1}}\right)^{k_1} - \frac{V}{V_0}\left(\frac{x - x_{02}}{x_{c2}}\right)^{k_2} \right], & x \geqslant \max\{x_{01}, x_{02}\} \\[2ex] 1 - \exp\left[-\frac{A}{A_0}\left(\frac{x - x_{01}}{x_{c1}}\right)^{k_1} \right], & x_{01} \leqslant x \leqslant x_{02} \\[2ex] 1 - \exp\left[-\frac{V}{V_0}\left(\frac{x - x_{02}}{x_{c2}}\right)^{k_2} \right], & x_{02} \leqslant x \leqslant x_{01} \end{cases} \quad (9.12)$$

Above the lower strength limit there is thus a stress range with only a volume or a surface area size effect and a stress range with a mixture of these two size effects.

An *inhomogeneous* stress distribution is considered for an ideal brittle solid. The stress $s(x,y,z)$ at a point (x,y,z) is taken as

$$s(x,y,z) = s\, w(x,y,z) \tag{9.13}$$

where s is a reference stress, e.g., the maximum stress in the body, and where $w(x,y,z)$ is a dimension-free function. The volume V of the solid is divided into k parts of volume $V_1, \ldots, V_k$. This division is close enough that the stress distribution within each volume V_i can be taken as homogeneous. On the other hand, the volumes are large enough that volume strengths can be assumed mutually independent and following a distribution function of type (9.6). The distribution function of the strength R corresponding to the reference stress s in (9.13) is

$$F_R(r) = 1 - \prod_{i=1}^{k}(1 - F_{R_i}(r)) \tag{9.14}$$

$$= 1 - \prod_{i=1}^{k}\exp\left|\frac{V_i}{V_0}\log(1 - F_0(r\,w(x,y,z)_i))\right|$$

$$= 1 - \exp\left|\frac{1}{V_0}\sum_{i=1}^{k}V_i\log(1 - F_0(r\,w(x,y,z)_i))\right|$$

The summation can be represented by an integral, whereby

$$F_R(r) = 1 - \exp\left|\frac{1}{V_0}\int_V \log(1 - F_0(r\,w(x,y,z)))\,dV\right| \tag{9.15}$$

For a limiting form (9.7) with $x_0=0$ this becomes

$$F_R(r) = 1 - \exp\left|-\frac{1}{V_0}\left(\frac{r}{r_c}\right)^k \int_V w(x,y,z)^k\,dV\right| \tag{9.16}$$

$$= 1 - \exp\left|-\frac{V^*}{V_0}\left(\frac{r}{r_c}\right)^k\right|$$

where the stress volume V^* has been introduced as

$$V^* = \int_V w(x,y,z)^k\,dV \tag{9.17}$$

The distribution function then becomes a Weibull distribution, (Weibull, 1939). Example 9.1 illustrates the computation of the stress volume for a beam with various support conditions and loading.

Example 9.1

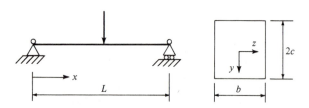

Figure 9.2 Simply supported beam of ideal brittle material.

Figure 9.2 shows a simply supported slender beam with rectangular cross section and loaded at the midpoint by a concentrated force. The beam material is linear elastic and ideal brittle. The tensile strength of a standard test specimen of volume V_0 has a Weibull distribution with mean value μ_0 and standard deviation σ_0. Equation (9.8) then determines the strength distribution with $x_0 = 0$ and the exponent k is determined by solving (7.12) with respect to k. x_c is determined as

$$x_c = \frac{\mu_0}{\Gamma\left(1+\frac{1}{k}\right)}$$

Shear stresses and compressive axial stresses are ignored. The axial tensile stress distribution $s(x,y,z)$ is

$$s(x,y,z) = \begin{vmatrix} \frac{2x}{l}\frac{y}{c}\,s, & 0 \leqslant x < \frac{l}{2}, & -c \leqslant y \leqslant c, & -\frac{b}{2} \leqslant z \leqslant \frac{b}{2} \\[2mm] \frac{2(l-x)}{l}\frac{y}{c}\,s, & \frac{l}{2} \leqslant x \leqslant l, & -c \leqslant y \leqslant c, & -\frac{b}{2} \leqslant z \leqslant \frac{b}{2} \end{vmatrix}$$

where s is the maximum tensile stress. The stress volume V^* is

$$V^* = 2\int_{-b/2}^{b/2} dz \int_0^c dy \int_0^{l/2} \left(\frac{2x}{l}\frac{y}{c}\right)^k dx = \frac{bcl}{(k+1)^2} = \frac{1}{2(k+1)^2}\,V$$

For pure bending the stress volume is similarly

$$V^* = \frac{1}{2(k+1)}\,V$$

Figure 9.3 shows a comparison of the mean strength of a beam with various support conditions and various loading resulting in the same maximum tensile stress. The mean strength is shown as a function of the standard deviation σ_0 or equivalently as a function of the exponent k. The simply supported beam under pure bending is the reference condition.

9.1.2 Ideal Plastic Material

A cross section of a homogeneous bar under uniaxial loading is considered. The force deformation relation for a particle is ideal

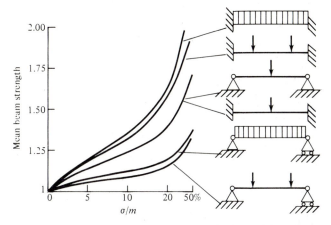

Figure 9.3 Comparison of mean strength of beam with various support and loading conditions and standard deviation of reference volume strength, from Johnson (1953).

plastic or elastic-ideal plastic with yield force R_i. The cross-section strength R is

$$R = \sum_i R_i \tag{9.18}$$

where the summation is over all particles. The mean value and variance of R are

$$E[R] = \sum_i E[R_i] \tag{9.19}$$

$$\text{Var}[R] = \sum_i \sum_j \text{Cov}[R_i, R_j] \tag{9.20}$$

When the number of particles is large and the dependence between particle strengths is sufficiently weak, the distribution of R tends to a normal distribution due to the central limit theorem.

The sum (9.18) can be approximated by an integral

$$R = \int_A \sigma(x,y)\, dA \tag{9.21}$$

where $\sigma(x,y)$ is the yield stress at (x,y). The mean value and variance of R are

$$E[R] = \int_A E[\sigma(x,y)]\, dA \tag{9.22}$$

$$\text{Var}[R] = \int_A \int_A \text{Cov}[\sigma(x_1,y_1), \sigma(x_2,y_2)]\, dA_1\, dA_2 \tag{9.23}$$

When the yield stress process is homogeneous, these expressions reduce to

$$E[R] = \mu_\sigma A \qquad (9.24)$$

$$Var[R] = \int_A \int_A C_\sigma(x_2-x_1, y_2-y_1)\, dA_1 dA_2 \qquad (9.25)$$

where μ_σ and $C_\sigma(,)$ are the mean value and covariance function, respectively. The mean strength per unit area is thus independent of the area, whereas the variance decreases with the area, except for $C_\sigma \equiv 1$.

Example 9.2 illustrates a combination of the ideal brittle model and the ideal plastic model.

Example 9.2

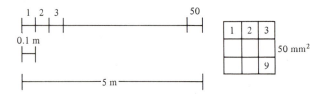

Figure 9.4   Modeling of bar.

A bar of ideal plastic material has the length 5 m and the cross-section area 450 mm^2. To compute the strength distribution test results for specimens of length 0.1 m and cross-section area 50 mm^2 are available. The test results show that the strength distribution of the test specimens can be taken as normal with mean value 10×10^3 N and standard deviation 0.9×10^3 N. To utilize the test results the bar is modeled as shown in Fig. 9.4. Each cross section is divided into 9 parts of equal area 50 mm^2. The bar is further divided into 50 sections of equal length 0.1 m. The bar therefore consists of $9 \times 50 = 450$ specimens with strength distribution as that of the test specimens. The strengths of the specimens composing the bar are assumed mutually independent. The distribution of a single section strength R' is normal with mean value and standard deviation from (9.19) and (9.20)

$$E[R'] = 9 \times 10 \times 10^3 \text{ N} = 90 \times 10^3 \text{ N}$$

$$D[R'] = \sqrt{9} \times 0.9 \times 10^3 \text{ N} = 2.7 \times 10^3 \text{ N}$$

The bar consists of 50 such sections and the distribution function for the bar strength is according to (9.1)

$$F_R(r) = 1 - \left| 1 - \Phi\left(\frac{r - 90 \times 10^3 \text{ N}}{2.7 \times 10^3 \text{ N}} \right) \right|^{50}$$

The mean value and standard deviation of R can be computed numerically as

$$E[R] = 90 \times 10^3 \text{ N} - 2.25 \times 2.7 \times 10^3 \text{ N} = 83.93 \times 10^3 \text{ N}$$

$$D[R] = 0.46 \times 2.7 \times 10^3 \text{ N} = 1.24 \times 10^3 \text{ N}$$

Per unit area the mean value and standard deviation of the maximum stress σ are thus

$$(E[\sigma], D[\sigma]) = \begin{cases} (200 \text{ MPa}, 6 \text{ MPa}) & \text{test specimen} \\ (186.5 \text{ MPa}, 2.8 \text{ MPa}) & \text{bar} \end{cases}$$

The mean value of the maximum stress is thus less for the bar than for the test specimens due to the series structure of the bar along its length. On the other hand, the standard deviation is reduced due to the averaging effect over the cross section.

9.1.3   Fiber Bundle

A *fiber bundle* is a material model in which a cross section is composed of identical fibers in a parallel arrangement (see Fig. 9.5).

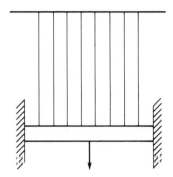

Figure 9.5   Fiber bundle model.

The force-displacement relation for a fiber is such that when a fiber breaks it is no longer capable of carrying any load. Figure 9.6 shows four material models for the force-displacement relations of a fiber bundle with three fibers. Failure of a fiber does not neces-sarily imply failure of the system, but the force is redistributed among the other fibers which may then be able to carry the increased loading. For the system in Fig. 9.5 the fibers all have the same displacement. Failure of a fiber then leads to the same dis-placement increase in all fibers. For the material model I this is tantamount to an equal increase in force, which is referred to as

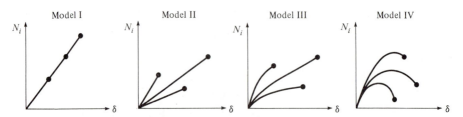

Figure 9.6 Examples of force-displacement relations for fibers.

equal load sharing. Fiber-reinforced materials have the fibers sur-
rounded by a matrix of another material. Failure of a fiber in a
cross section leads to a load sharing only between the nearest fibers
at that cross section. Through shear stresses the increased force is
also transmitted to other fibers for cross sections away from the
local failure. Many models for this *local load sharing* have been
proposed by Harlow and Phoenix (1981).

Daniels (1945) considered a system of n identical fibers with
equal load sharing and following the material model I. It was
further assumed that the random strengths X_i were mutually
independent with identical distribution function $F_X(\ )$. Taking the
fiber strengths in order of increasing magnitude gives a set of ran-
dom strengths $\hat{X}_1, \hat{X}_2, \ldots, \hat{X}_n$. The marginal distribution for $\hat{X}_m$ is
given in (7.1). The joint distribution of $(\hat{X}_1, \ldots, \hat{X}_n)$ can also be
expressed in terms of $F_X(\ )$. The strength R of the fiber bundle is
expressed as

$$R = \max\{n\hat{X}_1, (n-1)\hat{X}_2, \ldots, 2\hat{X}_{n-1}, \hat{X}_n\} \qquad (9.26)$$

Daniels (1945) determined an exact recursion formula for the distribution function of R. This formula is, however, not well suited for
numerical calculations. Daniels also showed that asymptotically as
$n \to \infty$ the distribution for R approaches a normal distribution with
expected value and variance:

$$E[R] = nx_0(1 - F_X(x_0)) \qquad (9.27)$$

$$\mathrm{Var}[R] = nx_0^2 F_X(x_0)(1 - F_X(x_0)) \qquad (9.28)$$

respectively, where x_0 maximizes the function $x(1 - F_X(x))$. The
result is valid under the conditions that the maximum is unique
and $\lim_{x \to \infty} x(1 - F_X(x)) = 0$.

It can be useful to derive the same asymptotic distribution
from a lower bound on the fiber bundle strength. A value x is
selected and the strengths of all fibers with strength less than x are
reduced to 0, while the strengths of all fibers with strength larger
than x are reduced to x. The reduced strength of the fiber bundle
is thus Mx, where M is the random number of fibers with strength

larger than x. M follows a binomial distribution. The reduced fiber bundle strength therefore also follows a binomial distribution and the mean value and variance are given as in (9.27) and (9.28) with x_0 replaced by x. The value of x can be selected freely and can, e.g., be selected as the value which maximizes the expected value. It then follows that x should be selected as x_0. When $n \to \infty$ the binomial distribution converges to a normal distribution and the distribution of the reduced fiber bundle strength converges to the exact asymptotic distribution.

The asymptotic normality of the distribution of the fiber bundle strength has been demonstrated under less strict assumptions than those used by Daniels, (see Hohenbichler, 1983, for a review). It has been shown that the asymptotic normality is also valid under certain weak dependencies between fiber strengths and for independent fibers but with general force-displacement relations.

Although the asymptotic distribution is known for many cases, this result is not always a good approximation for small to medium-size fiber bundles. This is because the convergence is generally slow. Based on the procedure of Hohenbichler and Rackwitz (1983), the exact distribution can be determined from the system reliability results given in Chapter 5. First, the material model II is considered with uniform deformation and mutually independent fiber strengths with identical distribution. Figure 9.7 illustrates this case with four fibers.

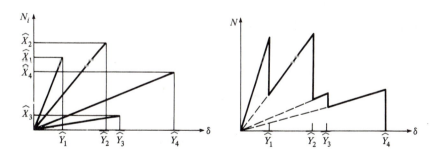

Figure 9.7   Force-deformation relation for fibers and fiber bundle.

At the deformation δ the total force $N(\delta)$ is

$$N(\delta) = \sum N_i(\delta) \tag{9.29}$$

where the summation is over all unbroken fibers. The strength of the fiber bundle is the maximum value of $N(\delta)$:

$$R = \max_{\delta} N(\delta) = \max_{\delta} \sum N_i(\delta) \tag{9.30}$$

The distribution function of R is expressed as

$$F_R(r) = P(\max_\delta \sum N_i(\delta) \leq r) = P(\sum N_i(\delta) \leq r \text{ for all } \delta) \quad (9.31)$$

$$= P(\cap_\delta \{\sum N_i(\delta) - r \leq 0\}) \leq \min_\delta \{P(\sum N_i(\delta) - r \leq 0)\}$$

The inequality follows from the reliability bound on parallel systems in (5.94).

For the material model II the maximum value of $N(\delta)$ occurs when the load in one fiber attains its maximum value. The intersection in (9.31) is therefore over n events. Taking the deformations at fiber failure in increasing magnitudes gives the ordered random variables $\hat{Y}_1, \hat{Y}_2, \ldots, \hat{Y}_n$. For $\delta = \hat{Y}_k$ the total force is

$$N(\hat{Y}_k) = \sum_{i=k}^{n} N_i(\hat{Y}_k) = \sum_{i=k}^{n} \hat{Y}_k \frac{\hat{X}_i}{\hat{Y}_i} \quad (9.32)$$

where $\hat{X}_i$ is the strength of the fiber with deformation $\hat{Y}_i$ at failure. The fiber bundle strength follows from (9.30) as

$$R = \max_{k=1}^{n} N(\hat{Y}_k) = \max_{k=1}^{n} \sum_{i=k}^{n} \hat{Y}_k \frac{\hat{X}_i}{\hat{Y}_i} \quad (9.33)$$

For the material model I with a constant ratio $\hat{X}_i/\hat{Y}_i$ this specializes to (9.26), which is then written as

$$R = \max_{k=1}^{n} \{(n-k+1)\hat{X}_k\} \quad (9.34)$$

The distribution function of the strength is

$$F_R(r) = P\left|\bigcap_{k=1}^{n} \left\{\sum_{i=k}^{n} \hat{Y}_k \frac{\hat{X}_k}{\hat{Y}_k} - r \leq 0\right\}\right| \quad (9.35)$$

which for the material model I reduces to

$$F_R(r) = P\left|\bigcap_{k=1}^{n} \{(n-k+1)\hat{X}_k - r \leq 0\}\right| \quad (9.36)$$

These distribution functions can be computed directly based on the theory of reliability of parallel systems outlined in Section 5.5. For material model II there are n independent pairs of basic variables $(X_1, Y_1), \ldots, (X_n, Y_n)$. The random variables X_i and Y_i within each pair may be dependent. In the expressions for $F_R(\ )$ the basic variables have have been ordered as $(\hat{X}_1, \hat{Y}_1), \ldots, (\hat{X}_n, \hat{Y}_n)$ and these pairs are no longer mutually independent. It is useful to demonstrate the first step in the reliability calculation of the parallel system, namely the Rosenblatt transformation (5.16) of the basic variables into a set of mutually independent and standardized normal variables.

The basic variables are collected in the basic variable vector $\mathbf{Z}$

$$\mathbf{Z} = (Z_1, \ldots, Z_n, Z_{n+1}, \ldots, Z_{2n}) \tag{9.37}$$

$$= (\hat{Y}_1, \ldots, \hat{Y}_n, \hat{X}_1, \ldots, \hat{X}_n)$$

The Rosenblatt transformation is defined from the conditional distribution functions $F_i(z_i \mid z_1, \ldots, z_{i-1})$. For $Z_1 = \hat{Y}_1$ the distribution function follows directly from the results of Section 9.1.1:

$$F_1(z_1) = F_{\hat{Y}_1}(z_1) = 1 - [1 - F_Y(z_1)]^n \tag{9.38}$$

The distribution of any Y_i, other than $\hat{Y}_1$, given that $\hat{Y}_1 = z_1$ is the original distribution truncated from below at z_1:

$$P(Y_i \leqslant y \mid \hat{Y}_1 = z_1) = \frac{F_Y(y) - F_Y(z_1)}{1 - F_Y(z_1)}, \quad y > z_1, \quad i = 2, \ldots, n \tag{9.39}$$

$\hat{Y}_2 \mid \hat{Y}_1 = z_1$ is the smallest of $n-1$ random variables with distribution function as in (9.39). The conditional distribution function needed in the Rosenblatt transformation is thus

$$F_2(z_2 \mid z_1) = 1 - \left| 1 - \frac{F_Y(z_2) - F_Y(z_1)}{1 - F_Y(z_1)} \right|^{n-1}, \quad z_2 > z_1 \tag{9.40}$$

Continuation of this procedure leads to the conditional distribution function $F_i(z_i \mid z_1, \ldots, z_{i-1})$ for $i \leqslant n$ and $z_i > z_{i-1} > \cdots > z_1$:

$$F_i(z_i \mid z_1, \ldots, z_{i-1}) = 1 - \left| 1 - \frac{F_Y(z_i) - F_Y(z_{i-1})}{1 - F_Y(z_{i-1})} \right|^{n-i+1} \tag{9.41}$$

For $i > n$ the conditional distributions are simpler since fiber properties are independent from fiber to fiber. The conditional distribution function of $Z_{n+i} = \hat{X}_i$ given $(Z_1, \ldots, Z_{n+i-1}) = (\hat{Y}_1, \ldots, \hat{Y}_n, \hat{X}_1, \ldots, \hat{X}_{i-1}) = (z_1, \ldots, z_n, z_{n+1}, \ldots, z_{n+i-1})$ therefore reduces to

$$F_{n+i}(z_{n+i} \mid z_1, \ldots, z_{n+i-1}) = F_{n+i}(z_{n+i} \mid z_i) \tag{9.42}$$

and the Rosenblatt transformation can be carried out.

A similar analysis can be carried out with material models III and IV. These force-deformation relations are typically described in terms of more than two random variables. For material model IV the deformations at which local maxima of the force occur are not directly determined. A lower bound on the bundle strength is found by considering only the deformations for which, e.g., one fiber has its maximum force or a fiber fails. The intersection in (9.31) is then over $2n$ values of δ.

9.2 FATIGUE

This section gives an introduction to models for accumulated damage due to fluctuating stresses and strains caused by time-varying external loading. Examples of such accumulated damages include crack initiation and subsequent growth in metals, successive crushing of concrete, and successive rupture of fibers in fiber-reinforced materials. When failure occurs after a relatively small number of load cycles (e.g., less than 100 cycles), it is common to use the name *low-cycle fatigue* for the damage accumulation. Otherwise, the term *high-cycle fatigue* is used.

In the analysis of damage accumulation the concept of damage indicators is conveniently introduced. The damage indicators, denoted $D_1, D_2, \ldots$, summarize at any time the state of damage. They do not describe the state at a microscopic level but are rather aggregated force deformation parameters describing the macroscopic appearance of damage. Damage indicators can have a specific physical meaning such as crack length or can be parameters in a mathematical model of the structural behavior. The damage indicators are ideally as few and as informative as possible. As observed by Veneziano (1981), the problem of deciding which damage indicators to use is conceptually similar to that of selecting the order of a regression analysis — the larger the number of damage indicators is, the better the damage accumulation can be described, but at the same time the more complex will the analysis be and the larger will the (statistical) uncertainty in the determination of the damage indicators be.

The damage indicators are usually taken as zero in the initial state and are thereafter nondecreasing functions of time until failure. The event of failure before time t is therefore related to the damage indicator functions, $D_1(\tau)$, $D_2(\tau)$, $\ldots$, during the time interval from 0 to t.

9.2.1 Damage Accumulation Laws

The damage accumulation laws considered in this section all operate with a simplified measure of damage in terms of a scalar damage indicator D. The value of D is taken as zero in the initial state and as one at failure. The damage accumulation is determined entirely by the stress variation and stresses are assumed to vary in cycles with stress ranges S_n. The damage indicator is a nondecreasing function of the number of stress cycles. The damage increment ΔD_n in the nth stress cycle depends on the damages accumulated at the end of each of the preceding $n-1$ stress cycles and on the stress range of the nth stress cycle

$$\Delta D_n = D_n - D_{n-1} = \xi(D_1, D_2, \ldots, D_{n-1}, S_n), \quad i = 1, 2, \ldots \quad (9.43)$$

where ξ is a nonnegative function which can also depend on variables such as the frequency of load cycles, the shape of the stress cycles, the environmental conditions, and the average stress in the jth stress cycle.

A damage accumulation theory is an *interaction-free* theory if the damage accumulated after n stress cycles is independent of the order in which the stress cycles occur. In that case the damage increment in a stress cycle depends only on the damage accumulated at the beginning of the stress cycle and on the stress cycle itself.

$$\Delta D_n = \xi(D_{n-1}, S_n), \quad n = 1, 2, \ldots \quad (9.44)$$

When the damage is a slowly varying function of the number of stress cycles the finite difference ΔD_n in (9.44) can be replaced with good accuracy by the derivative dD/dn. This leads to the "kinetic equation" for damage accumulation (Bolotin, 1969, 1981).

$$\frac{dD}{dn} = \xi(D, S) \quad (9.45)$$

The ξ-function can be related to the damage indicator D_S for a *homogeneous regime*, i.e., for S constant. The number of stress cycles with stress range S necessary to cause failure is denoted $N = N(S)$. The damage indicator is $D_S = D_S(n/N, S)$, since there is a one-to-one correspondence between N and S. Figure 9.8 shows three different examples of the function D_S

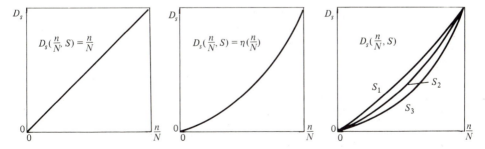

Figure 9.8   Damage accumulation under constant-amplitude loading.

In the first two examples the damage indicator D_S is not an explicit function of the stress range S but depends only on the ratio

n/N. Such a damage theory is called a *stress-independent* theory. In the first example the damage accumulation is further a linear function of n/N. For a stress-independent theory $D_S = \eta(n/N)$ and the derivative with respect to n is

$$\frac{dD_S}{dn} = \frac{1}{N(S)} \eta'(\frac{n}{N}) = \frac{1}{N(S)} \eta'(\eta^{-1}(D_S)) \qquad (9.46)$$

If this relation is assumed valid even for an *inhomogeneous regime*, the kinetic equation (9.45) is

$$dD = \frac{1}{N(S)} \eta'(\eta^{-1}(D)) \, dn \qquad (9.47)$$

where dn is the number of stress cycles with stress range between S and $S + dS$. The variables can be separated as

$$\frac{dn}{N(S)} = \frac{dD}{\eta'(\eta^{-1}(D))} \qquad (9.48)$$

Since $\eta(0) = 0$ and $\eta(1) = 1$ it follows by a variable substitution $D = \eta(x)$ that

$$\int_{D=0}^{1} \frac{dD}{\eta'(\eta^{-1}(D))} = \int_{x=0}^{1} dx = 1 \qquad (9.49)$$

Integrating both sides of (9.48), the following fundamental condition is found at failure:

$$\sum_i \frac{n_i}{N(S_i)} = 1 \qquad (9.50)$$

where the integration of the left-hand side of (9.48) has been replaced by a summation, and where n_i is the number of stress cycles with stress range S_i. The condition (9.50) is usually attributed to Palmgren (1924) and Miner (1945), who independently proposed the condition, but based on the assumption that $\eta(\ )$ is a linear function. This additional assumption is not necessary for the validity of (9.50). Application of (9.50) is generally referred to as Miner's rule. Without any detailed knowledge of the damage process the assumptions behind Miner's rule are the most obvious ones to be made. The only material characteristic entering Miner's rule is the number of stress cycles to failure under constant-amplitude loading. Miner's rule thus provides a method for extrapolation of this material characteristic to more complicated stress variations.

Experiments disclose variations from Miner's rule and a number of other damage accumulation laws have been proposed which are all variants of Miner's rule, but take into account some of the experimentally observed phenomena. Many reviews of these laws have been made, (see, e.g., Leve, 1969; Schijve, 1972; Stallmeyer and Walker, 1968). In a comparison of a number of cumulative damage theories Leve (1969) concludes that the theories cannot be relied on to produce fatigue life predictions of sufficient accuracy for many encountered circumstances, and because there is no indication at present that any cumulative damage theory will give predictions significantly better than all other proposed theories, the use of Miner's rule in the majority of applications is still recommended. This is true in particular when no mechanical model for the damage accumulation is available. Although Miner's rule has shortcomings, it provides a model which has at least some similarities with the real situation. Further, while the absolute validity of Miner's rule is in question, its comparative usefulness as a simple analytical criterion for comparing different designs of a system is of considerable value.

The relationship between S and N for a homogeneous regime has a very definite random character even under the most controlled and uniform test conditions. It is therefore meaningless to speak of a deterministic relation $N = N(S)$. The relation should be in terms of the conditional cumulative distribution function $F_{N \mid S}(n,s)$. Repeated testing at some stress level $S = s$ can give test results which allow an estimation of statistics for the random variable $N \mid S = s$. Such a statistical analysis of test data is performed in Example 5.3.

The relation between S and N used in Example 5.3 is

$$N = KS^{-m}, \quad S > 0 \tag{9.51}$$

in which K and m are random variables due to inherent physical and statistical uncertainty. The value of K can depend on the mean stress S_a in the stress cycles. The effect can be accounted for, e.g., by the Goodmann criterion

$$K = K_0 \left(1 - \frac{S_a}{S_u} \right)^m \tag{9.52}$$

where K_0 is the value of K from tests with zero mean stress and where S_u is the ultimate tensile strength. This is called the Basquin equation and is generally used for the high cycle range in which stresses are restricted to the elastic range.

In structural codes such as API (1982) and DnV (1982) a stress range threshold S_0 is included. For stress ranges below this threshold no damage is assumed to occur. The S-N relation is then

$$N = \begin{cases} KS^{-m}, & S > S_0 \\ \infty, & S \leqslant S_0 \end{cases} \tag{9.53}$$

In the recommendations from Department of Energy (1982) the S-N relation is proposed as

$$
N = \begin{cases} KS^{-m}, & S > S_0 \\ K_1 S^{-m_1}, & S \leqslant S_0 \end{cases}
\tag{9.54}
$$

where $KS_0^{-m} = K_1 S_0^{-m_1}$.

A more general model which also includes the low-cycle range was proposed independently by Coffin and Manson and is referred to in ASCE, Committee on Fatigue and Fracture Reliability (1982):

$$
\frac{\Delta \varepsilon}{2} = \frac{\sigma_f'}{E}(2N)^{c_1} + \varepsilon_f'(2N)^{c_2}
\tag{9.55}
$$

Here $\Delta \varepsilon$ is the strain range and E is the modulus of elasticity, while the other factors are empirical constants. The first term on the right-hand side dominates in the high-cycle range, and when set equal to the elastic strain range leads to the Basquin equation. The second term, which is equal to the plastic strain range, dominates in the high-strain low-cycle region.

When (9.51) is inserted into (9.50) the failure criterion becomes

$$
1 - \frac{1}{K}\sum_i S_i^m \leqslant 0
\tag{9.56}
$$

where the summation is over all stress ranges. The joint distribution of K and m is addressed in Example 5.3 and to compute the reliability the distribution of $\sum S_i^m$ for a fixed value of m must be computed. This problem is considered in the next sections.

9.2.2 Cycle Counting

Figure 9.9 shows the principal behavior of sample functions for three different stress processes. The first case corresponds to a constant-amplitude loading. Stress cycles are easily identified and stress ranges are the same for all stress cycles. The second case corresponds to a stationary ideal narrow-band Gaussian process. Each upcrossing of the mean level corresponds to exactly one local maximum and a stress cycle is defined as the stress process between two consecutive upcrossings of the mean level. Stress ranges varies from stress cycle to stress cycle and follow a Rayleigh distribution as described in Section 7.4. The third case corresponds to a stationary Gaussian process which is not ideal narrow-banded. No obvious definition of stress cycles can be given and the stress range distribution cannot be directly determined. To identify stress cycles, *cycle counting* methods are applied.

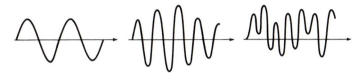

Figure 9.9   Three examples of stress variation around mean level.

Three different methods of cycle counting are included here: *peak counting, range counting,* and *rain-flow counting*. Other counting methods have been proposed in the literature, as explained in detail in Dowling (1972). All three counting methods give the same result for a pure sinusoidal stress history and for an ideal narrow-band stress history. The three methods are described below for stationary processes. For simplicity, the mean level is taken as zero in the description.

In the *peak counting method* all local maxima above zero are counted. For the stress history in Fig. 9.10 four such maxima are identified.

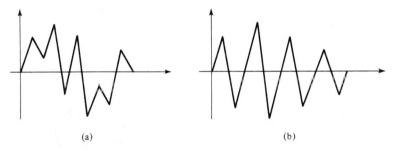

(a)   (b)

Figure 9.10   Illustration of peak counting method: (a) stress history; (b) equivalent stress history.

A local maximum is paired with a local minimum of the same size and an equivalent stress history is obtained, as shown in the figure. It is thus implicitly assumed that the distribution functions $F_{\max}(\ )$ and $F_{\min}(\ )$ of local maxima and local minima satisfy

$$F_{\min}(-x) = 1 - F_{\max}(x) \tag{9.57}$$

Furthermore, local maxima and minima are paired to form stress cycles independent of their relative location in the stress history. The distribution function for a stress range is

$$F_S(s) = F_{max}(s/2) \tag{9.58}$$

The mean number of stress cycles in a time period $[0,T]$, N_T, is equal to the mean number of local maxima above zero:

$$N_T = \frac{1+\alpha}{2} v_m T \tag{9.59}$$

where α is the regularity factor of the process and v_m is the mean rate of local maxima.

In the peak counting method a number of stress reversals are ignored corresponding to local maxima below zero. Furthermore, the method has the characteristic that all small stress reversals above zero are counted as much larger stress reversals. All effects of the sequence of stress cycles are ignored since a local maximum is combined with a local minimum independent of its location in time in relation to the maximum.

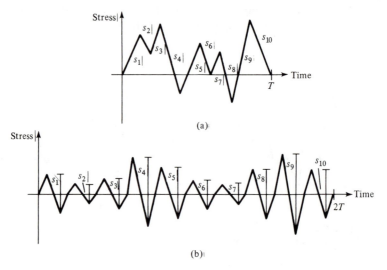

(a)

(b)

Figure 9.11 Illustration of range counting method: (a) stress history; (b) equivalent stress history of double length.

In the *range counting method* each stress range, i.e., the difference between two successive local extremes, is counted as one-half a stress cycle. For the stress history in Fig. 9.11, 10 half-cycles are thus identified. The distribution of stress ranges is addressed for Gaussian processes in the next section. The mean number of stress cycles in a time period $[0,T]$, N_T, is equal to the mean number of local maxima in that period:

$$N_T = \nu_m T \qquad (9.60)$$

The range counting method uses only local information about the stress process, as each local extreme is combined only with the preceding and the following local extreme. The range counting method has the characteristic that if all small stress reversals are counted, the larger ranges are broken up and counted as several smaller ones.

In the *rain-flow counting* method the stress history is first con- verted into a series of peaks and troughs as shown in Fig. 9.12 with the peaks evenly numbered. The time axis is oriented vertically with the positive direction downward. The time series is then viewed as a sequence of roofs with rain falling on them. The rain- flow paths are defined according to the following rules (Wirsching and Shehata, 1977):

1. A rain flow is started at each peak and trough.
2. When a rain-flow path started at a trough comes to the tip of the roof, the flow stops if the opposite trough is more nega- tive than that at the start of the path under consideration (e.g., path [1-8], path [9-10], etc.). A path started at a peak is stopped by a peak which is more positive than that at the start of the rain path (e.g., path [2-3], path [4-5], path [6-7], etc.).
3. If the rain flowing down a roof intercepts flow from a previ- ous path, the present path is stopped (e.g., path [3-3a], path [5-5a], etc.).
4. A new path is not started until the path under consideration is stopped.

Half-cycles of trough-originated stress range magnitudes S_i are pro- jected distances on the stress axis (e.g., [1-8], [3-3a], [5-5a], etc.). It should be noted that for time series sufficiently long, any trough- originated half-cycle will be followed by another peak-originated half-cycle of the same range. This is also the case for short stress histories if the stress history starts and ends at the same stress value.

The rain-flow method is not restricted to high-cycle fatigue but can also be used for low-cycle fatigue where strain range is the important parameter. Figure 9.12 shows a simple example. In this sequence four events that resemble constant-amplitude cycling are

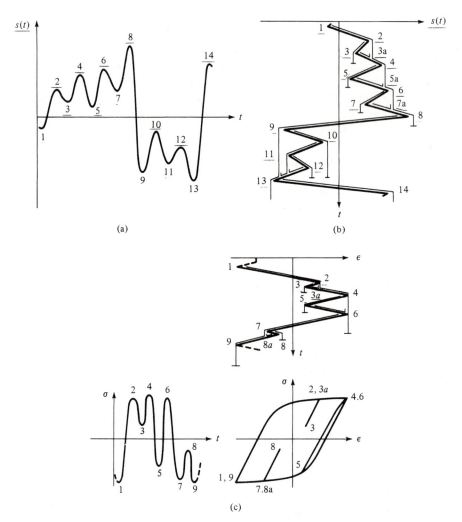

Figure 9.12 Illustration of the rain-flow counting method: (a), (b) application to a stress history, from Wirsching and Shehata (1977); (c) application to a low-cycle strain history.

recognized, 1-6-9, 2-3-3a, 4-5-6, and 7-8-8a. These events are closed hysteresis loops and each event is associated with a strain range and a mean strain. Each closed hysteresis loop can therefore be compared with constant-amplitude data in order to calculate the accumulated damage.

In the rain-flow method small reversals are treated as interruptions of the larger ranges and the rain-flow method also identifies a mean stress for each stress cycle.

A comparison between test results and results predicted by the three methods shows that the rain-flow counting method generally gives the best results. This method is the only one of the three that identifies both slowly varying stress cycles and more rapid stress reversals on top of these. The peak counting method will in general assign larger probabilities to larger stress ranges, while the range counting method will assign larger probabilities to smaller stress ranges. Compared to the rain-flow counting method, the peak counting will therefore result in larger estimates for the accumulated damage, while the range counting method will predict smaller values of the damage. Analytical results for the stress range distribution obtained through rain-flow counting are very difficult to obtain and the method is generally used only with measured or simulated stress histories.

9.2.3 Damage Statistics

The failure criterion (9.56) for fatigue failure in a period $[0,T]$ involves the random variable $\sum S_i^m$, where the summation is over all stress ranges. The distribution type of this random variable is generally difficult to obtain. If the number of stress cycles is large and the dependence between stress ranges is sufficiently weak, the distribution is well approximated by the normal distribution. The mean value and variance of the sum shall be derived for various stationary processes.

For a *Poisson spike process* defined in Section 7.3.4 and illustrated in Fig. 7.7, the stress ranges are equal to the heights of the spikes. The number of spikes in $[0,T]$ is denoted by N and it follows a Poisson distribution with expected value and variance νT. Spikes are mutually independent and identically distributed and the mean value and variance of the sum are (Ditlevsen, 1981, p. 162)

$$\mathrm{E}\left[\sum_{i=1}^{N} S_i^m\right] = \mathrm{E}[N]\mathrm{E}[S_i^m] = \nu T \mathrm{E}[S_i^m] \qquad (9.61)$$

$$\mathrm{Var}\left[\sum_{i=1}^{N} S_i^m\right] = \mathrm{E}[N]\mathrm{Var}[S_i^m] + \mathrm{Var}[N]\mathrm{E}[S_i^m]^2 \qquad (9.62)$$

$$= \nu T(\mathrm{Var}[S_i^m] + \mathrm{E}[S_i^m]^2)$$

These results are valid independent of the counting method.

For a narrow-band Gaussian process, stress ranges are Rayleigh-distributed in the limit as $\alpha \to 1$. The mean value of the sum then follows directly as

$$E[\textstyle\sum S_i^m] = v_0 T E[S_i^m] = v_0 T \int_0^\infty (2x)^m \frac{x}{\sigma_X^2} \exp\left|-\frac{1}{2}\left(\frac{x}{\sigma_X}\right)^2\right| dx \quad (9.63)$$

$$= v_0 T (2\sqrt{2})^m \sigma_X^m \Gamma(1 + m/2)$$

Assuming a deterministic number of stress cycles $N = v_0 T$ in the time period $[0,T]$, the variance of the sum can be written as

$$\text{Var}[\textstyle\sum S_i^m] = \sum_{i=1}^{N}\sum_{j=1}^{N} \text{Cov}[S_i^m, S_j^m] \quad (9.64)$$

$$= N\text{Var}[S_i^m] + 2\sum_{k=1}^{N-1}(N-k)\text{Cov}[S_i^m, S_{i+k}^m]$$

The variance of S_i^m is computed as

$$\text{Var}[S_i^m] = E[S_i^{2m}] - E[S_i^m]^2 \quad (9.65)$$

$$= (2\sqrt{2})^{2m}\sigma_X^{2m}\left|\Gamma(1+m) - \Gamma^2(1+m/2)\right|$$

To compute the covariance $\text{Cov}[S_i^m, S_{i+k}^m]$ an envelope process $R(t)$ is considered as illustrated in Fig. 9.13.

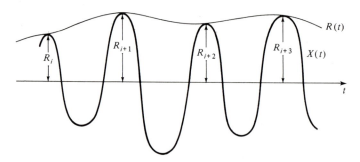

Figure 9.13   Definition of stress cycle amplitude from envelope process.

The mean distance between stress cycles i and $i+k$ is k/v_0. The covariance is therefore computed as

$$\text{Cov}[S_i^m, S_{i+k}^m] = 4^m \text{Cov}[R_i^m, R_{i+k}^m] \quad (9.66)$$

$$= 4^m \text{Cov}[R(t)^m, R(t+k/v_0)^m] = 4^m C_{R^m}(k/v_0)$$

For integer values of m the covariance function $C_{R^m}(\ )$ is simple. Using the envelope definition (7.99), it follows from (7.101) that for $m = 2$

$$C_{R^2}(\tau) = \text{Var}\,[R(t)^2]k(\tau)^2 = 2\sigma_X^4 \left| \rho(\tau)^2 + 2\frac{\rho'(\tau)^2}{\omega_0^2} + \frac{\rho''(\tau)^2}{\omega_0^4} \right| \quad (9.67)$$

$$\approx 4\sigma_X^4 \left| \rho(\tau)^2 + \frac{\rho'(\tau)^2}{\omega_0^2} \right|$$

For noninteger values the approximation to the joint distribution of $R(t_1)$ and $R(t_2)$ given in (7.103) can be applied. Using this distribution, moments $E[R(t)^m R(t+\tau)^m]$ are

$$E[R(t)^m R(t+\tau)^m] = 2^m \sigma_X^{2m} \Gamma^2 \left(1+\frac{m}{2}\right) {}_2F_1\left(-\frac{m}{2}, -\frac{m}{2}; 1; k(\tau)^2\right) \quad (9.68)$$

where ${}_2F_1(\ )$ is the hypergeometric function. The covariance function is then

$$C_{R^m}(\tau) = E[R(t)^m R(t+\tau)^m] - E[R(t)^m]^2 \quad (9.69)$$

$$= 2^m \sigma_X^{2m} \Gamma^2 \left(1+\frac{m}{2}\right) \left| {}_2F_1\left(-\frac{m}{2}, -\frac{m}{2}; 1; k(\tau)^2\right) - 1 \right|$$

From these expressions the variance of the sum (9.64) can then be computed.

In connection with the peak counting method a general result can be given for the expected value of the sum. Since all information about the sequence of stress cycles is ignored, it makes little sense to consider any variance of the sum in this case. Use is made of a general result for the mean-upcrossing rate. The number of upcrossings of a level x is equal to the number of maxima greater than x minus the number of minima greater than x. This leads to the following expression for $v(x)$:

$$v(x) = v_m\,(F_{\min}(x) - F_{\max}(x)) = \frac{v_0}{\alpha}[F_{\min}(x) - F_{\max}(x)] \quad (9.70)$$

Upon differentiation this gives

$$f_{\max}(x) - f_{\min}(x) = -\alpha \frac{1}{v_0}\frac{dv(x)}{dx} \quad (9.71)$$

The peak counting method is based on the idea of obtaining stress ranges by pairing a positive maximum with a negative minimum of the same numerical magnitude independent of the relative position of the two local extremes in the stress history. Figure 9.14 illustrates this idea. Only positive maxima in the cross-hatched area are combined with corresponding minima. The number of stress cycles is thereby reduced but larger weights are given to large stress ranges, so although no rigorous proof is given, the procedure is expected to result in estimates of damage larger than the actual

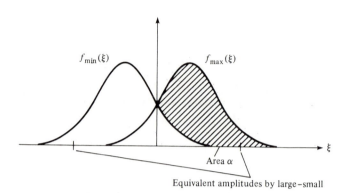

$$f_{min}(\xi) \qquad\qquad f_{max}(\xi)$$

Area α

Equivalent amplitudes by large–small

Figure 9.14 Upper bound on damage from stress cycle counting.

values. The number of stress cycles in a time interval $[0,T]$ becomes $\alpha v_m T = v_0 T$. The stress amplitude distribution $f_m(x)$ to be used in the damage calculation is given from the figure as

$$f_m(x) = \frac{1}{\alpha} [f_{max}(x) - f_{min}(x)] \qquad (9.72)$$

$$= -\frac{1}{v_0} \frac{dv(x)}{dx}, \quad x > 0$$

and the expected value of the sum becomes

$$E[\textstyle\sum S_i^m] = v_0 TE[S_i^m] = -T \int_0^\infty (2x)^m \frac{dv(x)}{dx}\, dx \qquad (9.73)$$

The right-hand side of (9.73) has been computed in Section 8.4 in connection with Markov vector methods for nonlinear random vibration and in Section 10.2 in connection with the Morison load model. For a stationary Gaussian process

$$-\frac{dv(x)}{dx} = v_0 \frac{x}{\sigma_{\bar X}^2} \exp\left[-\frac{1}{2}\left(\frac{x}{\sigma_X}\right)^2\right] \qquad (9.74)$$

and (9.73) becomes identical to (9.63). For a stationary Gaussian process damage is thus computed as for an equivalent ideal narrow-band process with the same values of v_0 and σ_X, i.e., with the same spectral moments λ_0 and λ_2.

Stationary Gaussian processes can be studied more closely with respect to range counting. Analogous with Fig. 9.13 stress ranges are defined from the double envelope process (7.118) and (7.119) (see Fig. 9.15), The stress amplitudes follows a Rayleigh distribution (7.120) and the expected value of the sum is

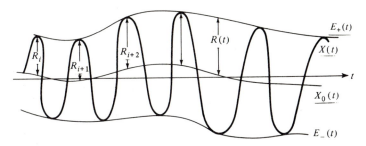

Figure 9.15 Definition of stress cycle amplitudes from double envelope process.

$$E[\sum S_i^m] \;=\; \nu_m T E[(2R_i)^m] \;=\; \frac{\nu_0}{\alpha} T(2\sqrt{2})^m (\alpha\sigma_X)^m \Gamma(1+m/2) \quad (9.75)$$

$$= \alpha^{m-1}\nu_0 T(2\sqrt{2})^m \sigma_X^m \Gamma(1+m/2)$$

This result deviates by a factor of α^{m-1} from the result for the limit $\alpha \to 1$ in (9.63). The variance of the sum follows from (9.64) to (9.69) with ν_0 replaced by ν_m, σ_X replaced by $\alpha\sigma_X$, and $k(\tau)^2$ defined as in (7.101) with $R(t)$ defined in (7.119).

In Fig. 9.15 a stress range is defined as the difference between an upper and a lower envelope process at a single point in time. In reality the difference should refer to different times of the two envelopes

$$S_i \;=\; X(t_i) - X(t_i + T_i) \;=\; E_+(t_i) - E_-(t_i + T_i) \qquad (9.76)$$

where T_i is the distance between the local maximum and the following local minimum. The expected value of S_i defined in this way is given in (7.91) and the distribution of S_i is here analyzed somewhat further.

Figure 9.16 illustrates the definition of stress range S and half wavelength T_1 after a local maximum of height U. The stress range depends on the curvature at the maximum, which again is correlated to the value of the maximum. It is therefore convenient to consider a new random process $Z(t)$ which is the original stress process $X(t)$ conditioned on the value of $\mathbf{Y}=(X(\tau), \dot{X}(\tau), \ddot{X}(\tau))$ at time τ. Without loss of generality, $\tau = 0$.

$$Z(t) \;=\; X(t) \mid \mathbf{Y}=\mathbf{y}=(u,v,w) \qquad (9.77)$$

$Z(t)$ is a (nonstationary) normal process and therefore completely described by its mean-value function and covariance function. These

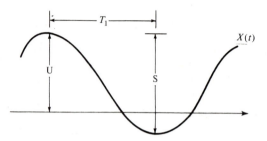

Figure 9.16 Definition of stress range S and half-wave length T_1.

two functions are determined by application of (7.137) and (7.138) for the random vector $(\mathbf{X},\mathbf{Y})=(X(t),X(s),X(0),\dot{X}(0),\ddot{X}(0))$. The mean value of this vector is $\mathbf{0}$ and the covariance matrix is

$$
\begin{vmatrix} c_{\mathbf{XX}} & c_{\mathbf{XY}} \\ c_{\mathbf{YX}} & c_{\mathbf{YY}} \end{vmatrix} = \begin{vmatrix} \lambda_0 & c(t-s) & c(t) & -c'(t) & -c''(t) \\ c(t-s) & \lambda_0 & c(s) & -c'(s) & -c''(s) \\ c(t) & c(s) & \lambda_0 & 0 & -\lambda_2 \\ -c'(t) & -c'(s) & 0 & \lambda_2 & 0 \\ -c''(t) & -c''(s) & -\lambda_2 & 0 & \lambda_4 \end{vmatrix} \quad (9.78)
$$

where $c(\ )$ is the covariance function for $X(t)$ and where $\lambda_0 = c(0)$, $\lambda_2 = -c''(0)$, and $\lambda_4 = c^{iv}(0)$ have been introduced. The mean-value function of $Z(t)$ is linear in u, v, and w.

$$
E[Z(t)] = uA(t) + vB(t) + wC(t) \quad (9.79)
$$

with the functions $A(t)$, $B(t)$, and $C(t)$ defined as

$$
A(t) = \frac{1}{\lambda_0\lambda_4-\lambda_2^2}(\lambda_4 c(t) + \lambda_2 c''(t)) \quad (9.80)
$$

$$
B(t) = -\frac{1}{\lambda_2}c'(t) \quad (9.81)
$$

$$
C(t) = -\frac{1}{\lambda_0\lambda_4-\lambda_2^2}(\lambda_2 c(t) + \lambda_0 c''(t)) \quad (9.82)
$$

The covariance function is independent of u, v, and w

$$
\mathrm{Cov}[Z(t), Z(s)] = c(t-s) \quad (9.83)
$$

$$
-\frac{1}{\lambda_2(\lambda_0\lambda_4-\lambda_2^2)}\{\lambda_2\lambda_4 c(t)c(s) + \lambda_2^2 c(t)c''(s)
$$

$$
+ (\lambda_0\lambda_4-\lambda_2^2)c'(t)c'(s)+\lambda_2^2 c''(t)c(s)+\lambda_0\lambda_2 c''(t)c''(s)\}
$$

When the interest is on the stress range, the conditioned process is considered after a local maximum and therefore $v = 0$ and $w < 0$. The half-wavelength T_1 is the time until the first downcrossing of zero level by the derivative process $\dot{Z}(t)$. The first-passage-time probability density function $f_{T_1 \mid \mathbf{y}}(t_1)$ can be bounded or approximated as described in Section 7.4.5. The unconditioned probability density function of the stress range can now be expressed in analogy with (7.73) as

$$f_S(s) = \frac{1}{v_m} \int_{-\infty}^{\infty} du \int_{-\infty}^{0} dw \int_{0}^{\infty} |w| \, f_{S \mid t_1, \mathbf{y}}(s, t_1, u, 0, w) \qquad (9.84)$$

$$\times f_{T_1 \mid \mathbf{y}}(t_1, u, 0, w) \, f_{\mathbf{Y}}(u, 0, w) \, dt_1$$

Here $f_{S \mid t_1, \mathbf{y}}(s, t_1, u, 0, w)$ is the probability density function of S conditioned on the values of the maximum and curvature and on the time to the first minimum. This distribution is approximated by the probability density function of S conditioned on the values of the maximum and curvature and on the existence of a minimum — but not necessarily the first — at $T_1 = t_1$. This has been found to be a good approximation for this application for several example processes with unimodal spectral densities (Lindgren and Rychlik, 1982). Conditional on a minimum for $T_1 = t_1$, the stress range is

$$S = u - Z(t_1) \mid \dot{Z}(t_1) = 0, \ddot{Z}(t_1) > 0 \qquad (9.85)$$

Similar to (7.73) the conditional distribution of S is then

$$f_{S \mid t_1, \mathbf{y}}(s, t_1, u, 0, w) = \frac{1}{v_m^Z(t_1)} \int_{0}^{\infty} z \, f_{\mathbf{Y}_1}(u - s, 0, z) \, dz \qquad (9.86)$$

where $\mathbf{Y}_1 = (Z(t_1), \dot{Z}(t_1), \ddot{Z}(t_1))$ and $v_m^Z(t_1)$ is the mean rate of local minima at time t_1 for the process $Z(t)$.

Computation of the distribution of S in this way therefore requires significant numerical efforts. Furthermore, the analysis uses the complete covariance function and its derivatives of up to and including fourth order. Only in special cases is such information available with sufficient confidence.

A simpler approximation to the distribution S is obtained by considering only the mean value of $Z(t)$ in (9.78) with $v = 0$ and $w < 0$ (Lindgren and Rychlik, 1982). The complete fourth-order derivative of the correlation function is then no longer used, only its value at zero λ_4. After a local maximum the stress process is approximated by

$$U A(t) + W C(t) = W\left(\frac{U}{W}A(t) + C(t)\right) \qquad (9.87)$$

where U and W are the (random) values of the maximum and the second derivative at the maximum. The time to the first minimum satisfies

$$\frac{U}{W} A'(t) + C'(t) = 0 \qquad (9.88)$$

Excluding the possibility $A'(t) = C'(t) = 0$ for $t > 0$, this leads to

$$\frac{U}{W} = -\frac{C'(T_1)}{A'(T_1)} = p(T_1) \qquad (9.89)$$

The stress range is

$$S = U - (U A(T_1) + W C(T_1)) \qquad (9.90)$$

W can now be expressed as

$$W = S\frac{-A'(T_1)}{C'(T_1)(1 - A(T_1)) + A'(T_1)C(T_1)} = Sq(T_1) \qquad (9.91)$$

The joint probability density function of (U,W) at a local maximum is [see also (7.73)]

$$f_{UW}(u,w) = \frac{1}{v_m} \mid w \mid f_{UVW}(u,0,w) \qquad (9.92)$$

$$= \frac{1}{\sqrt{2\pi\lambda_0(1-\alpha^2)\lambda_4}} \mid w \mid$$

$$\times \exp\left|-\frac{1}{2(\lambda_0\lambda_4 - \lambda_2^2)}(\lambda_4 u^2 + 2\lambda_2 u\, w + \lambda_0 w^2)\right|, \quad w < 0$$

Through a transformation of variables the joint probability density function of (T_1,S) is obtained as

$$f_{T_1 S}(t,s) = \frac{1}{\sqrt{2\pi\lambda_0(1-\alpha^2)\lambda_4}} \, s^2 \mid q(t)^3 p'(t) \mid \qquad (9.93)$$

$$\times \exp\left|-\frac{1}{2(1-\alpha^2)}s^2 q(t)^2\frac{\lambda_4 p(t)^2 + 2\lambda_2 p(t) + \lambda_0}{\lambda_0\lambda_4}\right|, \quad t > 0, \; s > 0$$

From this expression the marginal distribution of S can be obtained by integration over t.

The two approximations for the distribution of S have been compared for several processes with unimodal spectral density (Lindgren and Rychlik, 1982). The agreement is generally very close. Characteristic properties of a normal process with a bimodal spectral density have also been reproduced by the first approximation technique.

9.3 STOCHASTIC MODELING OF FATIGUE CRACK GROWTH

The prediction of fatigue crack growth in metals and other materials is very important for many structures, such as aircrafts, components in power plants, transportation vehicles, transmission towers, and offshore structures. For these structures the crack growth stage of fatigue is often the most important one because cracks are present for a major fraction of the useful fatigue lifetime. There are many possible origins for fatigue crack initiation, such as inclusions in welds, voids in polycrystals, and notches, which are present in many components owing to the fabrication process. In other cases fatigue cracks are developed from microcracks after a nucleation period.

Fatigue crack growth is a repetitive process that is unknown at the atomic level, but for which several macroscopic observations have been made. Fatigue cracks initiated in slip bands begin their growth along the slip planes with the highest resolved shear stress. The shear mode is called stage I propagation and is normally very short before a critical crack length is reached and stage II propagation takes over. In stage II propagation the crack direction turns and follows a course perpendicular to the maximum principal stress at least as long as the crack growth rate is small, (Schijve, 1979). At higher crack growth rates the growth direction remains perpendicular to the main principal stress but the plane of the fatigue fracture will for several materials be under an angle of 45° with that stress. Stage II growth is better understood than stage I growth. The repetitive nature of stage II growth follows from the observation of striations on the fracture surface, where each striation has been shown to represent the crack advance for one cycle of the load. The observation has been made both for ductile and brittle materials and is very important, since the understanding of the mechanism of advance for one cycle is then sufficient to describe the growth mechanism for the entire stage. One such mechanism of advance is the plastic blunting mechanism suggested by Laird (1967). This mechanism is applicable to ductile materials under moderate to large fatigue load amplitudes. In brittle materials there is often an additional static fracture besides the striations. This is probably due to small size brittle fracture in the material ahead of the crack tip, which is then connected with the fatigue fracture by the striation mechanism.

The stress in a cracked element can be calculated within the theory of elasticity, assuming linear elastic behavior. The stress field near crack tips can be divided into three basic types, each associated with a local mode of deformation. In Paris and Sih (1965) a comprehensive collection of stress and displacement field solutions is given. The near-field stresses all contain a singularity as exemplified by the stress field solution for the mode I, plane strain case:

$$\sigma_x = \frac{K_I}{(2\pi r)^{1/2}} \cos \frac{\theta}{2} \left| 1 - \sin \frac{\theta}{2} \sin \frac{3\theta}{2} \right|$$

$$\sigma_y = \frac{K_I}{(2\pi r)^{1/2}} \cos \frac{\theta}{2} \left| 1 + \sin \frac{\theta}{2} \sin \frac{3\theta}{2} \right|$$

$$\sigma_z = \nu(\sigma_x + \sigma_y) \tag{9.94}$$

$$\tau_{xy} = \frac{K_I}{(2\pi r)^{1/2}} \sin \frac{\theta}{2} \cos \frac{\theta}{2} \cos \frac{3\theta}{2}$$

$$\tau_{xz} = \tau_{yz} = 0$$

The parameters K_I, K_{II}, and K_{III} are stress intensity factors which are independent of the coordinates (r,θ) and hence control the intensity but not the distribution of the stress field. The stress intensity factors contain the magnitude of loading forces linearly and also depends on geometrical variables. Physically, the stress intensity factors can be interpreted as parameters which reflect the redistribution of stresses in a body due to the introduction of a crack. It follows from dimensional considerations that K is of the form

$$K = Y\sigma\sqrt{\pi a} \tag{9.95}$$

where σ represents the far-field stress resulting from the applied load, and where $Y = Y(a)$ is a geometry function accounting for the shape of the specimen and the crack geometry.

In (9.94) higher-order terms in r are neglected, so the formulas are valid in the region where r is small compared to other planar dimensions of the body, such as crack length. The formulas show a square-root stress singularity, implying that stresses tend to infinity as r tends to zero. Ductile materials will therefore always show some plasticity at the tip of the crack. As long as the plastic zone is sufficiently small, the stress field solution (9.94) can still be applied outside the plastic zone.

The crack growth rate, Δa or da/dn, is understood as the crack extension of a crack of length a during one stress cycle. From a description of crack growth mechanisms it follows that da/dn depends on the cyclic stress in the crack tip area, the elastic-plastic response of the material in the same area, and on a fracture criterion. da/dn also depends on environmental conditions, such as temperature and the presence of a corrosive aggregate. These dependencies are, however, not considered here. A wide variety of formulas for the fatigue crack growth rate have been suggested. Paris and Erdogan (1963) have reviewed crack growth equations which were introduced before 1963, while a review of equations introduced later than 1963 was conducted by Irving and McCartney (1977).

The most simple and generally applicable crack growth equation was proposed in Paris and Erdogan (1963).

$$\frac{da}{dN} = C(\Delta K)^{m}, \quad \Delta K > 0 \tag{9.96}$$

where ΔK is the stress intensity factor range in a stress cycle, and C and m are regarded as material constants. C can depend on the stress-strain and fracture relation parameters mentioned earlier and also on the mean stress often characterized by the stress ratio $R = K_{min}/K_{max}$. m, on the other hand, is almost insensitive to R. If the Paris and Erdogan equation was correct, then a plot of $\log da/dn$ against $\log \Delta K$ would show a straight line. In Fig. 9.17 a schematic plot of typical fatigue crack growth data is shown.

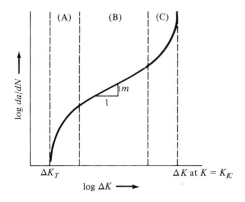

Figure 9.17   Schematic plot of typical fatigue crack growth data.

The agreement between the model (9.96) and the experiments is good for the mid-range of growth rates (B), typically 10^{-5} to 10^{-3} mm/cycle. At higher growth rates (C), when K_{max} approaches the fracture toughness K_{IC}, the equation underestimates the propagation rate, whereas for lower values of K (A) it overestimates the propagation rate. The figure shows a threshold stress intensity ΔK_{T} below which crack propagation cannot be detected. The value of ΔK_{T} is found to depend on the mean stress intensity. The existence and an explanation of the threshold still remain a subject of controversy. Based on a comparison of experimental data spanning four orders of magnitude in growth rate a value of $m = 3$ was suggested by

Paris and Erdogan. Many other crack growth equations can be written in a form similar to (9.96) and the value of m is usually predicted as $m = 2$ or as $m = 4$. In practical analysis the value of m is often selected based on a statistical analysis of crack growth data.

For variable-amplitude loading additional difficulties arise concerning the definition and counting of stress cycles as discussed in Section 9.2.2, and the effect of the sequence of load cycles on acceleration and retardation of crack growth caused by high compressive and tensile load peaks. The simplest models are the *noninteraction models,* which simply ignore sequence effects and assume that the crack extension in any load cycle depends only on the value of ΔK and possibly also R for that cycle. Crack retardation and acceleration effects are thus completely ignored and the increment in crack size during one load cycle is computed from an equation for constant-amplitude loading. Only such models are considered here.

The crack growth equation of Paris and Erdogan is randomized in the next section to provide crack growth curves similar to curves observed experimentally. A section then describes the application of reliability methods in connection with two failure criteria related to crack growth. The last section describes reliability updating, and updating of the distribution of the basic variables as information, in the form of measured crack sizes, becomes available.

9.3.1 Probabilistic Crack Growth Equations

The Paris and Erdogan equation (9.96) combined with the expression for the stress intensity factor (9.95) leads to the following equation for the crack growth rate for *constant-amplitude loading.*

$$\frac{da}{dn} = CY(a)^m S^m (\sqrt{\pi a})^m \tag{9.97}$$

where $S = \Delta\sigma$ is the constant stress range. The solution to the differential equation is obtained by separating the variables and integrating.

$$\psi(a) = CS^m n \tag{9.98}$$

where the function $\psi(a)$ is

$$\psi(a) = \int_{a_0}^{a} \frac{dz}{Y(z)^m (\sqrt{z\pi})^m} \tag{9.99}$$

and a_0 is the initial crack size. $\psi(a)$ is an increasing function, and the crack length after n stress cycles, $a(n)$, is obtained by solving

(9.98) with respect to a. As an example, let the geometry factor be taken corresponding to an infinite panel, i.e., $Y(a) \equiv 1$. The crack length after n stress cycles is obtained from (9.98) as

$$a(n) = \begin{cases} \left(a_0^{(2-m)/2} + \dfrac{2-m}{2} C\pi^{m/2} S^m N \right)^{2/(2-m)}, & m \neq 2 \\ a_0 \exp(C\pi S^2 N), & m = 2 \end{cases} \tag{9.100}$$

It follows from (9.98) that if failure is defined by the crack size exceeding a critical value a_c, then the following equation is valid at failure

$$NS^m = \frac{\psi(a_c)}{C} = \text{constant} \tag{9.101}$$

This relation is in agreement with the S-N relation presented in Section 9.2.1. One way to define a damage indicator, D, in terms of the crack size is

$$D = \frac{\psi(a)}{\psi(a_c)} \tag{9.102}$$

Using this definition it follows from (9.98) that damage increases linearly with the number of stress cycles from 0 to 1.

Numerous experimental results exist for crack growth under constant-amplitude loading. Figure 9.18 shows experimental results for 68 center cracked specimens made of 2024-T3 aluminum, (Virkler et al., 1979). The specimens were from the same lot, and the experiments were highly controlled and performed by the same laboratory. The initial half crack length of each specimen was $a_0 = 9$ mm. It is observed that the sample curves are all different, they are irregular and not very smooth, they become more smooth for larger crack sizes, and they intermingle, in particular for small crack sizes. The experiments therefore indicate that the material is inhomogeneous and that the crack therefore grows through zones of varying resistance. To achieve sample curves as those of Fig. 9.18, the material parameters C and/or m must vary, as the initial crack size and the loading were the same for all specimens. Here m is assumed constant and the same for all specimens, while C is randomized.

First, let C be a random variable varying independently from specimen to specimen. The general behavior of sample curves is then as shown in Fig. 9.18 and different sample curves are obtained. The sample curves are, however, quite smooth and no intermingling takes place.

Second, C is written as (Ortiz, 1984),

$$C = C(a) = \frac{C_1}{C_2(a)} \tag{9.103}$$

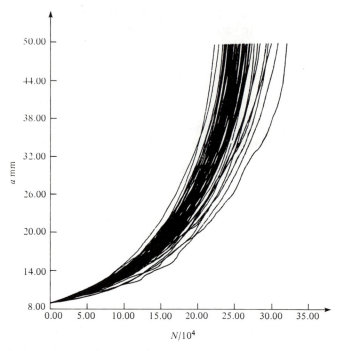

Figure 9.18   Experimental results, data from Virkler et al. (1979).

Here C_1 is a random variable describing variations between mean values in different specimens, while $C_2(a)$ is a positive random process describing variations from the mean value along the crack path within each specimen. $C_2(a)$ is assumed to be stationary and, as an example, $C_2(a)$ can be taken as a lognormal process, i.e., $\log C_2(a)$ is a normal process. The corresponding stochastic differential equation for a follows from (9.97) as

$$\frac{da}{dn} = \frac{C_1}{C_2(a)} Y(a)^m S^m (\sqrt{\pi a})^m \tag{9.104}$$

Separation of the variables and integration leads to the solution

$$\psi_1(a) = C_1 S^m n \tag{9.105}$$

where the random function $\psi_1(a)$ is

$$\psi_1(a) = \int_{a_0}^{a} C_2(z) \frac{dz}{Y(z)^m (\sqrt{z\pi})^m} \tag{9.106}$$

The expected value and covariance function for $\psi_1(a)$ are

$$E[\psi_1(a)] = \psi(a) \tag{9.107}$$

$$\text{Cov}[\psi_1(a_1), \psi_1(a_2)] = \int_{a_0}^{a_1}\int_{a_0}^{a_2}\text{Cov}[C_2(z_1), C_2(z_2)] \tag{9.108}$$

$$\times \frac{dz_1}{Y(z_1)^m(\sqrt{z_1\pi})^m} \frac{dz_2}{Y(z_2)^m(\sqrt{z_2\pi})^m}$$

$\psi_1(a)$ is a stochastic integral and the distribution of $\psi_1(a)$ is not readily determined from the integral and the distribution of the random process $C_2(a)$. The distribution of the crack length $a(n)$ is even more complicated to determine. Sample curves for the number of cycles $n = n(a)$ to reach a certain crack length can be obtained from a simulation of $C_2(a)$ and integration of (9.106). With a suitable choice for the process $C_2(a)$ these sample curves are similar to those of Fig. 9.18.

Ortiz (1984) has computed estimates for the statistical properties for $C_2(a)$ based on the experimental results of Fig. 9.18, i.e., for 2024-T3 aluminum. The correlation function of $C_2(a)$ is found to decrease to zero very rapidly, and $C_2(a)$ can therefore be approximated by a white noise process without introducing a large error. Furthermore, the distribution of $\psi_1(a)$ is well approximated by a normal distribution for large crack increments $a - a_0$.

The model in (9.104) can be directly extrapolated to variable-amplitude loading when the appropriate value of S is inserted for each stress cycle. It must be emphasized that this is an extrapolation beyond experimental experience and that possible sequence effects are neglected by use of such a non-interaction model.

The crack growth equation is written as

$$\Delta a_i = \frac{C_1}{C_2(a_i)} Y(a_i)^m S_i^m (\sqrt{a_i\pi})^m \tag{9.109}$$

where S_i is the stress range, and a_i is the crack length in the ith stress cycle. The corresponding increment $\Delta\psi_{1,i}$ is analogous

$$\Delta\psi_{1,i} = C_1 S_i^m \tag{9.110}$$

and the crack growth equation is solved as

$$\psi_1(a) = C_1 \sum_{i=1}^{n} S_i^m \tag{9.111}$$

The statistics for S_i^m are computed or approximated for stationary Gaussian processes and for Poisson spike processes in Section 9.2.3.

The increase in the damage indicator (9.102) is

$$\Delta D_i = \frac{\Delta\psi_{1i}}{\psi(a_c)} = \frac{C_1 S_i^m}{C_1 N_i S_i^m} = \frac{1}{N_i} \tag{9.112}$$

where N_i is the number of stress cycles of stress range S_i necessary to cause failure. The damage accumulation thus follows Miner's rule.

9.3.2 Reliability Analysis for Fatigue Crack Growth

Reliability problems for fatigue crack growth can be formulated as limit state problems and both serviceability and ultimate limit states can be defined. Two separate types of failure criteria can be envisioned (ASCE, Committee on Fatigue and Fracture Reliability, 1982).

$$a_c - a \leq 0 \tag{9.113}$$

$$K_{IC} - K \leq 0 \tag{9.114}$$

In the first case a critical crack size a_c is selected, perhaps based on serviceability considerations. In the second case failure occurs when the stress intensity factor K exceeds the fracture toughness K_{IC}. Then the crack growth becomes unstable and rapid failure occurs. Four cases should be considered corresponding to the two failure criteria and constant- or variable-amplitude loading.

Case 1 : Critical crack size under constant-amplitude loading

$$g(\mathbf{z}) = \psi_1(a_c) - \psi_1(a) = \int_{a_0}^{a_c} \frac{dz}{C_2(z)Y(z)^m \sqrt{\pi z}\,)^m} - C_1 S^m N \tag{9.115}$$

Case 2 : Brittle fracture under constant-amplitude loading

$$g(\mathbf{z}) = K_{IC} - Y(a(N))(\sigma_a + \frac{1}{2}S)\sqrt{\pi a(N)} \tag{9.116}$$

where σ_a is the average far-field stress and where $a(N)$ is obtained by solving (9.98) with respect to a.

Case 3 : Critical crack size under variable-amplitude loading

$$g(\mathbf{z}) = \psi_1(a_c) - \psi_1(a) = \int_{a_0}^{a_c} \frac{dz}{C_2(z)Y(z)^m \sqrt{\pi z}\,)^m} - C_1 \sum_{i=1}^{N} S_i^m \tag{9.117}$$

Case 4 : Brittle fracture under variable-amplitude loading

Failure occurs if

$$\sigma > \frac{K_{IC}}{Y(a)\sqrt{\pi a}} \tag{9.118}$$

where $\sigma = \sigma(t)$ is the far-field stress. This is illustrated in Fig. 9.19. No failure occurs in the time period $[0,T]$ if the stress process $\sigma(t)$ does not cross over the time-varying threshold $\xi(t) = K_{IC}/Y(a(t))\sqrt{\pi a(t)}$ in $[0,T]$. This probability is approximated by [see (7.131)]

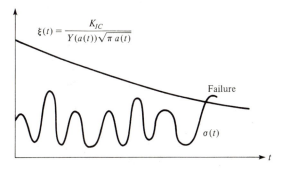

Figure 9.19 Brittle fracture under variable-amplitude loading analyzed as a first-passage problem.

$$P_R = 1 - F_{T_f}(T) \approx F_{\sigma(0)}\left(\frac{K_{IC}}{Y(a_0)\sqrt{\pi a_0}}\right) \exp\left|-\int_0^T v_\sigma(\xi(t))dt\right| \quad (9.119)$$

T_f is the random lifetime and $v_\sigma(\xi(t))$ is the mean-upcrossing rate of the level $\xi(t)$ by the process $\sigma(t)$ at time t. This mean-upcrossing rate is computed by Rice's formula (7.59). In this application the time derivative $\dot{\xi}$ can be neglected. Equation (7.59) therefore reduces to

$$v_\sigma(\xi(t)) = \int_0^\infty \dot{\sigma} f_{\sigma\dot{\sigma}}(\xi(t),\dot{\sigma})d\dot{\sigma} \quad (9.120)$$

For a stationary Gaussian process the result follows from (7.66)

$$v_\sigma(\xi(t)) = v_0 \exp\left|-\frac{1}{2}\left(\frac{\xi(t)-\mu_X}{\sigma_X}\right)^2\right| \quad (9.121)$$

The limit state function can be stated as

$$g(\mathbf{z}) = T_f - T \quad (9.122)$$

The reliability is computed by a first-order or second-order reliability method, FORM or SORM, as described in Section 5.1. The first step is to transform the set of basic variables into a set of independent and standardized normal variables U_i. The Rosenblatt transformation can be used directly, when the distribution type of ψ_1 conditioned on a_0, m, and possible random parameters in the geometry function is selected.

Example 9.3

The simple structure shown in Fig. 9.20 is considered. It consists of a large panel with a center crack and loaded uniaxially perpendicular to the crack. The loading is a constant-amplitude loading leading to a far-field stress range S. The geometry function increases with the ratio between the

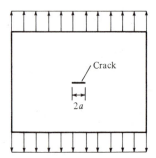

Figure 9.20   Panel with center crack.

crack length and the panel width.   The geometry function is identical 1 for an infinite panel and is modeled as

$$Y(a) \;=\; \exp(Y_1(\frac{a}{50})^{Y_2})$$

where Y_1 and Y_2 are parameters which are assumed random.   For simplicity $C_2(a)$ is taken as 1.   Lengths are measured in mm and stresses in N/mm^2. The distribution of the basic variables are taken as

$$S \in N(60, 10^2)$$
$$Y_1 \in LN(1, 0.2^2) \quad \Leftrightarrow \quad \ln Y_1 \in N(-0.020, 0.198^2)$$
$$Y_2 \in LN(2, 0.1^2) \quad \Leftrightarrow \quad \ln Y_2 \in N(0.692, 0.050^2)$$
$$a_0 \in EX(1)$$
$$(\ln C_1, m) \in N_2(-33.00, 0.47^2, 3.5, 0.3^2; -0.9)$$

$N(\alpha, \beta^2)$ denotes a normal distribution with mean value α and variance β^2. Similarly $LN(\alpha, \beta^2)$ denotes a log-normal distribution with mean value α and variance β^2. $N_2(\alpha, \beta^2, \gamma, \delta^2; \rho)$ denotes a bi-normal distribution with mean values α and γ, variances β^2 and δ^2 and correlation coefficient ρ, and $EX(\lambda)$ denotes an exponential distribution with mean value λ.

The example has six basic variables and the corresponding standardized and independent normal variables $U_1 - U_6$ are defined by the Rosenblatt transformation as

$$U_1 \;=\; \Phi^{-1}(F_S(S)) \;=\; \frac{S - 60}{10}$$

$$U_2 \;=\; \Phi^{-1}(F_{Y_1}(Y_1)) \;=\; \frac{\ln Y_1 + 0.020}{0.198}$$

$$U_3 \;=\; \Phi^{-1}(F_{Y_2}(Y_2)) \;=\; \frac{\ln Y_2 - 0.692}{0.050}$$

$$U_4 \;=\; \Phi^{-1}(F_{a_0}(a_0)) \;=\; \Phi^{-1}(1 - \exp(-a_0))$$

$$U_5 = \Phi^{-1}(F_m(m)) = \frac{m - 3.5}{0.3}$$

$$U_6 = \Phi^{-1}(F_{\ln C_1}(\ln C_1 \mid m)) = \frac{1}{\sqrt{1 - 0.9^2}} \left(\frac{\ln C_1 + 33.00}{0.47} + 0.9\frac{m - 3.5}{0.3} \right)$$

a_C is taken as 50 mm. The first-order reliability index is shown in Table 9.1 together with the vector of sensitivity factors $\boldsymbol{\alpha}$ and the first- and second-order approximation to the failure probability for various values of n. The first- and second-order approximations to the failure probability are close in all cases.

TABLE 9.1 Example Results of FORM and SORM					
N	10^5	2×10^5	5×10^5	10^6	2×10^6	5×10^6
β	3.430	3.024	2.480	2.063	1.641	1.076
α	0.3473	0.3498	0.3532	0.3560	0.3590	0.3632
	0.0140	0.0124	0.0104	0.0092	0.0081	0.0068
	-0.0099	-0.0088	-0.0074	-0.0065	-0.0057	-0.0047
	0.5300	0.5336	0.5403	0.5469	0.5549	0.5683
	0.7642	0.7603	0.7537	0.7475	0.7399	0.7272
	0.1193	0.1207	0.1225	0.1239	0.1254	0.1273
$P_{F,FORM}$	3.0×10^{-4}	1.2×10^{-3}	6.6×10^{-3}	2.0×10^{-2}	5.0×10^{-2}	1.4×10^{-1}
$P_{F,SORM}$	2.6×10^{-4}	0.9×10^{-3}	5.3×10^{-3}	1.7×10^{-2}	4.0×10^{-2}	1.2×10^{-1}

The first-order approximation to the expected value of n at failure is

$$\mathrm{E}[N] = \int_0^\infty (1 - P(N \le n)) \, dn \approx \int_0^\infty \Phi(\beta(n)) \, dn$$

The sensitivity of the reliability index to changes in the distribution parameters can be computed from the results of Section 5.6. For the mean value of S and $N = 10^6$ stress cycles, one obtains:

$$\frac{\partial \beta}{\partial \mu_S} \sim -\frac{\alpha_1}{\sigma_S} = -\frac{0.3560}{10} = -0.0356$$

A change in μ_S from 60 MPa to 50 MPa is thus expected to lead to a first-order reliability index of

$$2.063 + 0.0356 \times 10 = 2.419$$

The exact value is computed as 2.442. Similarly, for the standard deviation of m and $N = 10^6$ stress cycles, one obtains:

$$\frac{\partial \beta}{\partial \sigma_m} \sim -\frac{\beta \alpha_5^2}{\sigma_m} + \frac{0.9}{\sqrt{1 - 0.9^2}} \left(-\frac{\beta \alpha_5 \alpha_6}{\sigma_m} \right) = -5.133$$

A reduction of the standard deviation to 0.15 is thus expected to lead to a first-order reliability index of

$$2.063 + 5.133 \times 0.15 = 2.833$$

The exact value is computed as 2.815.

9.3.3 Updating Through Inspection

Two types of observation are considered (Madsen, 1985):

$$a(N_i) = A_i \qquad (9.123)$$

$$a(N_i) \leqslant a_d \qquad (9.124)$$

In the first case a crack size A_i is observed after N_i stress cycles. A_i may be random due to measurement error and due to uncertainties in the interpretation of a measured signal as a crack size. Measurements of the type (9.123) can be envisioned for several times corresponding to several values of N_i.

For each measurement a safety margin can be defined in analogy with (9.115) or (9.117) as

$$g_i(\mathbf{Z}) = \begin{vmatrix} \psi_1(A_i) - C_1 S^m N_i, & \text{constant-amplitude loading} \\ \\ \psi_1(A_i) - C_1 \sum_{r=1}^{N_i} S_r^m, & \text{variable-amplitude loading} \end{vmatrix} \qquad (9.125)$$

The corresponding first-order safety margin M_i in u-space is defined according to (4.17) and (4.29) as

$$M_i = \beta_i - \boldsymbol{\alpha}_i^T \mathbf{U} \qquad (9.126)$$

Compared to the dimension for the safety margin corresponding to (9.115) or (9.117), the dimension of the U-vector is increased by the number of random variables describing the inspection uncertainty. The first-order reliability index $\beta_{\,|\,i}$ after the inspection is

$$\beta_{\,|\,i} = \frac{E[M \mid M_i = 0]}{D[M \mid M_i = 0]} \qquad (9.127)$$

When more observations are included this generalizes to

$$\beta_{\,|\,1,2,...,k} = \frac{E[M \mid M_1 = M_2 = \cdots = M_k = 0]}{D[M \mid M_1 = M_2 = \cdots = M_k = 0]} \qquad (9.128)$$

Because the joint distribution of the set of all first-order safety margins is normal the conditional reliability index follows from linear regression results as (see, e.g., Ditlevsen, 1981, p. 126)

$$\beta_{\,|\,1,2,...,k} = \frac{\hat{E}[M \mid M_1 = M_2 = \cdots = M_k = 0]}{D[M - \hat{E}[M \mid M_1 = M_2 = \cdots = M_k = 0]]} \qquad (9.129)$$

$$= \frac{\beta - \boldsymbol{\rho}_{M,\mathbf{M}}^T \boldsymbol{\rho}_{\mathbf{MM}}^{-1} \boldsymbol{\beta}}{\sqrt{1 - \boldsymbol{\rho}_{M,\mathbf{M}}^T \boldsymbol{\rho}_{\mathbf{MM}}^{-1} \boldsymbol{\rho}_{M,\mathbf{M}}}}$$

where $\hat{E}[\ \mid\ ]$ denotes the linear regression, $\mathbf{M} = (M_1, M_2, ..., M_k)$,

$\rho_{M,M} = \{\rho[M, M_i]\} = \{\boldsymbol{\alpha}^T \boldsymbol{\alpha}_i\}$, $\rho_{MM} = \{\rho[M_i, M_j]\} = \{\boldsymbol{\alpha}_i^T \boldsymbol{\alpha}_j\}$, and $\boldsymbol{\beta} = \{\beta_i\}$. For one observation (9.129) reduces to

$$\beta_{|1} = \frac{\beta - \rho_1\beta_1}{\sqrt{1 - \rho_1^2}} \tag{9.130}$$

If the measurement uncertainty is large, then ρ_1 is small and it follows directly from (9.130) that the reliability index is only changed by a small amount.

Example 9.4

Example 9.3 is continued. The design life-time is taken as $N = 10^6$ stress cycles and the first-order reliability index is required to be larger than 2. This is satisfied at the design stage as shown from the value $\beta = 2.063$ for $N = 10^6$ cycles in Table 9.1. In service, the observation $a = 3.9$ mm is made after 10^5 stress cycles. The design point for the safety margin (9.125) becomes

$$\mathbf{u}_1^* = \beta_1\boldsymbol{\alpha}_1 = 2.022\,(0.0168, 0.0001, -0.0002, 0.9993, 0.0340, 0.0059)$$

With a critical crack length $a_C = 50$ mm after 10^6 stress cycles the correlation coefficient ρ_1 becomes

$$\rho_1 = \boldsymbol{\alpha}^T\boldsymbol{\alpha}_1 = 0.5786$$

with $\boldsymbol{\alpha}$ from Table 9.1. The updated reliability index from $N = 10^6$ cycles follows from (9.130)

$$\beta_{|1} = \frac{\beta - \rho_1\beta_1}{\sqrt{1 - \rho_1^2}} = \frac{2.063 - 0.5786 \times 2.022}{\sqrt{1 - 0.5786^2}} = 1.095$$

This reliability index is less than 2. Corresponding to $N = 2 \times 10^5$ cycles the updated reliability index can be computed as

$$\beta_{|1} = \frac{3.024 - 0.5657 \times 2.022}{\sqrt{1 - 0.5657^2}} = 2.280$$

This value is larger than the required minimum level and no repair is needed before $N = 2 \times 10^5$. With an additional observation $a = 4.0$ mm after 2×10^5 stress cycles the reliability index can be further updated. The design point for this observation is

$$\mathbf{u}_2^* = \beta_2\boldsymbol{\alpha}_2 = 2.031\,(0.0384, 0.0003, -0.0004, 0.9961, 0.0781, 0.0135)$$

and the correlation coefficients ρ_2 and ρ_{12} are

$$\rho_2 = \boldsymbol{\alpha}^T\boldsymbol{\alpha}_2 = 0.6184, \quad \rho_{12} = \boldsymbol{\alpha}_1^T\boldsymbol{\alpha}_2 = 0.9988$$

The updated value of the reliability index is given by (9.129)

$$\beta_{|12} = \frac{\beta - [\rho_1 \ \ \rho_2]\dfrac{1}{1-\rho_{12}^2}\begin{vmatrix} 1 & -\rho_{12} \\ -\rho_{12} & 1 \end{vmatrix}\begin{vmatrix} \beta_1 \\ \beta_2 \end{vmatrix}}{\left| 1 - [\rho_1 \ \ \rho_2]\dfrac{1}{1-\rho_{12}^2}\begin{vmatrix} 1 & -\rho_{12} \\ -\rho_{12} & 1 \end{vmatrix}\begin{vmatrix} \rho_1 \\ \rho_2 \end{vmatrix}\right|^{1/2}} = 56.41$$

The nominator is very small and a very high degree of accuracy in the reliability indices and correlation coefficients was required to achieve this result.

The small reliability index after the first inspection is explained by a large initial crack size. With the second inspection it is, however, predicted that the crack grows slowly enough that it does not become critical before $N = 10^6$ cycles.

The fraction of the variation which has been removed after the inspections can be measured by the multiple correlation coefficient. The multiple correlation coefficient is denoted ρ^2 and is equal to (see, e.g., Ditlevsen, 1981, p. 127)

$$\rho^2 = \rho_{M,M}^T \rho_{MM}^{-1} \rho_{M,M}$$

With the two inspection results above, the multiple correlation coefficient is as large as 0.9999. Essentially all uncertainty in thus covered by two combinations of the basic variables and with two observations the further development can almost be computed deterministically.

It is next assumed that the inspection results are subjected to uncertainty. This is modeled by assigning a normal distribution to the measured crack size. The mean values are taken as above and the variation of the updated reliability indices with standard deviation σ_A is studied. The two inspection results are assumed independent. For the first inspection the design point is found as

$$\mathbf{u}_1^* = \begin{vmatrix} 2.022\,(0.0168, 0.0001, -0.0002, 0.9992, 0.0340, 0.0059, -0.0081) \\ 2.020\,(0.0167, 0.0001, -0.0002, 0.9985, 0.0339, 0.0059, -0.0405) \\ 2.015\,(0.0165, 0.0001, -0.0002, 0.9960, 0.0335, 0.0058, -0.0812) \\ 1.980\,(0.0152, 0.0001, -0.0001, 0.9782, 0.0307, 0.0054, -0.2049) \end{vmatrix}$$

for $\sigma_A = 0.02$ mm, 0.1 mm, 0.2 mm, and 0.5 mm, respectively; and where the last coordinate corresponds to the uncertainty in the first inspection result. For the second inspection the design point is similarly:

$$\mathbf{u}_2^* = \begin{vmatrix} 2.031\,(0.0384, 0.0003, -0.0004, 0.9961, 0.0781, 0.0135, 0, -0.0077) \\ 2.029\,(0.0382, 0.0003, -0.0004, 0.9954, 0.0777, 0.0134, 0, -0.0386) \\ 2.025\,(0.0377, 0.0003, -0.0004, 0.9932, 0.0767, 0.0132, 0, -0.0774) \\ 1.993\,(0.0345, 0.0002, -0.0003, 0.9775, 0.0698, 0.0121, 0, -0.1959) \end{vmatrix}$$

for the same four choices for σ_A. The inspection results are assumed independent and the coordinate at the design point is 0 for A_1. The updated reliability indices and the multiple correlation coefficient are:

$$\beta_{|1} = \begin{vmatrix} 1.095, & \sigma_A = 0.02 \text{ mm}, \\ 1.097, & \sigma_A = 0.1 \text{ mm}, \\ 1.103, & \sigma_A = 0.2 \text{ mm}, \\ 1.146, & \sigma_A = 0.5 \text{ mm}, \end{vmatrix} \qquad \beta_{|12} = \begin{vmatrix} 4.054, & \sigma_A = 0.02 \text{ mm} \\ 1.311, & \sigma_A = 0.1 \text{ mm} \\ 1.129, & \sigma_A = 0.2 \text{ mm} \\ 1.098, & \sigma_A = 0.5 \text{ mm} \end{vmatrix}$$

$$\rho^2 = \begin{vmatrix} 0.97, & \sigma_A = 0.02 \text{ mm} \\ 0.64, & \sigma_A = 0.1 \text{ mm} \\ 0.46, & \sigma_A = 0.2 \text{ mm} \\ 0.36, & \sigma_A = 0.5 \text{ mm} \end{vmatrix}$$

It follows that the inspection uncertainty has a large effect on the reliability index after the second inspection. To increase the multiple correlation coefficient further, more inspections can be included.

The distribution of the original basic variables can also be updated at each observation of the crack length. Since $(\mathbf{U}, M_1)$ has a normal distribution, it follows directly from (9.126),

$$\begin{aligned} E[\mathbf{U} \mid M_1 = 0] &= -\boldsymbol{\alpha}_1 \beta_1 \\ \mathbf{C}_{\mathbf{U} \mid M_1 = 0} &= \mathbf{I} - \boldsymbol{\alpha}_1^T \boldsymbol{\alpha}_1 \end{aligned} \qquad (9.131)$$

where $\mathbf{C}$ denotes the covariance matrix and $\mathbf{I}$ the unity matrix. $\mathbf{U} \mid M_1 = 0$ is normal and the updated distribution of the basic variable vector $\mathbf{Z}$ follows directly from the Rosenblatt transformation with the updated distribution for $\mathbf{U}$ being inserted. In particular, for independent basic variables, the relation between Z_j and U_j before the inspection is

$$F_{Z_j}(z_j) = \Phi(u_j) \qquad (9.132)$$

After the inspection the relation is

$$F_{Z_j}(z_j \mid M_1 = 0) = \Phi(u_j \mid M_1 = 0) \qquad (9.133)$$

$$= \Phi(\frac{u_j + \alpha_{1j}\beta_1}{\sqrt{1-\alpha_{1j}^2}}) = \Phi(\frac{\Phi^{-1}(F_{Z_j}(z_j)) + \alpha_{1j}\beta_1}{\sqrt{1-\alpha_{1j}^2}})$$

The basic variables are dependent after the inspection, even when they are independent before the inspection. In case a repair takes place after the inspection and the reliability of the repaired structure is sought, the updated distributions of the basic variables which are not influenced by the repair should be used. The updating of the original basic variables can also be carried out based on several inspections.

In the second case, (9.124), a_d represents a lower limit of detectability. A safety margin is defined as

$$g_d(\mathbf{Z}) = \begin{vmatrix} C_1 S^m N_i - \psi_1(a_d), & \text{constant} - \text{amplitude loading} \\ C_1 \sum_{r=1}^{N_i} S_r^m - \psi_1(a_d), & \text{variable} - \text{amplitude loading} \end{vmatrix} \qquad (9.134)$$

and the corresponding first-order safety margin M_d in u-space becomes

$$M_d = \beta_d - \boldsymbol{\alpha}_d^T \mathbf{U} \qquad (9.135)$$

The reliability index after the inspection is

$$\beta_{\mid d} = -\Phi^{-1}(P(M \leqslant 0 \mid M_d \leqslant 0)) \qquad (9.136)$$

$$= -\Phi^{-1}\left(\frac{P(M \leqslant 0 \cap M_d \leqslant 0)}{P(M_d \leqslant 0)}\right)$$

An asymptotically correct approximation to the probability $P(M \leqslant 0 \cap M_d \leqslant 0)$ is obtained by determination of the joint design point for the event, as it corresponds to failure of a parallel system (see Section 5.5). A first-order approximation is

$$P(M \leqslant 0 \mid M_d \leqslant 0) \approx \frac{\Phi_2(-\beta, -\beta_d ; \boldsymbol{\alpha}^T \boldsymbol{\alpha}_d)}{\Phi(-\beta_d)} \qquad (9.137)$$

where $\Phi_2(\ ,\ ;\rho)$ is the bi-normal standardized distribution function for a correlation coefficient ρ. When the observation (9.124) is made for several successive numbers of stress cycles only the largest number of stress cycles is used in the updating procedure. This is, however, only true for an inspection procedure which is perfect, i.e., a procedure that does not miss a detectable crack.

The distributions of the basic variables are also updated through the inspection within the same framework. For independent basic variables the distribution function is updated as

$$F_{Z_j}(z_j \mid M_d \leqslant 0) = \frac{P(Z_j - z_j \leqslant 0 \cap M_d \leqslant 0)}{P(M_d \leqslant 0)} \qquad (9.138)$$

A first-order approximation is

$$F_{Z_j}(z_j \mid M_d \leqslant 0) \approx \frac{\Phi_2(-\beta(z_j), -\beta_d ; \alpha_{dj})}{\Phi(-\beta_d)} \qquad (9.139)$$

where the correlation coefficient α_{dj} is the jth component in $\boldsymbol{\alpha}_d$ and the reliability index $\beta(z_j)$ is

$$\beta(z_j) = -\Phi^{-1}(F_{Z_j}(z_j)) \qquad (9.140)$$

9.4 BOGDANOFF'S CUMULATIVE DAMAGE MODEL

One way to introduce a probabilistic structure in a cumulative damage model is to start with some deterministic law for damage accumulation, and then introduce random variables or random processes in place of the model parameters. Another way consists in assuming an evolutionary probabilistic structure from the start. In the first approach, the accumulated damage as a function of time is described, whereas in the second approach, the probabilistic distribution of the accumulated damage as a function of time is described. In Section 9.2 the first approach is exemplified in connection with Miner's rule and S-N curves. In Section 9.3 the same approach is exemplified for crack growth in metals. The second approach is elaborated on here with the model of Bogdanoff et al. (1978, 1980) as a starting point. In Bogdanoff's model use is made of various Markovian processes (see, e.g., Parzen, 1962). Such processes have a large domain of applicability and are widely used in probabilistic modeling, partly because the processes have a number of mathematically attractive properties.

A basic element in the model is a *duty cycle* which is a repetitive period of operation in the life of a component in which damage can accumulate. Under constant-amplitude loading a duty cycle can correspond to a certain number of load cycles. Another example is the operation of an aircraft for which each mission can be divided into duty cycles for taxiing, takeoff, cruise, descent, and landing. The loading during a duty cycle can thus also be random in nature. The state of damage is considered only at the end of each duty cycle. The damage accumulation during a duty cycle is assumed nonnegative, but the mechanism of the accumulation is not considered. The probability distribution of damage after a duty cycle is assumed to depend, in a probabilistic manner, only on the duty cycle itself and on the damage accumulated at the start of the duty cycle. It is, however, independent of how the damage was accumulated up to the start of the duty cycle. These assumptions are the Markov assumptions and the damage process is viewed as a discrete-time Markov process. The time t is measured in units of duty cycles, hence $t = 1, 2, \ldots$. Equal increments in t need not correspond to equal increments in chronological time, as different duty cycles can be present in the lifetime, such as in the aircraft example already mentioned.

For reasons of simplicity a number of additional assumptions are now made, some of which are relaxed later. A discretization of damage into the set of states $i = 1, 2, \ldots, b$ is made. Here state b denotes a state of failure in some sense. The damage accumulation process is then a discrete-time, discrete-state Markov process and as such can be viewed as a Markov chain. The probability distribution of damage is completely determined by the transition matrix for each duty cycle and by the initial damage present at $t = 0$. The initial state of damage is specified by the vector $\mathbf{p}_0 = \{\pi_i\}$, where π_i is the probability of damage being in state i at $t = 0$.

$$\mathbf{p}_0 = [\pi_1 \ \pi_2 \cdots \pi_b]; \quad \pi_i \geq 0; \quad \sum_{i=1}^{b} \pi_i = 1 \qquad (9.141)$$

The type of initial damage is not specified and may arise from material defects, manufacturing defects, or other types of initial defects.

The transition matrix for a duty cycle is $\mathbf{P} = \{P_{ij}\}$, where P_{ij} is the probability of the damage being in state j after the duty cycle given that the damage is in state i at the beginning of the duty cycle. It is first assumed that damage accumulation during a duty cycle can only be of one unit. The transition matrix is then of the form

$$\mathbf{P} = \begin{bmatrix} p_1 & q_1 & 0 & 0 & \cdot & 0 & 0 \\ 0 & p_2 & q_2 & 0 & \cdot & 0 & 0 \\ 0 & 0 & p_3 & q_3 & \cdot & 0 & 0 \\ \cdot & \cdot & \cdot & \cdot & \cdot & & \cdot \\ 0 & 0 & 0 & 0 & \cdot & p_{b-1} & q_{b-1} \\ 0 & 0 & 0 & 0 & \cdot & 0 & 1 \end{bmatrix} \qquad (9.142)$$

where $p_i + q_i = 1$, $p_i \geq 0$, and $q_i \geq 0$ for all i. The corresponding Markov chain is pictured in Fig. 9.21. The Markov chain has $b-1$ transient states and one absorbing state $i = b$.

Figure 9.21 Markov chain with transition probabilities.

The state of damage at time t is given by the vector $\mathbf{p}_t = \{p_t(i)\}$, where $p_t(i)$ is the probability that damage is in state i at time t.

$$\mathbf{p}_t = [p_t(1) \ p_t(2) \cdots p_t(b)]; \quad p_t(i) \geq 0; \quad \sum_{i=1}^{b} p_t(i) = 1 \qquad (9.143)$$

It follows from Markov chain theory that

$$\mathbf{p}_t = \mathbf{p}_0 \mathbf{P}_1 \mathbf{P}_2 \cdots \mathbf{P}_{t-1} \mathbf{P}_t \qquad (9.144)$$

where $\mathbf{P}_j$ is the transition matrix for the jth duty cycle. By (9.144) the probability distribution of damage is completely specified at any time and is calculated by simple matrix operations. If the duty cycles are of the same severity, (9.144) reduces to

$$\mathbf{p}_t = \mathbf{p}_0 \mathbf{P}^t \qquad (9.145)$$

where $\mathbf{P}$ is the common transition matrix. Since matrix multiplication is generally not commutative, it follows from (9.144) that not only the severity but also the order of the duty cycles influence the damage accumulation.

A generalization of the model consists in filling out the upper portion of the transition matrix. Then damage increases by more than one unit during a duty cycle are possible. Equation (9.144) still holds, so the calculation of the probability distribution of damage is just as simple. The number of parameters in the model is, however, increased, and the statistical uncertainty resulting from parameter estimation based on a limited number of test results is consequently increased.

The probability distribution of various random variables associated with the damage accumulation process can now be determined. The time W_b to failure, i.e., the time to absorption at state b, has the cumulative distribution function $F_W(t,b)$ given by

$$F_W(t,b) = P(W_b \leq t) = p_t(b), \quad t=1,2,\ldots \qquad (9.146)$$

The time W_i to reach state i has the cumulative distribution function of (9.146) with $p_t(b)$ replaced by $\sum_{j=i}^{b} p_t(j)$. The hazard function $\rho(t)$ defined in Chapter 1 becomes

$$\rho(t) = \frac{F_W(t,b) - F_W(t-1,b)}{1 - F_W(t-1,b)}, \quad t=1,2,\ldots \qquad (9.147)$$

The mean value and variance of the lifetime W_b are

$$E[W_b] = \sum_{t=1}^{b} (1 - F_W(t,b)) \qquad (9.148)$$

$$\text{Var}[W_b] = 2\sum_{t=1}^{b} t(1 - F_W(t,b)) + E[W_b] - E[W_b]^2 \qquad (9.149)$$

respectively, and similar results hold for W_i. The time T_j spent in state j has a geometric distribution

$$P(T_j = k) = p_j^{k-1}(1 - p_j), \quad k=1,2,\ldots \qquad (9.150)$$

where p_j is the jth diagonal element in $\mathbf{P}$. The mean value and variance are

$$E[T_j] = \frac{1}{1-p_j} \qquad (9.151)$$

$$\text{Var}[T_j] = \frac{p_j}{(1-p_j)^2} \qquad (9.152)$$

The state of damage at time t is denoted $D(t)$. It has the probability density function

$$P(D(t)=i) = p_t(i), \quad i=1,2,\ldots \tag{9.153}$$

from which the cumulative distribution function, the mean value and variance, and other statistics are computed directly.

The results above are all functions of the π_i's and the P_{ij}'s. In some cases analytical results can be calculated in closed form. To obtain these results it is often convenient to use geometric transforms. This is illustrated in Bogdanoff (1978, part 3). A numerical evaluation is always possible with the values $p_t(i)$ obtained from (9.144). If sample function behavior is of particular interest, e.g., in connection with model identification, a simple simulation can yield such sample functions.

Some further generalizations of the model are next discussed. Continuous time can be introduced by subordinating the model to a renewal process. The damage in a duty cycle can then depend in a probabilistic manner on the duration of the duty cycle. The damage process thus belongs to the class of semi-Markov processes as defined in Cinlar (1975). The computations become considerably more complicated when semi-Markov processes are introduced, so the value of this generalization is doubtful.

The state b at failure can be randomized into a finite number of failure states $b-a$, $b-a+1,\ldots,b$. In Bogdanoff (1978, part 1) this is done by introducing the random vector $\mathbf{\rho}=\{\rho_i\}$, where ρ_i is the probability of state $i\leqslant b$ being the failure state.

$$\mathbf{\rho} = [\rho_{b-a}\ \rho_{b-a+1}\ \cdots\ \rho_b]; \quad \rho_i\geqslant0; \quad \sum_{i=b-a}^{b}\rho_i=1 \tag{9.154}$$

Equation (9.144) is still valid when the transition matrix $\mathbf{P}$ in (9.142) is slightly modified.

Another generalization is the generalization to a vector-valued damage (Madsen, 1982). The case of a two-dimensional damage vector (D_1,D_2) is described here. Similar to Fig. 9.21 a two-dimensional Markov chain is then defined. Figure 9.22 shows such a chain with six states for D_1 and four states for D_2 and with the possible state changes and the failure states. The initial damage is specified by the matrix $\{\pi_{ij}\}$, where π_{ij} is the probability that the initial damage is in state (i,j). All probabilities π_{ij} are nonnegative and their sum is equal to 1. The transition matrix is $\{P_{ijkl}\}$, where P_{ijkl} is the probability that the damage (D_1,D_2) is in state (k,l) after a duty cycle given that it is in state (i,j) at the beginning of the duty cycle. The P_{ijkl}'s comply with $P_{ijkl}\geqslant0$ and $\sum_{k,l}P_{ijkl}=1$. Figure 9.22 which is a direct generalization of Fig. 9.21 corresponds

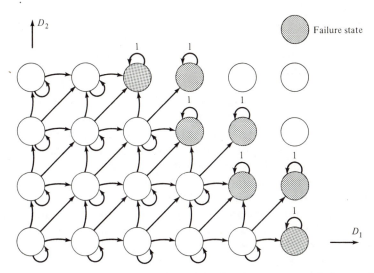

Figure 9.22   Two dimensional Markov chain for damage accumulation model.

to a situation where P_{ijkl} is zero except in cases where $k - i = 1$ or $l - j = 1$. The probability density function for the damage vector at time t is $p_t(k,l)$, which is determined as

$$p_t(k,l) = \sum_{i,j} P_{ijkl}\, p_{t-1}(i,j) \qquad (9.155)$$

from which the probability distribution of the damage vector can be completely determined recursively with $p_0(i,j) = \pi_{ij}$. The waiting time W_b to failure has the cumulative distribution function

$$F_{W}(t,0) = P(W_b \leq t) = \sum_{\text{failure states}} p_t(i,j) \qquad (9.156)$$

and distributions and statistics of other relevant random variables are calculated as easily as in the one-dimensional model.

Two aspects of the model which are thoroughly described in Bogdanoff (1978, part 1) are the effects of inspection and replacement strategies. These aspects are not dealt with here; it is only stated that they are easily incorporated into the model.

Two crucial points for the applicability of the model are the possibilities of model identification and parameter estimation. These points are not directly solvable since from the outset of model formulation there is no immediate relation between damage and measurable physical quantities. The identification and estimation therefore rely totally on test data. The type of test data reported is usually lifetime data, i.e., data on W_b. Such data are, however, not sufficient in the model identification since many completely different models can result in the same distribution for W_b, or at least in distributions which are very close and which cannot be rejected from the available data.

These shortcomings do not present a problem if the interest is solely on the ultimate situation tested, but the main purpose of the damage model is to predict lifetimes accurately under conditions not covered by the data and also to take inspection data into account. Additional information on damage accumulation is therefore needed. This information must necessarily be obtained by non-destructive testing methods and it must be decided how model damage states can be related to the measured inspection quantities. The additional information can, e.g., consist of information on sample curve behavior or, as illustrated in Bogdanoff and Kozin (1980), of statistics for the time to reach damage levels below failure. In the model identification phase experience with the model plays a significant role, but consistent and systematic model identification procedures can be developed. This has, however, not been done yet and it presents a weak point for the application of the model.

The estimation of the model parameters is a major statistical problem in itself. In the examples of Bogdanoff et al. (1978, 1980), standard methods based on maximum likelihood or second-order moment fit are used. The statistical uncertainty is very large if the number of parameters is large for the data available. The statistical uncertainty must be balanced against the model uncertainty, which generally decreases with the number of parameters. How to evaluate the statistical uncertainty is not yet clear.

Given the model and the model parameters estimated from the data available, the next step is to relate the model parameters to test parameters, such as stress conditions, material properties, geometry, environment, etc.

The main advantages of the model are summarized in the following points:

The model is very flexible and can be used for almost all types of continuous damage accumulation problems.

It is possible separately to assess the major sources of uncertainty in manufacturing standards, severity and order of duty cycles, specification of failure, and the effect of inspection and replacement strategies.

The model is simple to use and statistics for random variables such as lifetime and state of damage after a given time are easily determined.

In the model identification and parameter estimation phases, the model is well suited to suggest further test programs for model verification.   Also, consequences of not having a uniquely defined process are easily checked.

Among the shortcomings of the model, the following can be mentioned:

The number of model parameters easily becomes very large and the statistical uncertainty resulting from their estimation from test data becomes large.

Damage states are not uniquely related to measurable physical quantities.

No systematic methods for system identification and parameter estimation have yet been developed.

9.5 SUMMARY

Models for material strength are presented based on the stochastic models from the previous chapters. Reliability problems involving the stochastic strength models are solved by the reliability methods developed in Chapter 5.   The updating of the computed reliability and the distribution of basic variables from experienced performance is also discussed.

The first section describes classical strength theories for brittle materials, ductile materials, and fiber bundles. The size effect for brittle materials is explained in detail and the strength analysis of fiber bundles by the first-order reliability method is demonstrated.

The second section is devoted to damage accumulation theories. It is shown that an interaction-free and stress-independent theory leads to the celebrated Miner's rule for damage accumulation under variable-amplitude loading, based on a fatigue strength analysis under constant-amplitude loading.   Cycle counting methods are used to identify stress cycles in a variable-amplitude loading history. The stress cycles appear in the form of a sum of powers in the damage accumulation equations. Statistics for this sum are determined for the Poisson spike process and the stationary Gaussian process defined in Chapter 8.

The third section presents stochastic models for fatigue crack growth based on a randomization of the crack growth equation of Paris and Erdogan.   Random fluctuations of the material properties as the crack propagates are given particular attention. Failure criteria for failure due to crack growth to a critical crack size and to brittle fracture are analyzed by the first-order reliability method. Updating of the reliability and of the distribution of the basic variables from inspection, i.e., from measured crack sizes, is performed by a simple extension of the first-order reliability methods. Inspection results can be deterministic or random, and one or several inspection results can be available.

The chapter concludes with a presentation of a damage accumulation model due to Bogdanoff. The damage process is modeled as a discrete-time, discrete-state Markov process. The initial state of damage is specified by a random vector and the transition matrix determines the accumulation of damage.

REFERENCES

API (American Petroleum Institute), "Recommended Practice for Planning, Designing and Constructing Fixed Offshore Platforms," API RP2A, 13th ed., January 1982.

ASCE (American Society of Civil Engineers), Committee on Fatigue and Fracture Reliability of the Committee on Structural Safety and Reliability of the Structural Division, "Fatigue Reliability 1-4," *Journal of the Structural Division,* ASCE, Vol. 108, 1982, pp. 3-88.

BOGDANOFF, J. L., "A New Cumulative Damage Model, Part 1," *Journal of Applied Mechanics,* Vol. 45, 1978, pp. 246-250.

BOGDANOFF, J. L., "A New Cumulative Damage Model, Part 3," *Journal of Applied Mechanics,* Vol. 45, 1978, pp. 733-739.

BOGDANOFF, J. L. and F. KOZIN, "A New Cumulative Damage Model, Part 4," *Journal of Applied Mechanics,* Vol. 47, 1980, pp. 40-44.

BOGDANOFF, J. L. and W. KRIEGER, "A New Cumulative Damage Model, Part 2," *Journal of Applied Mechanics,* Vol. 45, 1978, pp. 251-257.

BOLOTIN, V. V., *Statistical Methods in Structural Mechanics,* Holden-Day, San Francisco, 1969.

BOLOTIN, V. V., *Warscheinlichkeitsmethoden zur Berechnung von Konstruktionen,* VEB Verlag für Bauwesen, Berlin, 1981.

CINLAR, E., "Markov Renewal Theory: A Survey," *Management Science,* Vol. 21, 1975, pp. 727-752.

DANIELS, H. E., "The Statistical Theory of the Strength of Bundles of Threads," *Proceedings of the Royal Society,* Vol. A183, 1945, pp. 405-435.

DEPARTMENT OF ENERGY, "New Fatigue Design Guidance for Steel Welded Joints in Offshore Structures," Recommendation, UK, 1982.

DITLEVSEN, O., *Uncertainty Modeling with Applications to Multidimensional Civil Engineering Systems,* McGraw-Hill, New York, 1981.

DnV (Det norske Veritas), "Rules for the Design, Construction and Inspection of Offshore Structures, Appendix C Steel Structures," Reprint with Corrections, 1982.

DOWLING, N. E., "Fatigue Failure Prediction Methods for Complicated Stress-Strain Histories," *Journal of Materials,* Vol. 7, 1972, pp. 71-84.

GEORGE, K. P. and A. A. BASMA, "An Extreme-Value Model for Strength of Stiff Clay," in *Probabilistic Characterization of Soil Properties: Bridge Between Theory and Practice,* eds. D. S. Bowles and H.-Y. Ko, ASCE, 1984, pp. 157-169.

GOODING, E. J., "Investigation of the Tensile Strength of Glass," *Journal of the Society of Glass Technology,* 16, 1932.

GRIFFITH, A. A., "The Phenomenon of Rupture and Flow in Solids," *Philosophical Transactions of the Royal Society,* A221, 1920, pp. 163-168.

HARLOW, D. G. and S. L. PHOENIX, "Probability Distributions for the Strength of Composite Materials I: Two-Level Bounds," *International Journal of Fracture,* Vol. 17, 1981, pp. 347-371.

HOHENBICHLER, M., "Resistance of Large Brittle Parallel Systems," in *Proceedings,* Fourth International Conference on Application of Statistics and Probability in Soil and Structural Engineering, ICASP4, University of Firenze, Italy, June 1983, pp. 1301-1312.

HOHENBICHLER, M. and R. RACKWITZ, "Reliability of Parallel Systems under Imposed Uniform Strain," *Journal of the Engineering Mechanics Division,* ASCE, Vol. 109, 1983, pp. 896-907.

IRVING, P. E. and L. N. MCCARTNEY, "Prediction of Fatigue Crack Growth Rates: Theory, Mechanisms, and Experimental Results," Fatigue 77 Conference, Cambridge University, *Metal Science,* August/September 1977, pp. 351-361.

JOHNSON, A. I., *Strength, Safety and Economical Dimensions of Structures,* Statens Kommitte för Byggnadsforskning, Meddelanden No. 22, Stockholm, 1953.

LAIRD, C., "The Influence of Metallurgical Structure on the Mechanism of Fatigue Crack Propagation," in *Fatigue Crack Propagation,* ASTM STP 415, 1967, pp. 131-181.

LEVE, H. L., "Cumulative Damage Theories," in *Metal Fatigue Theory and Design,* ed. A. F. Madayag, John Wiley, New York, 1969.

LINDGREN, G. and I. RYCHLIK, "Wave Characteristic Distributions for Gaussian Waves – Wave Length, Amplitude and Steepness," *Ocean Engineering,* Vol. 9, 1982, pp. 411-432.

MADSEN, H. O., "Deterministic and Probabilistic Models for Damage Accumulation Due to Time Varying Loading," DIALOG 5-82, Danish Engineering Academy, Lyngby, Denmark, 1982.

MADSEN, H. O., "Random Fatigue Crack Growth and Inspection," in *Proceedings ICOSSAR'85,* Kobe, Japan, May 1985.

MINER, M. A., "Cumulative Damage in Fatigue," *Journal of Applied Mechanics,* ASME, Vol. 12, 1945, pp. A159-A164.

ORTIZ, K., "Stochastic Modeling of Fatigue Crack Growth," Ph.D. dissertation, Stanford University, Stanford, California, 1984.

PALMGREN, A., "Die Lebensdauer von Kugellagern," *Zeitschrift der Vereines Deutches Ingenieure,* Vol. 68, No. 14, 1924, pp. 339-341.

PARIS, P. C. and F. ERDOGAN, "A Critical Analysis of Crack Propagation Laws," *Journal of Basic Engineering,* ASME, Vol. 85, 1963, pp. 528-534.

PARIS, P. C. and G. C. SIH, "Stress Analysis of Cracks," in *Fracture Toughness Testing and Its Applications,* ASTM STP 381, 1965, pp. 30-82.

PARZEN, E., *Stochastic Processes,* Holden-Day, San Francisco, 1962.

RICHARDS, C. W., "Size Effect in the Tension Test of Mild Steel," *Proceedings,* ASTM, Vol. 54, 1954, pp. 995-1002.

SCHIJVE, J., "The Accumulation of Fatigue Damage in Aircraft Materials and Structures," *AGARD Conference Proceedings No. 118,* Symposium on Random Load Fatigue, Lyngby, Denmark, April 1972, pp. 3.1-3.120.

SCHIJVE, J., "Four Lectures on Fatigue Crack Growth," *Engineering Fracture Mechanics,* Vol. 11, 1979, pp. 167-221.

STALLMEYER, J. E. and W. H. WALKER, "Cumulative Damage Theories and Application," *Journal of the Structural Division,* ASCE, Vol. 94, 1968, pp. 2739-2750.

VENEZIANO, D., "Probabilistic Seismic Resistance of Reinforced Concrete Frames," in *Structural Safety and Reliability,* ed. T. Moan and M. Shinozuka, Proceedings ICOSSAR'81, Trondheim, Norway, June 1981, pp. 241-258.

VIRKLER, D. A., B. M. HILBERRY and P. K. GOEL, "The Statistical Nature of Fatigue Crack Propagation," *Journal of Engineering Materials and Technology,* ASME, Vol. 101, 1979, pp. 148-153.

WEIBULL, W., "A Statistical Theory of the Strength of Materials," *Proceedings, Royal Swedish Institute of Engineering Research,* No. 151, Stockholm, Sweden, 1939.

WEIBULL, W., "The Phenomenon of Rupture in Solids," *Proceedings, Royal Swedish Institute of Engineering Research,* No. 153, Stockholm, Sweden, 1939.

WIRSCHING, P. H. and A. M. SHEHATA, "Fatigue under Wide Band Random Stresses Using the Rain-Flow Method," *Journal of Engineering Materials and Technology,* ASME, July 1977, pp. 205-211.

ZECH, B. and F. H. WITTMAN, "Probabilistic Approach to Describe the Behavior of Materials," *Transactions of SMIRT 4,* 1977, pp. 575-584.

10

STOCHASTIC MODELS FOR LOADS

10.1 GUST WIND LOADS

Structures exposed to gusty wind will experience a fluctuating wind load created by turbulence. The magnitude and character of the load fluctuations depend on the shape of the structure and its orientation with respect to the mean wind. The following presentation is limited to cases where the effective wind pressure can be determined by a quasi-stationary wind pressure coefficient C in the form

$$p(t) = \frac{1}{2} \rho C U(t)^2 \tag{10.1}$$

ρ is the density of air — typically $\rho \sim 1.2 \, \text{kg}/\text{m}^3$ — and $U(t)$ is the fluctuating wind speed. This type of wind load has been treated by, e.g., Davenport (1967, 1977), while a description of other wind-load phenomena, such as vortex shedding, galloping, and flutter, may be found in Simiu and Scanlan (1978).

10.1.1 Response Analysis

Let the response of the structure be resolved into eigenmodes of discrete or continuous type (see, e.g., Clough and Penzien, 1975).

The time dependence of each of these modes can then be described by functions $T_i(t)$ satisfying the differential equations

$$\ddot{T}_i(t) + 2\zeta_i\omega_i \dot{T}_i(t) + \omega_i^2 T_i(t) = \frac{q_i(t)}{m_i} \tag{10.2}$$

where ω_i and ζ_i are the eigenfrequency and the corresponding damping ratio, and $q_i(t)$ and m_i are the generalized load and mass of mode i, i.e., weighted averages of the load and the mass.

The wind velocity $U(t)$ is resolved into a time-independent mean value $\bar{U}$ and a stationary fluctuating part $u(t)$:

$$U(t) = \bar{U} + u(t) \tag{10.3}$$

When the fluctuations are small compared with the mean wind, formula (10.1) for the wind pressure can be linearized:

$$p(t) \sim \frac{1}{2}\rho C(U^2 + 2\bar{U}^T u(t)) \tag{10.4}$$

Thus the fluctuating part of the wind pressure is determined by the component $u(t)$ in the direction of the mean wind.

Formula (10.4) implicitly assumes the structure to be fixed. When this is not the case, the fluctuating wind-speed component $u(t)$ should be replaced by the relative wind speed. In terms of the normalized mode-shape functions $\mathbf{v}_j(\mathbf{x})$ with the components $v_j(\mathbf{x})$ in the direction of the mean wind, the generalized force corresponding to the wind fluctuations is

$$q_i(t) = \rho\int_A C\,U\,(u(t) - \sum_j v_j \dot{T}_j(t))\,v_i\,dA \tag{10.5}$$

where A is the projected area of the structure. The terms in (10.5) containing the modal velocities $\dot{T}_j(t)$ correspond to additional damping, called aerodynamic damping. It is therefore advantageous to rewrite (10.2) in the form

$$\ddot{T}_i(t) + 2\zeta_i\omega_i \dot{T}_i(t) + \frac{1}{m_i}\sum_j b_{ij} \dot{T}_j(t) + \omega_i^2 T_i(t) = \frac{q_i^e(t)}{m_i} \tag{10.6}$$

where the matrix

$$b_{ij} = \rho\int_A CUv_i v_j\,dA \tag{10.7}$$

contains the aerodynamic damping and

$$q_i^e(t) = \rho\int_A CUu(t)v_i\,dA \tag{10.8}$$

is the effective generalized load. The aerodynamic damping matrix b_{ij} may not be diagonal and it will then introduce modal coupling.

If a mode is excited alone, the total damping ratio is the sum of the structural damping ratio ζ_i and the aerodynamic damping ratio:

$$\zeta_i^a = \frac{\rho}{2\omega_i m_i} \int_A CU v_i^2 dA \qquad (10.9)$$

For structures with low natural frequency the aerodynamic damping may constitute a substantial addition to the structural damping, which may be of the order 0.01 to 0.02 for concrete structures and 0.005 to 0.01 for steel structures.

Within limited time intervals the wind-speed fluctuations $u(t)$ can be considered as a zero-mean stationary stochastic process, and the structural response may then be described by its spectral density $S^u(\omega)$. The fluctuating wind component $u(t)$ depends on position as well as on time and therefore constitutes a stochastic field. A full frequency spectrum description therefore requires the cross spectrum $S^u(\mathbf{x}_1, \mathbf{x}_2, \omega)$ of the components $u(\mathbf{x}_1, t_1)$ and $u(\mathbf{x}_2, t_2)$ at the arbitrary positions $\mathbf{x}_1$ and $\mathbf{x}_2$. It is convenient to treat the spatial dependence in normalized form by introducing the dimensionless function

$$\chi(\mathbf{x}_1, \mathbf{x}_2, \omega) = \frac{S^u(\mathbf{x}_1, \mathbf{x}_2, \omega)}{\sqrt{S^u(\mathbf{x}_1, \mathbf{x}_1, \omega) S^u(\mathbf{x}_2, \mathbf{x}_2, \omega)}} \qquad (10.10)$$

Obviously, $\chi = 1$ for $\mathbf{x}_1 = \mathbf{x}_2$, and it may be shown that $|\chi| \leqslant 1$ for all $\mathbf{x}_1$ and $\mathbf{x}_2$. The function χ is a measure of the spatial structure of the stochastic field $\mathbf{u}(\mathbf{x}, t)$, and the square of its magnitude, $|\chi|^2$, is called the *coherence*. A small coherence for a frequency ω at the points $\mathbf{x}_1$ and $\mathbf{x}_2$ implies that the corresponding frequency components of $u(\mathbf{x}_1, t)$ and $u(\mathbf{x}_2, t)$ are nearly uncorrelated.

The spectral matrix of the generalized loads (10.8) can be expressed as

$$S_{ij}^q(\omega) = \rho^2 \int_A \int_A (CU v_i)_1 (CU v_j)_2 S^u(\mathbf{x}_1, \mathbf{x}_2, \omega) dA_1 dA_2 \qquad (10.11)$$

If the spectral density is independent of the position $\mathbf{x}$, the influence of the geometry is conveniently extracted in dimensionless form by introducing reference values U_0 and C_0 and the aerodynamic admittance

$$F_{ij}(\omega) = \frac{1}{(C_0 U_0 A)^2} \int_A \int_A (CU v_i)_1 (CU v_j)_2 \chi(\mathbf{x}_1, \mathbf{x}_2, \omega) dA_1 dA_2 \qquad (10.12)$$

The spectral matrix of the generalized loads then is

$$S_{ij}^q(\omega) = (\rho C_0 U_0 A)^2 F_{ij}(\omega) S^u(\omega) \qquad (10.13)$$

The effects of the spatial structure of the turbulent wind and the geometry have now been concentrated in the aerodynamic admittance $F_{ij}(\omega)$ expressed in terms of an integral of $\chi(\mathbf{x}_1, \mathbf{x}_2, \omega)$.

Before proceeding to specific calculations of structural response properties, some theoretical considerations are presented concerning the functions $\chi(\mathbf{x}_1,\mathbf{x}_2,\omega)$ and $S^u(\omega)$.

10.1.2 The Turbulent Wind

Several features of turbulent wind can be understood and described to a reasonable degree of approximation by Taylor's hypothesis of "frozen turbulence" (see Batchelor, 1959). The idea consists of the assumption that the structure of the eddies making up the turbulence is sufficiently stable to be considered fixed while the mean wind translates it across the field of observation. Under this assumption the whole structure of the turbulence can be inferred from an instantaneous observation of all fluctuation velocities $\mathbf{u}(\mathbf{x})$. The temporal variation is obtained via a translation with the mean wind velocity $\mathbf{U}$. In the following the turbulence is assumed to be homogeneous in space.

Introduce the spatial covariance matrix

$$\mathbf{C}(\mathbf{x}_1,\mathbf{x}_2) \;=\; \text{Cov}\,[\mathbf{u}(\mathbf{x}_1),\mathbf{u}(\mathbf{x}_2)] \;=\; E\,[\mathbf{u}(\mathbf{x}_1)\mathbf{u}(\mathbf{x}_2)^T] \qquad (10.14)$$

In the case of homogeneous turbulence, $\mathbf{C}(\mathbf{x}_1,\mathbf{x}_2)$ depends only on the spatial separation $\mathbf{r}=\mathbf{x}_1-\mathbf{x}_2$, and in complete analogy with stochastic processes of a single scalar variable in Section 8.1 the spectral representation is

$$\mathbf{C}(\mathbf{r}) \;=\; \int \mathbf{\Phi}(\mathbf{k})e^{i\mathbf{k}^T\mathbf{r}}d\mathbf{k} \qquad (10.15)$$

The integration is over the full three-dimensional wave number space (k_1,k_2,k_3). The turbulent energy per unit mass of air is

$$E\left|\frac{1}{2}\mathbf{u}^T\mathbf{u}\right| \;=\; \frac{1}{2}\sum_{j=1}^{3} C_{jj}(0) \;=\; \frac{1}{2}\int\sum_{j=1}^{3}\Phi_{jj}(\mathbf{k})d\mathbf{k} \qquad (10.16)$$

As in the case of one independent variable, the integral (10.16) may be considered as a decomposition of the energy into contributions associated with the wave number vector $\mathbf{k}$. In the case of isotropic turbulence, which is the main concern here, the integrand $\sum\Phi_{jj}(\mathbf{k})$ depends only on the length k of the vector $\mathbf{k}$, and the integral (10.16) is conveniently expressed in terms of spherical coordinates.

$$E\left|\frac{1}{2}\mathbf{u}^T\mathbf{u}\right| \;=\; \int_0^{\infty} E(k)dk \qquad (10.17)$$

The energy density $E(k)$ is seen to be

$$E(k) \;=\; 2\pi k^2 \sum_{j=1}^{3}\Phi_{jj}(\mathbf{k}) \qquad (10.18)$$

where $4\pi k^2$ is the area of a spherical shell of radius k. The tensor $\Phi(\mathbf{k})$ must be isotropic, and it can then be shown that in the case of incompressibility it must be of the form

$$\Phi_{ij}(\mathbf{k}) = \frac{E(k)}{4\pi k^4}(k^2\delta_{ij} - k_i k_j) \tag{10.19}$$

where δ_{ij} is Kronecker's delta. Thus the spectral density of a homogeneous isotropic incompressible field is determined by a single scalar function, $E(k)$.

These fragments of the theory of homogeneous turbulence lead to important results concerning the asymptotic behavior of the spectrum for large k (and large ω), the relative magnitude of the spectral components, and the form of the normalized cross spectra such as the function χ in (10.10).

The full, time-dependent covariance matrix for $\mathbf{u}(\mathbf{x}_1,t_1)$ and $\mathbf{u}(\mathbf{x}_2,t_2)$ follows from the hypothesis of "frozen turbulence":

$$\mathbf{C}(\mathbf{x}_1,\mathbf{x}_2,t_1,t_2) = \mathbf{C}(\mathbf{r}-\mathbf{U}\tau) \tag{10.20}$$

where $\tau = t_1 - t_2$. By substitution of (10.19) into (10.15) the components of the covariance matrix are found to be

$$C_{ij}(\mathbf{r}-\mathbf{U}\tau) = \int \frac{E(k)}{4\pi k^4}(k^2\delta_{ij} - k_i k_j)\, e^{i\mathbf{k}^T(\mathbf{r}-\mathbf{U}\tau)}\, d\mathbf{k} \tag{10.21}$$

In particular this expression can be used to calculate the frequency spectra for each of the fluctuating components u_1, u_2, and u_3 at a point $\mathbf{x}_1 = \mathbf{x}_2$. Let u_1 be the component in the direction of the mean wind velocity. The integral is therefore symmetric in k_2 and k_3, permitting the introduction of polar coordinates. The resulting spectrum for the longitudinal component u_1 is

$$S_{11}^u(\omega) = \frac{1}{2\pi}\int_{-\infty}^{\infty} C_{11}(-\mathbf{U}\tau)e^{-i\omega\tau}d\tau \tag{10.22}$$

$$= \int\int \frac{E(k)}{4\pi k^4}(k^2 - k_1^2)\,dk_2 dk_3$$

$$= \frac{1}{2}\int_{k_1}^{\infty}\left(1 - \frac{k_1^2}{k^2}\right)\frac{E(k)}{k}\,dk, \quad k_1 = \frac{\omega}{U}$$

The spectra for the transverse components u_2 and u_3 are identical.

$$S_{22}^u(\omega) = S_{33}^u(\omega) = \frac{1}{4}\int_{k_1}^{\infty}\left(1 + \frac{k_1^2}{k^2}\right)\frac{E(k)}{k}\,dk, \quad k_1 = \frac{\omega}{U} \tag{10.23}$$

Although these relations involve integration, it is seen that if $E(k)$ is known for large values of k, all three spectra are known for large values of ω.

The asymptotic behavior of $E(k)$ for large values of k follows from the energy cascade theory of Kolmogorov (see Batchelor, 1959). It is based on the observation that if turbulent energy is generated in the large eddies (small k) and dissipated in the small eddies (large k), there is an intermediate region in which $E(k)$ is independent of the specific mechanisms of generation and dissipation. In this region

$$E(k) \sim k^{-5/3} \qquad (10.24)$$

and it then follows from (10.22) and (10.23) that

$$S_{11}^u(\omega) \approx \frac{3}{4} S_{22}^u(\omega) = \frac{3}{4} S_{33}^u(\omega) \sim \left(\frac{\omega}{U}\right)^{-5/3} \qquad (10.25)$$

There is thus a theoretical basis — as well as experimental verification — of the asymptotic behavior $\omega^{-5/3}$ of the frequency spectra. This also explains why different empirical expressions for the spectra are compared by first adjusting the asymptotic behavior for large ω.

Two spectral forms that are often used are those of Davenport (1967):

$$\frac{\omega S_{11}^u(\omega)}{\kappa U^2} = 4 \frac{\xi^2}{(1+\xi^2)^{4/3}} \qquad (10.26)$$

and Harris (1971):

$$\frac{\omega S_{11}^u(\omega)}{\kappa U^2} = 4 \frac{\xi}{(2+\xi^2)^{5/3}} \qquad (10.27)$$

The dimensionless variable ξ is

$$\xi = \frac{\omega \lambda}{2\pi U}, \qquad 0 < \xi < \infty \qquad (10.28)$$

where λ is a length scale proposed to be 1200 m in (10.26) and 1800 m in (10.27). κ is a drag coefficient that relates the dissipated energy to the reference mean wind speed U. Representative values are given in Table 10.1 for the reference speed U_{10} at 10 m height.

Table 10.1 Drag Coefficient κ and Exponent α		
Terrain	κ	α
City center	0.050	0.40
Forest and suburbs	0.015	0.30
Open grassland	0.005	0.16
Rough sea	0.001	0.12

The formula (10.21) also enables a theoretical approximation of the coherence of the turbulence. Here only the function $\chi(x_1, x_2, \omega)$ for transverse separation is considered, while the general theoretical problem and the experimental verification has been dealt with by Kristensen and Jensen (1979). For transverse separation in the 2-direction the cross spectrum is

$$S_{11}^u(r, \omega) = \frac{1}{2\pi} \int_{-\infty}^{\infty} C_{11}(\mathbf{r} - \mathbf{U}\tau) e^{-i\omega\tau} d\tau \tag{10.29}$$

$$= \int\int \frac{E(k)}{4\pi k^4} (k^2 - k_1^2) e^{ik_2 r} dk_2 dk_3, \quad k_1 = \frac{\omega}{U}$$

By introducing polar coordinates in the (k_2, k_3) plane and carrying out the angular integration, the following expression is obtained:

$$S_{11}^u(r, \omega) = \frac{1}{2} \int_0^\infty \frac{E(\sqrt{k_1^2 + K^2})}{(k_1^2 + K^2)^2} K^3 J_0(Kr) dK, \quad k_1 = \frac{\omega}{U} \tag{10.30}$$

$J_0(\ )$ is the Bessel function of order zero and it follows that $S_{11}^u(r, \omega)$, and thereby $\chi(r, \omega)$, is real for transverse separation. Furthermore, only values of $E(k)$ with $k > \omega/U$ contribute to the integral, and therefore an asymptotic estimate can be obtained by use of (10.24). The integral can then be evaluated explicitly (Abramowitz and Stegun, 1965, formula 11.4.44). The normalized result is

$$\chi(r, \omega) = \left(\frac{r\omega}{2U}\right)^{5/6} \left| 2K_{5/6}\left(\frac{r\omega}{U}\right) - \left(\frac{r\omega}{U}\right) K_{1/6}\left(\frac{r\omega}{u}\right) \right| \frac{1}{\Gamma(5/6)} \tag{10.31}$$

$K_\nu(\ )$ is the modified Bessel function of order v. χ appears to be a function of the single variable $r\omega/U$. However, this property is of asymptotic nature as it depends on (10.24), which is valid only for large ω. The theoretical result (10.31) shows fair agreement with experimental results for separations substantially smaller than the length scale λ, but is too complicated for most practical calculations. It is often replaced by a simple exponential relation

$$\chi(r, \omega) = \exp\left(-a \frac{r\omega}{U}\right) \tag{10.32}$$

A typical value of the parameter a is 1.3, but larger values are sometimes used for horizontal separation.

In the natural wind the mean wind speed increases with height, and it is therefore necessary to include this variation in the load (10.8) and damping (10.9). It is also necessary to replace the uniquely defined U from homogeneous turbulence with suitable reference wind speeds in the spectra (10.26) to (10.28) and in the spatial correlation (10.32). Dimensional analysis leads to a logarithmic dependence of U upon height, but in engineering calculations, in particular those of analytic nature, it is customary to represent the variation with height in power form with the reference value at 10 m height,

$$U = U_{10}\left(\frac{z}{z_{10}}\right)^{\alpha} \tag{10.33}$$

The theoretical shortcoming of this relation leads to different exponents for different types of terrain. Representative values for α are given in Table 10.1.

A suitable reference wind speed is also necessary in (10.28) in terms of U/λ and in (10.32) in terms of U/a. In both cases there is considerable uncertainty regarding the optimal formulation. Thus Kaimal et al. (1972) propose the spectrum

$$\frac{\omega S_{11}^{u}(\omega)}{\kappa U_{10}^{2}} = \frac{200 f}{(1 + 50 f)^{5/3}} \tag{10.34}$$

with the dimensionless frequency

$$f = \frac{\omega z}{2\pi U(z)}, \quad 0 < f < \infty \tag{10.35}$$

This corresponds to a length scale that increases somewhat less than proportionally with z, while Davenport (1967) proposed a fixed length scale λ used in connection with the reference wind speed U_{10} in (10.28). There is experimental evidence supporting a spectral length scale increasing with height, and also a variable ratio U/a producing a similar effect (Simiu and Scanlan, 1978, pp. 53-61).

10.1.3 Slender Structures

The influence of the size and shape of the structure is contained in the aerodynamic admittance $F_{ij}(\omega)$ given in (10.12). In general it requires evaluation of a double area integral, but for slender structures such as masts, chimneys, and some bridges full correlation may be assumed over each cross section, thereby reducing the computation to evaluation of a double line integral. In addition to the immediate usefulness of these results, they illustrate some general features of interest.

First introduce a dimensionless length coordinate $\xi = x/L$. With the notation $\mu(\xi) = (CUH)/C_0 U_0 h_0$, where h is the transverse dimension of the cross section, the aerodynamic admittance (10.12) takes the form

$$F_{ij}(\omega) = \int_0^1 \int_0^1 \mu(\xi_1)\,\mu(\xi_2)\,v_i(\xi_1)\,v_2(\xi_2)\,\chi(\xi_1,\xi_2,\omega)\,d\xi_1 d\xi_2 \tag{10.36}$$

In particular, for exponential coherence $\chi(r,\omega)$ is written as

$$\chi(\xi_1,\xi_2,\omega) = \exp(-\varphi \mid \xi_1 - \xi_2 \mid) \qquad (10.37)$$

where size, wind speed, and frequency are combined in the dimensionless parameter

$$\varphi = \frac{a \omega L}{U} \qquad (10.38)$$

Under this assumption $F_{ij} = F_{ij}(\varphi)$.

The asymptotic behavior of $F_{ij}(\varphi)$ for small and large values of φ can be determined directly as follows. For $\varphi \to 0$, χ is approximately constant on the full area of integration, giving the limit value

$$F_{ij}(0) = \int_0^1 \mu(\xi_1) v_i(\xi_1) d\xi_1 \int_0^1 \mu(\xi_2) v_j(\xi_2) d\xi_2 \qquad (10.39)$$

For large values of φ, χ only deviates appreciably from zero around $\xi_1 = \xi_2$, i.e., around the diagonal of the square of integration. The asymptotic contribution from the diagonal is determined by the integral

$$\int_{-\infty}^{\infty} \exp(-\varphi \Delta\xi) \, d\Delta\xi = \frac{2}{\varphi} \qquad (10.40)$$

The asymptotic behavior of $F_{ij}(\varphi)$ for large φ then is

$$F_{ij} \sim \frac{2}{\varphi} \int_0^1 \mu(\xi)^2 v_i(\xi) v_j(\xi) d\xi, \quad \varphi \gg 1 \qquad (10.41)$$

For $i = j$ the integral in (10.41) is always positive, while $F_{ij}(0)$ may vanish if either of the shape functions $v_i(\xi)$ or $v_j(\xi)$ changes sign.

For mode shapes without change of sign, such as the fundamental mode of cantilevers, $F_{ii}(\varphi)$ decreases monotonically, and Davenport (1977) has suggested use of the approximation

$$F_{ii}(\omega) \approx \frac{1}{A + B\varphi} \qquad (10.42)$$

where the parameters A and B are determined from (10.39) and (10.41), respectively.

Consider next two simple mode shapes, appropriate for the along wind and torsional vibrations of a symmetric cantilever bridge in the construction period (Fig. 10.1). When only the wind load on the bridge deck is considered,

$$v_1(\xi) = 1, \quad v_2(\xi) = 2\xi \qquad (10.43)$$

It follows from symmetry that $F_{12}(\varphi) = 0$, and for $i = j$ the integral is conveniently written as

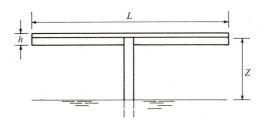

Figure 10.1   Cantilever bridge during construction phase.

$$F_{ii}(\varphi) = 2\int_{-1/2}^{1/2}\mu(\xi_2)v_i(\xi_2)e^{-\varphi\xi_2}\int_{-1/2}^{\xi_2}\mu(\xi_1)v_i(\xi_1)e^{\varphi\xi_1}d\xi_1 \quad (10.44)$$

For $\mu(\xi)=1$ the following expressions are obtained:

$$F_{11}(\varphi) = \frac{2}{\varphi^2}(e^{-\varphi} - 1 + \varphi) \quad (10.45)$$

$$F_{22}(\varphi) = \frac{2}{\varphi^4}\left(4 - \varphi^2 + \frac{1}{3}\varphi^3 - (2+\varphi)^2 e^{-\varphi}\right) \quad (10.46)$$

The aerodynamic admittances $F_{11}(\varphi)$ and $F_{22}(\varphi)$ are shown in Fig. 10.2. This figure also contains the approximation

$$F_{11} \approx \frac{1}{1+0.5\varphi} \quad (10.47)$$

In view of the other approximations involved in the evaluation of wind-induced response, the accuracy of (10.47) is good, and this type of approximation is easily extended to cases of variable $\mu(\xi)$. The curves clearly illustrate the different effects of size in the two cases. While the relative uniformity of the wind for small φ leads to maximum response of mode 1, the torsional moment becomes increasingly balanced, leading to decreased response of mode 2.

Example 10.1

Many bridges are constructed from prefabricated elements mounted on the piers as shown in Fig. 10.1. During the construction period the structure is subject to torsional vibrations that are not characteristic of the completed bridge. The quantity of primary interest is the torsional moment M in the pier. It may be expressed in the form

$$M(t) = \bar{P}L\psi(t)$$

where $\bar{P}$ is the mean wind load on the bridge deck — possibly including construction equipment — and $\psi(t)$ is a dimensionless stochastic process, the intensity of which is to be determined.

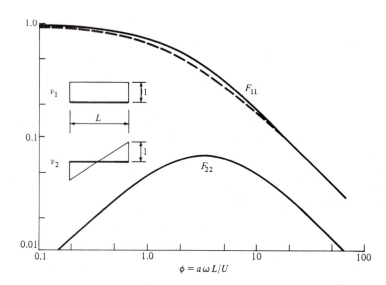

Figure 10.2 Asymptotic admittance functions.

Let the displacement be given by the angle of twist θ. When the height of the cross section h is constant, the generalized load is

$$q(t) = \int_{-L/2}^{L/2} p(t)hx\,dx = \rho CUhL^2 \int_{-1/2}^{1/2} u(t)\xi\,d\xi$$

$$= 2\bar{P}L \int_{-1/2}^{1/2} \frac{u(t)}{U}\xi\,d\xi$$

The generalized mass is

$$m = \int_{-L/2}^{L/2} \rho_s hx^2\,dx = I$$

and the aerodynamic damping ratio then follows from (10.9):

$$\zeta^a = \frac{\rho CU}{2\omega_0 I} \int_{-L/2}^{L/2} hx^2\,dx = \frac{\rho CUhL^3}{24\omega_0 I} = \frac{\bar{P}L}{12\omega_0 UI}$$

The equation of motion (10.6) then is

$$\ddot{\theta}(t) + 2(\zeta^s + \zeta^a)\omega_0\dot{\theta}(t) + \omega_0^2\theta(t) = \frac{2\bar{P}L}{I} \int_{-1/2}^{1/2} \frac{u(t)}{U}\xi\,d\xi$$

The main concern is with the torsional moment M, and the angle θ is therefore eliminated by use of the relation

$$M = k\theta = \omega_0^2 I\theta$$

where k is the stiffness. Upon multiplication of the equation of motion with the factor $\omega_0^2 I/\bar{P}L$, the following equation is obtained for the dimensionless gust response process $\psi(t)$:

$$\ddot{\psi}(t) + 2(\zeta^s + \zeta^a)\omega_0\dot{\psi}(t) + \omega_0^2\psi(t) = 2\omega_0^2\int_{-1/2}^{1/2}\frac{u(t)}{U}\xi d\xi$$

This equation illustrates that the loading term depends only on the stiffness and mass of the structure through the natural frequency, while the aerodynamic damping by its direct relation to displacements requires knowledge of stiffness and mass.

The spectral density of $\psi(t)$ is determined from the transfer function $H(\omega)$ of the structure and the spectral density of the excitation. The latter follows from suitable normalization of (10.13), whereby

$$S^{\psi}(\omega) = \left(\frac{2\omega_0^2}{U}\right)^2 S^u(\omega)\mid H(\omega)\mid^2 F(\omega)$$

The aerodynamic admittance $F(\omega)$ for the present problem is given by (10.46) and is shown as the lower curve in Fig. 10.2. Different representations of the turbulence spectrum $S^u(\omega)$ are given in (10.26), (10.27) and (10.34). The variance of ψ follows from integration of the spectrum. An estimate can be obtained by use of the wide-band approximation discussed in Section 8.2.2. According to this approximation the value of the excitation spectrum at the resonance frequency ω_0 is decisive. In the present context the one-sided spectrum is used, and the result is

$$\sigma_{\psi}^2 \approx \frac{\pi}{\zeta}\frac{\omega_0 S^u(\omega_0)}{U^2}F(\varphi_0)$$

Note that for aerodynamic admittance functions with a finite limit for the low frequencies such as $F_{11}(\varphi)$ in Fig. 10.2, the low frequency components of the excitation may contribute significantly to the response. This effect may be included in an approximate way (see, e.g., Davenport, 1967).

The mean wind speed refers to a specific averaging time — often 10 minutes — and the expected maximum response within this time can be found as described in Section 7.4.3. What is needed is an expression for how much the expected maximum in N periods exceeds the mean value. This quantity is measured in units of a standard deviation and is called the *gust factor*. An asymptotic expression for the gust factor was derived in (7.88):

$$\mu_{max} \approx \sqrt{2\log N} + \frac{0.577}{\sqrt{2\log N}}$$

In the present context $N = 2\pi T/\omega_0$, where T is the averaging time of the mean wind. In the present problem the expected maximum moment then is

$$M \approx \mu_{max}\bar{P}L\sigma_{\psi} = \mu_{max}\bar{P}L\left(\frac{\pi}{\zeta}\frac{\omega_0 S^u(\omega)}{U^2}F(\varphi_0)\right)^{1/2}$$

Consider a uniform bridge deck of total length 80 m placed symmetrically on a pier of height $z = 26$ m as shown in Fig. 10.1. The natural frequency in torsion is $\nu_0 = 0.55$ Hz. When the 10 minute design wind speed is $U_{10} = 32$ m/s, the mean wind at $z = 26$ m follows from (10.33):

$$U = 32\left(\frac{26}{10}\right)^{0.12} = 36 \, \text{m/s}$$

The reduced frequency ξ with length scale $\lambda = 1200$ m is

$$\xi = 0.55 \times 1200 / 36 = 18.3$$

and with coherence parameter $2\pi a = 12$,

$$\varphi = 0.55 \times 12 / 36 = 0.183$$

The variance σ_ψ^2 can now be evaluated by use of the spectrum (10.26) with $\kappa = 0.001$ from Table 10.1 and the aerodynamic admittance function (10.46). For an estimated damping $\zeta \approx 0.015$,

$$\sigma_\psi^2 \approx \frac{\pi\kappa}{\zeta} \frac{U_{10}^2}{U^2} \frac{\omega_0 S^u(\omega_0)}{\kappa U_{10}^2} F(\varphi_0)$$

$$= \frac{\pi 0.001}{0.015} \left(\frac{32}{36}\right)^2 0.576 \times 0.0305 = 0.00291$$

The gust factor is calculated from $N = 600 \times 0.55 = 330$, whereby

$$\mu_{max} = 3.58$$

Thus the expected maximum moment is

$$M = 3.58 \times 0.054 \bar{P} L = 0.19 \bar{P} L$$

The factor 0.19 represents the effect of gusts, and the expected maximum torsional moment is seen to be only slightly less than if the mean load on each arm was applied in opposite direction. In particular the role of the gust factor should be noted.

10.2 WAVE LOADS

Waves as relating to wave loads on structures are usually described in terms of sea states. A sea state is an approximately stationary condition described by parameters with long-term fluctuations. These parameters include but are not necessarily limited to the significant wave height H_S, the mean wave period T_Z, and the wave direction θ_0. Precise definitions of these quantities are given in the following. For some purposes, e.g., equipment installation, it is desirable to know the properties of the processes describing the long-term fluctuations of the sea-state parameters – see, e.g., Example 5.4 on design wave determination from the joint distribution of H_S and T_Z.

The sea states are characterized by properties determined by visual observation or measurement. In addition to parameters such as H_S, T_Z, and θ_0, time records are used to determine the spectral density of the sea elevation $\eta(t)$ observed at a fixed location. More detailed measurements can also be used to determine the spread of the direction of wave propagation around the mean value θ_0. Typical spectra are described in Section 10.2.1.

The particle velocities in the water below the surface can be related to the surface elevation $\eta(x,y,t)$ by use of a wave theory. Several theories with various degrees of refinement are available, but statistical descriptions of wave forces are mainly limited to the linear theory of gravity waves, introduced briefly in Section 10.2.2. An important result of this theory is the so-called "dispersion relation," which gives the wavelength associated with any wave frequency and thereby provides a link between the time history at a specific position and the instantaneous form of the full water surface.

An important parameter for wave loads on structures is the size of the structural member compared with the wavelength λ. For large wavelengths the structural member experiences a flow that in spite of its transient character is generally in the same direction at any instant of time. Given sufficient time separation develops, and the wave load therefore consists of two distinct contributions: a drag force due to separation, and an inertial – or mass – force created by the transient.   The drag force increases approximately with the square of the incident particle velocity, while the inertial force is linear in the particle accelerations. Thus the relative importance of the two terms depends on the wave height. Generally, drag forces are most important near the surface and for high waves. The general features of this type of load is captured by the Morison formula discussed in Section 10.2.3.

Large structures – e.g., storage tanks – experience a different kind of wave load. When the wavelength is compatible with a typical cross-sectional dimension – say $\lambda < 5D$ – the dominating mechanism is wave scattering. An illustration of this and a discussion of the size effect is presented in Section 10.2.4.   An extensive account of wave forces has been given by Sarpkaya and Isaacson (1981), where much additional material can be found.

10.2.1 Wave Spectra

At a specific position the sea elevation $\eta(t)$ constitutes a stochastic process. For reasons of analysis it is convenient to divide the time into intervals of typically 5 to 10 hours and to assume the process $\eta(t)$ to be stationary in each of these intervals. A second-moment description of the process $\eta(t)$ in any particular sea state can then be given in terms of a one-sided wave spectrum $S_\eta(\omega)$:

$$\mathrm{Cov}[\eta(t),\eta(t+\tau)] = \int_0^\infty S_\eta(\omega) \cos(\omega\tau)\, d\omega \qquad (10.48)$$

Phillips (1958) used dimensional analysis combined with a hypothesis involving wave steepness limitation of a fully developed sea to obtain the asymptotic high-frequency relation

$$S_\eta(\omega) \sim \alpha g^2 \omega^{-5} \qquad (10.49)$$

where α would be a universal constant. On the basis of this result and observed spectra Pierson and Moskowitz (1964) proposed the spectrum

$$S_\eta(\omega) = \alpha g^2 \omega^{-5} \exp\left|-\beta\left(\frac{\omega}{\omega_0}\right)^{-4}\right|, \qquad \omega > 0 \qquad (10.50)$$

with the constants $\alpha = 0.0081$ and $\beta = 0.74$. The reference frequency $\omega_0 = g/U_{19.5}$ is given in terms of the mean wind speed at a height of 19.5 m above mean still water level. The spectrum is very sensitive to the value of $U_{19.5}$, and it is therefore practical to introduce parameters directly related to the observed waves.

Two important parameters of any wave spectrum are associated with the time scale and the wave height. The time scale T_Z is defined as the upcrossings of the mean still water level, and the significant wave height H_S is defined as the expected value of the highest one-third of the waves. If the wave process $\eta(t)$ is assumed to be Gaussian, the mean wave period T_Z is determined from (7.94) as

$$T_Z = 2\pi \sqrt{\lambda_0/\lambda_2} \qquad (10.51)$$

where λ_j is the (one-sided) moment of the wave spectrum $S_\eta(\omega)$. If furthermore the process $\eta(t)$ is assumed to be narrow-banded, the wave height is associated with the envelope, $H = 2R$, and the expected value of the largest one-third of the waves follows from (7.100) as

$$\frac{H_S}{\sigma_\eta} = 3\int_{r_{1/3}}^\infty 2r\, f_R(r)\, dr = 4 \qquad (10.52)$$

i.e., $H_S = 4\sigma_\eta$, where $\sigma_\eta = \sqrt{\lambda_0}$ is the standard deviation of $\eta(t)$. Thus H_S and T_Z follow from the first two moments of the wave spectrum $S_\eta(\omega)$. For the Pierson and Moskowitz spectrum (10.50) both H_S and T_Z are uniquely determined by the wind speed $U_{19.5}$:

$$H_S = 2\sqrt{\alpha/\beta}\,\frac{U_{19.5}^2}{g}, \qquad T_Z = \frac{2\pi}{(\beta\pi)^{1/4}}\frac{U_{19.5}}{g} \qquad (10.53)$$

The Pierson and Moskowitz spectrum is a special case of a gamma spectrum, i.e., a spectrum of the form

$$S_\eta(\omega) = A\,\omega^{-\xi}\exp(-B\omega^{-\zeta}), \qquad \omega > 0 \qquad (10.54)$$

The parameters A and B are related uniquely to H_S and T_Z by

$$A = \frac{\zeta}{16} H_S^2 \left(\frac{2\pi}{T_Z}\right)^{\xi-1} \frac{\Gamma\left(\frac{\xi-1}{\zeta}\right)^{(\xi-3)/2}}{\Gamma\left(\frac{\xi-3}{\zeta}\right)^{(\xi-1)/2}} \tag{10.55}$$

$$B = \left(\frac{2\pi}{T_Z}\right)^{\zeta} \frac{\Gamma\left(\frac{\xi-1}{\zeta}\right)^{\zeta/2}}{\Gamma\left(\frac{\xi-3}{\zeta}\right)^{\zeta/2}} \tag{10.56}$$

The Pierson and Moskowitz spectrum corresponds to $\xi = 5$ and $\zeta = 4$. The renormalized form in which the prior relation between H_S and T_Z is suspended was proposed by ISSC (1964). In this form

$$S_\eta(\omega) = \frac{H_S^2}{4\pi} \left(\frac{2\pi}{T_Z}\right)^4 \omega^{-5} \exp\left|-\frac{1}{\pi}\left(\frac{2\pi}{T_Z}\right)^4 \omega^{-4}\right| \tag{10.57}$$

While the parameter choice $\xi = 5$ has been justified by the asymptotic result (10.49), the exponent ζ can be used to adjust the width of the spectrum. In Section 7.4 the bandwidth was characterized by either of the dimensionless parameters $\alpha = \lambda_2 / \sqrt{\lambda_0 \lambda_4}$ or $\delta = \lambda_1 / \sqrt{\lambda_0 \lambda_2}$. Wave spectra satisfying the asymptotic condition (10.49) only allow use of the bandwidth measure δ. For a gamma spectrum with $\xi = 5$, the bandwidth measure becomes

$$\delta = \frac{\Gamma(3/\zeta)}{\sqrt{\Gamma(2/\zeta)\Gamma(4/\zeta)}} \tag{10.58}$$

For the Pierson and Moskowitz spectrum $\delta = \Gamma(3/4)/\pi^{1/4} = 0.92$, and δ increases for increasing values of the exponent ζ. The bandwidth as expressed by δ is a significant parameter in the joint distribution of the individual wave heights H and their associated periods T.

Often the development of waves is limited by the presence of coastlines. The corresponding wave pattern is said to be fetch-limited, and observations indicate that the spectrum is more peaked than the Pierson and Moskowitz spectrum. Although this feature in principle could be included in an appropriate gamma spectrum, a different and more detailed spectrum finds wider application. The JONSWAP spectrum resulted from the Joint North Sea Wave Project (Hasselmann et al., 1973) and it incorporates the fetch directly by means of the nondimensional parameter $\tilde{x} = g\,x/U_{10}^2$. x is the fetch and U_{10} is the mean wind speed at a height of 10 m above still water level. When the peak frequency ω_p is introduced, the JONSWAP spectrum is

$$S_\eta(\omega) = \alpha g^2 \omega^{-5} \exp\left[-\frac{5}{4}\left(\frac{\omega}{\omega_p}\right)^{-4}\right]\gamma^a \qquad (10.59)$$

making use of the functions

$$\alpha = 0.076\,\tilde{x}^{-0.22}, \quad \omega_p = 20\,\tilde{x}^{-0.33} \qquad (10.60)$$

$$a = \exp\left[-\frac{1}{2}\left(\frac{\omega - \omega_p}{\sigma\omega_p}\right)^2\right], \quad \sigma = \begin{cases}\sigma_-, & \omega < \omega_p \\ \sigma_+, & \omega > \omega_p\end{cases} \qquad (10.61)$$

and the parameters γ, σ_-, and σ_+. There is some scatter in the values for γ, σ_-, and σ_+ but no systematic trend with $\tilde{x}$. Their mean values are

$$\gamma = 3.3, \quad \sigma_- = 0.07, \quad \sigma_+ = 0.09 \qquad (10.62)$$

γ is a peak enhancement factor, while σ_- and σ_+ serve to determine the steepness of the flanks. For $\gamma = 1$ (10.59) has the same form as (10.50) expressed in terms of the peak spectral frequency ω_p.

Typically, wave spectra are unimodal with rather limited bandwidth. On the assumption of a normal process a simple approximation for the joint probability density of the wave amplitude R and the zero-crossing period T can then be obtained (Longuet-Higgins, 1975, 1983). It is a direct consequence of the asymptotic behavior (10.49) that the wave height cannot be determined by means of the local extrema, as this would require the spectral moment λ_4. The wave is therefore represented by its Hilbert transform envelope and a suitable phase angle θ. It is convenient to introduce the linear mean frequency $\bar{\omega} = \lambda_1/\lambda_0$ and use the representation

$$\eta(t) = \rho \cos(\bar{\omega}t + \theta) \qquad (10.63)$$

$$\hat{\eta}(t) = \rho \sin(\bar{\omega}t + \theta) \qquad (10.64)$$

where $\rho = \rho(t)$ and $\theta = \theta(t)$ change slowly with time. Upon differentiation and rearrangement of terms, one obtains

$$\xi = \dot{\eta} + \bar{\omega}\hat{\eta} = \dot{\rho}\cos(\bar{\omega}t + \theta) - \rho\dot{\theta}\sin(\bar{\omega}t + \theta) \qquad (10.65)$$

$$\hat{\xi} = \dot{\hat{\eta}} - \bar{\omega}\eta = \dot{\rho}\sin(\bar{\omega}t + \theta) + \rho\dot{\theta}\cos(\bar{\omega}t + \theta) \qquad (10.66)$$

The variables $(\eta,\hat{\eta},\xi,\hat{\xi})$ are joint normal, and as a consequence of the particular choice of $\bar{\omega}$ they are uncorrelated. The variances are $\text{Var}[\eta] = \text{Var}[\hat{\eta}] = \lambda_0$, $\text{Var}[\xi] = \text{Var}[\hat{\xi}] = \lambda_2(1-\delta^2) = \mu_2$, whereby

$$f(\eta,\hat{\eta},\xi,\hat{\xi}) = \frac{1}{(2\pi)^2\lambda_0\mu_2}\exp\left[-\frac{\eta^2 + \hat{\eta}^2}{2\lambda_0} - \frac{\xi^2 + \hat{\xi}^2}{2\mu_2}\right] \qquad (10.67)$$

The Jacobi determinant of the variable transformation from $(\eta,\hat{\eta},\xi,\hat{\xi})$ to $(\rho,\dot{\rho},\theta,\dot{\theta})$ is ρ^2, and upon a change of variables in (10.67), one finds

$$f(\rho,\dot{\rho},\theta,\dot{\theta}) = \frac{\rho^2}{(2\pi)^2\lambda_0\mu_2} \exp\left|-\frac{\rho^2}{2\lambda_0} - \frac{\dot{\rho}^2 + \rho^2\dot{\theta}^2}{2\mu_2}\right| \quad (10.68)$$

Integration of θ from 0 to 2π and $\dot{\rho}$ from $-\infty$ to ∞ gives

$$f(\rho,\dot{\theta}) = \frac{\rho^2}{\lambda_0\sqrt{2\pi\mu_2}} \exp\left|-\frac{\rho^2}{2\lambda_0} - \frac{\rho^2\dot{\theta}^2}{2\mu_2}\right| \quad (10.69)$$

In the final transformation, normalized variables R and T are introduced based on the assumption that $\theta + \bar{\omega}$ changes slowly, whereby

$$R \approx \frac{\rho}{\sqrt{2\lambda_0}}, \quad T \approx \frac{\bar{\omega}}{\bar{\omega} + \dot{\theta}} \quad (10.70)$$

The desired density then follows from (10.69) in the form

$$f(R,T) = \frac{2}{\sqrt{\pi}v} \frac{R^2}{T^2} \exp\left|-R^2\left(1 + (1-\frac{1}{T})^2v^{-2}\right)\right| L(v) \quad (10.71)$$

where $v^2 = \delta^{-2} - 1 = \lambda_0\lambda_2/\lambda_1^2 - 1$ and the normalizing factor $L(v)$ has been introduced to make the integral of $f(R,T)$ over the positive quarter plane equal to 1. Integration gives

$$\frac{1}{L(v)} = \frac{1}{2}[1 + (1+v^2)^{-1/2}] = \frac{1}{2}(1+\delta) \quad (10.72)$$

This formula bears a remarkable similarity to the fraction of positive maxima $1/2(1+\alpha)$ in terms of the regularity factor α. For the Pierson and Moskowitz spectrum, which is generally considered relatively wide, $\delta = 0.92$ and $v = 0.42$. Contour curves of $f(R,T)$ are shown for this case in Fig. 10.3, displaying a strong correlation between R and T. The results show good qualitative agreement with observations. Analyses of the same problem have been carried out by Canavie et al. (1976) and Lindgren and Rychlik (1982). Although their results show qualitative agreement, they require at least the fourth-moment not generally available for wave spectra.

10.2.2 Gravity Waves

The sea waves of interest for loads on structures are dominated by gravity forces (Lighthill, 1978). A coordinate system $\{x,y,z\}$ is introduced in the undisturbed water surface with the z-

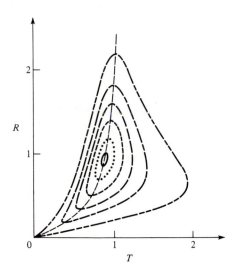

Figure 10.3 Contour curves of $f(R,T)/f_{max}$ for the Pierson and Moskowitz spectrum, $f/f_{max} = 0.99, 0.9, 0.7, 0.5, 0.3,$ and 0.1, from Longuet-Higgins (1983).

axis oriented upward. The particle velocities (u,v,w) are derivable from a potential $\psi(x,y,z,t)$,

$$(u,v,w) = \left(\frac{\partial}{\partial x}, \frac{\partial}{\partial y}, \frac{\partial}{\partial z}\right)\psi \tag{10.73}$$

The water can be considered as incompressible, and this implies that the potential ψ is harmonic,

$$\left(\frac{\partial^2}{\partial x^2} + \frac{\partial^2}{\partial y^2} + \frac{\partial^2}{\partial z^2}\right)\psi = 0 \tag{10.74}$$

The wave character of the solution arises from the existence of a free surface of elevation $\eta(x,y,t)$. At (x,y,η) the excess pressure is

$$p_e(x,y,\eta) = \rho g \eta(x,y) \tag{10.75}$$

In a linearized theory, where the convective terms are omitted and the boundary condition is exhausted at $z=0$, the momentum equation gives the excess pressure as $p_e = -\rho\,\partial\psi/\partial t$. Furthermore, a particle on the surface remains on the surface, and in a linearized theory this implies that $\partial\eta/\partial t = \partial\psi/\partial z$. With these linearized relations the boundary condition (10.75) takes the form

$$\left(\frac{\partial^2}{\partial t^2} + g\frac{\partial}{\partial z}\right)\psi(x,y,0,t) = 0 \tag{10.76}$$

Solutions of wave type to (10.74) and (10.76) are found by

introducing a harmonic time factor and separating the dependence on the depth variable z, i.e., by considering solutions of the form

$$\psi = \text{Re}[e^{i\omega t}\,\Psi(x,y)\,\zeta(z)] \tag{10.77}$$

For surface waves in uniform depth d the solution is

$$\psi = \text{Re}[e^{i\omega t}\,\Psi(x,y)\cosh k(z+d)] \tag{10.78}$$

where $\Psi(x,y)$ satisfies the two-dimensional Helmholz equation

$$\left(\frac{\partial^2}{\partial x^2} + \frac{\partial^2}{\partial y^2} + k^2\right)\Psi(x,y) = 0 \tag{10.79}$$

k is the wave number, and substitution into the boundary condition (10.76) gives a so-called dispersion relation between k and ω:

$$\omega^2 = gk\tanh(kd) \tag{10.80}$$

The most important solution is the plane − or long crested − wave. In this type of wave $\Psi(x,y)$ depends only on the arguments x and y through the linear combination $\mathbf{k}^T\mathbf{x}=k_x x + k_y y$, where $\mathbf{k}$ is a vector in the direction of wave propagation of length $|\mathbf{k}|=k$. When $\mathbf{k}=(k,0)$ the potential is independent of y:

$$\psi = \frac{\omega a}{k}\frac{\cosh k(z+d)}{\sinh(kd)}\,\text{Re}[e^{i(\omega t - kx)}] \tag{10.81}$$

The surface elevation is a sine wave with amplitude a progressing with phase velocity $c=\omega/k$.

$$\eta(x,t) = a\sin(\omega t - kx) \tag{10.82}$$

The particle velocities follow by differentiation of (10.81):

$$\begin{vmatrix} u \\ w \end{vmatrix} = \frac{\omega a}{\sinh(kd)}\begin{vmatrix} \cosh k(z+d)\sin(\omega t - kx) \\ \sinh k(z+d)\cos(\omega t - kx) \end{vmatrix} \tag{10.83}$$

The particle accelerations are

$$\begin{vmatrix} \dot{u} \\ \dot{w} \end{vmatrix} = \frac{\omega^2 a}{\sinh(kd)}\begin{vmatrix} \cosh k(z+d)\cos(\omega t - kx) \\ -\sinh k(z+d)\sin(\omega t - kx) \end{vmatrix} \tag{10.84}$$

The particle trajectories are ellipses with the ratio $\tanh k(z+d)$ between the vertical minor axis and the horizontal major axis.

For depths exceeding approximately half the wavelength, the solution may be replaced by the simpler deep-water approximation, in which the dispersion relation (10.80) is replaced by

$$\omega^2 = gk \tag{10.85}$$

and the particle velocity is

$$\begin{vmatrix} u \\ w \end{vmatrix} = \omega\, a\, e^{kz} \begin{vmatrix} \sin(\omega t - kx) \\ \cos(\omega t - kx) \end{vmatrix} \qquad (10.86)$$

More general wave forms can be generated by superposition of plane waves with different values of a and $\mathbf{k}$. The square of the amplitude corresponding to a particular value of $\mathbf{k}$ may be considered as the spectral density in a two-dimensional wave-number representation. In the case of plane waves in a particular direction the spectral description can be obtained either in terms of k by transforming an instantaneous space record or in terms of ω by transforming the time record obtained at a fixed position. The two spectral densities are related through the dispersion relation defining the group velocity $c_g = d\omega/dk$:

$$S_\eta(k) = c_g(\omega) S_\eta(\omega) \qquad (10.87)$$

A simple directional effect can be obtained by assuming the waves to be composed of independent plane waves propagating in different directions. The waves are then described by a spectrum $S_\eta(\omega)$ and an associated wave-energy spreading function. A common choice of spreading function is a cosine power, e.g.,

$$w(\theta - \theta_0) = \frac{2^{2n-1}}{\pi} \frac{\Gamma(n+1)^2}{\Gamma(2n+1)} \cos^{2n}\left(\frac{\theta - \theta_0}{2}\right), \qquad |\theta - \theta_0| < \pi \quad (10.88)$$

The normalizing factor allows for use of the standard unidirectional frequency spectrum.

10.2.3 The Morison Wave Force

When the characteristic dimension of a structure is small compared with the wavelength the load contains two important components: a drag force roughly proportional to the square of the normal component of the incident particle velocity and an inertia — or mass — force associated with the normal component of the particle acceleration. In addition, there may be a transverse force component associated with vortex shedding, but it is difficult to model and incorporate into a probabilistic analysis. The drag and mass terms are combined in the Morison formula for the force per unit length of a fixed cylinder (Morison et al., 1950):

$$\mathbf{p} = k_d \mathbf{u} |\mathbf{u}| + k_m \dot{\mathbf{u}} \qquad (10.89)$$

where $\mathbf{u}$ is the incident particle velocity normal to the cylinder. The parameters k_d and k_m are often given in terms of the dimensionless drag and mass coefficients C_d and C_m. In the particular case of a circular cylinder of diameter D,

$$k_d = C_d \rho D/2, \quad k_m = C_m \rho \pi D^2/4 \tag{10.90}$$

where ρ is the mass density of the water. For oscillating flow the ratio between the two terms is described by the Keulegan-Carpenter number $K = uT/D$, where u is the velocity amplitude and T the period of oscillation. It is important to note that the Morison formula as originally proposed relates to oscillatory flow with a definite frequency. For random waves the spectrum typically extends to infinity and according to the dispersion relation this implies the existence of short wave components for which (10.89) is not immediately applicable. Thus it is to be expected that the Morison formula should be used in connection with a suitable frequency cutoff or with a more general frequency dependence of the parameters k_d and k_m. The mass term is linear, and a theoretical estimate of the frequency dependence of k_m can therefore be obtained as outlined in the next section.

In the case of shallow waves the particle velocity $\mathbf{u}(t)$ and acceleration $\dot{\mathbf{u}}(t)$ can be represented by normal processes. The nonlinear form of the Morison equation then gives a force $\mathbf{p}(t)$ with a systematic deviation from the normal distribution. This problem was investigated without current by Pierson and Holmes (1965), while Borgman (1967) included a steady current. In the absence of current the relevant components $u(t)$ and $\dot{u}(t)$ are considered as independent, zero-mean normal processes with standard deviations σ_u and $\sigma_{\dot{u}}$. A measure of the relative importance of drag and inertia is

$$K = \frac{k_d}{k_m} \frac{\sigma_u^2}{\sigma_{\dot{u}}} \tag{10.91}$$

replacing the Keulegan-Carpenter number of harmonic flow. Here k_d and k_m are considered as representative constants, but in principle their frequency dependence could be included in the processes $u(t)$ and $\dot{u}(t)$. u/σ_u is a standard normal variable, and

$$y = \frac{Ku \mid u \mid}{\sigma_u^2} \tag{10.92}$$

then has the probability density

$$f_y(y) = f_u(u)\frac{du}{dy} = \frac{1}{2\sqrt{2\pi K \mid y \mid}} \exp\left(-\frac{\mid y \mid}{2K}\right) \tag{10.93}$$

The probability density of the normalized force,

$$q = \frac{p}{k_m \sigma_{\dot{u}}} = K \frac{u \mid u \mid}{\sigma_u^2} + \frac{\dot{u}}{\sigma_{\dot{u}}} \tag{10.94}$$

is then found by convolution of $f_y(y)$ with the standard normal density,

$$f_q(q) = \frac{1}{4\pi} \int_{-\infty}^{\infty} \exp\left| -\frac{1}{2}(q-y)^2 - \frac{|y|}{2K} \right| \frac{dy}{\sqrt{K|y|}} \quad (10.95)$$

The integral depends on q and on the parameter K. It is convenient for numerical evaluation to divide the interval of integration at $y = 0$ and to change sign in the negative interval. The probability density function of the normalized force q then is

$$f_q(q) = \frac{1}{4\pi\sqrt{K}} \exp\left(-\frac{q^2}{2}\right) \left| I\left(\frac{1}{2K}+q\right) + I\left(\frac{1}{2K}-q\right) \right| \quad (10.96)$$

where the function $I(\ )$ is the integral

$$I(z) = \sqrt{\pi} \, e^{z^2/4} D_{-1/2}(z) = \int_0^{-\infty} s^{-1/2} \exp\left(-\frac{s^2}{2} - zs\right) ds \quad (10.97)$$

$I(z)$ can be calculated by direct numerical integration or from tables of the parabolic cylinder function $D_{-1/2}(\ )$ (Abramowitz and Stegun, 1965, Chapter 19). The distribution of the wave force p — relating to a specific direction — is characterized by the parameter K and the variance σ_p^2. σ_p^2 follows directly from formula (10.94) in terms of normal variables and the moment formula given in Example 8.2:

$$\sigma_p^2 = (k_m \sigma_{\dot{u}})^2 \sigma_q^2 = 3(k_d \sigma_u^2)^2 + (k_m \sigma_{\dot{u}})^2 \quad (10.98)$$

Figure 10.4 shows the cumulative distribution of the normalized wave force p/σ_p for $K = 0.8$ and gives a comparison with the corresponding normal distribution. The distribution of the wave force displays more pronounced tails, and approximation by a normal distribution would underestimate the high loads.

The probability density of the force $p(t)$ at a given cross section depends on the parameters determining the scale and the shape. These parameters contain the coefficients k_d and k_m. Conversely, if records of $(p(t), u(t), \dot{u}(t))$ are available, the coefficients k_d and k_m can be determined from the second- and fourth-order moments. When $u(t)$ and $\dot{u}(t)$ are assumed normally distributed and independent, the second- and fourth-order moments of (10.89) provide equations for k_d and k_m in terms of σ_u, $\sigma_{\dot{u}}$, σ_p, and the kurtosis $\beta = E[p^4]/E[p^2]^2$. The solution is

$$k_d \sigma_u^2 = \sigma_p \left(\frac{\beta-3}{78}\right)^{1/4} \quad (10.99)$$

$$k_m \sigma_{\dot{u}} = \sigma_p \left(1 - 3\left(\frac{\beta-3}{78}\right)^{1/2}\right)^{1/2} \quad (10.100)$$

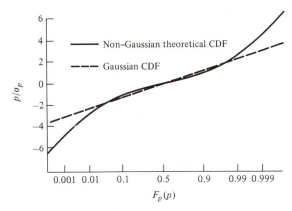

Figure 10.4 Cumulative distribution of the normalized wave force p/σ_p, from Tickell (1977).

This method for determination of k_d and k_m was proposed by Pierson and Holmes (1965). The ratio between the two expressions provides the following formula for K in terms of β:

$$K = \left| \frac{\left(\dfrac{\beta-3}{78}\right)^{1/2}}{1-3\left(\dfrac{\beta-3}{78}\right)^{1/2}} \right|^{1/2} \tag{10.101}$$

Thus the kurtosis β gives the shape of the distribution of $p(t)$. The applicability of the same distribution function to structural response parameters such as stresses has been investigated by Tickell (1977) and Holmes and Tickell (1979). They found good agreement with a suitably calibrated distribution of the form indicated in (10.96). Indeed, for quasi-static response of a structure to a very long wave, where the particle velocity field around the whole structure is in phase, this result would follow immediately from the theory. Dynamic effects, lack of correlation, and nonlinearity may lead to deviations.

The local extremes of the wave force $p(t)$ are closely related to the crossing rate of $Q(t)$, and for a narrow-band process an approximate distribution of the ranges can be deduced from the crossing rates. The upcrossing rate of the normalized wave force process $Q(t)$ can be found by use of Rice's formula (7.59), but consideration of a differentiable upcrossing of $Q(t)$ requires the second derivative of the particle velocity $U(t)$ and thereby the existence of the moment λ_4 for the velocity spectrum. A less restrictive result can be obtained for a modified form of the Morison equation, in which the inertia term is replaced by $-k_m \bar{\omega} \hat{U}(t)$, where $\bar{\omega} = \lambda_1 / \lambda_0$ is the mean frequency and $\hat{U}(t)$ is the Hilbert transform of $U(t)$. In that case only the moments λ_0, λ_1, and λ_2 of the velocity spectrum are needed.

The computation is complicated by the dependence between $Q(t)$ and $\dot{Q}(t)$. It is convenient to introduce the normalized variables

$$v_1 = \frac{u}{\sigma_u}, \quad v_2 = \frac{\dot{u}}{\sigma_{\dot{u}}}, \quad v_3 = \frac{\ddot{u}}{\sigma_{\ddot{u}}} \tag{10.102}$$

with the joint probability density

$$f_v(\mathbf{v}) = \frac{1}{2\pi} \frac{\varphi(v_2)}{\sqrt{1-\alpha^2}} \exp\left[-\frac{1}{2} \frac{v_1^2 + 2\alpha v_1 v_3 + v_3^2}{1-\alpha^2} \right] \tag{10.103}$$

$$= \varphi(v_1)\varphi(v_2) \frac{1}{\sqrt{1-\alpha^2}} \varphi\left[\frac{v_3 + \alpha v_1}{\sqrt{1-\alpha^2}} \right]$$

The parameter α is the regularity factor of the process $U(t)$ given by

$$\alpha = -\frac{\sigma_{\dot{u}}^2}{\sigma_u \sigma_{\ddot{u}}} = -\frac{\lambda_2}{\sqrt{\lambda_0 \lambda_4}} \tag{10.104}$$

When an auxiliary variable v is considered together with q and $\dot{q}$, Rice's formula is

$$v(q) = \int_0^\infty \dot{q} \, f_{q\dot{q}}(q,\dot{q}) d\dot{q} = \int_0^\infty \int_{-\infty}^\infty \dot{q} \, f_{q\dot{q}v}(q,\dot{q},v) d\dot{q}dv \tag{10.105}$$

In terms of the normalized variables v_1, v_2, and v_3,

$$q = Kv_1 |v_1| + v_2 \tag{10.106}$$

$$\dot{q} = \frac{\sigma_{\ddot{u}}}{\sigma_{\dot{u}}} (2\alpha K v_2 |v_1| + v_3) \tag{10.107}$$

and by choosing $v = v_1$ the Jacobian of the transformation is $\sigma_{\ddot{u}}/\sigma_{\dot{u}}$. After introducing the variable

$$z = \frac{v_3 + \alpha v_1}{\sqrt{1-\alpha^2}} = \frac{1}{\sqrt{1-\alpha^2}} \left| \frac{\sigma_{\dot{u}}}{\sigma_{\ddot{u}}} \dot{q} + \alpha(v_1 - 2Kv_2 |v_1|) \right| \tag{10.108}$$

$$= \frac{1}{\sqrt{1-\alpha^2}} \left| \frac{\sigma_{\dot{u}}}{\sigma_{\ddot{u}}} \dot{q} + \alpha(v + 2K^2 v^3 - 2K |v| q) \right|$$

instead of $\dot{q}$ as integration variable, the integral (10.105) takes the form

$$v(q) = \sqrt{1-\alpha^2}\frac{\sigma_{\ddot{u}}}{\sigma_{\dot{u}}}\int_{-\infty}^{\infty}\varphi(v)\varphi(q-Kv\mid v\mid)\int_{z_0}^{\infty}(z-z_0)\varphi(z)dzdv \quad (10.109)$$

$$= \frac{\sqrt{1-\alpha^2}}{2\pi}\frac{\sigma_{\ddot{u}}}{\sigma_{\dot{u}}}\int_{-\infty}^{\infty}\exp\left|-\frac{1}{2}v^2-\frac{1}{2}(q-Kv\mid v\mid)^2\right|\Psi(z_0)dv$$

where the function $\Psi(\ )$ was introduced in (7.67), and z_0 corresponds to $\dot{q}=0$ in (10.108). The integral (10.109) can be integrated by parts, when it is observed that

$$z_0 = \frac{\alpha}{\sqrt{1-\alpha^2}}(v+2K^2v^3-2K\mid v\mid q) \quad (10.110)$$

$$= \frac{\alpha}{\sqrt{1-\alpha^2}}\frac{1}{2}\frac{\partial}{\partial v}(v^2+(q-Kv\mid v\mid)^2)$$

Upon substitution of $\Psi(z_0)$ from (7.67) and integration by parts of the last term, the final form is

$$v(q) = \frac{\sigma_{\ddot{u}}}{\sigma_{\dot{u}}}\int_{-\infty}^{\infty}\left(\sqrt{1-\alpha^2}+\alpha\frac{\partial z_0}{\partial v}\right)\varphi(z_0)\varphi(v)\varphi(q-Kv\mid v\mid)\,dv \quad (10.111)$$

This expression is suited for numerical integration. However, it is difficult to evaluate a representative value of the parameter α theoretically. It is therefore of particular interest to consider the narrow-band limit $\alpha\to 1$. In this limit the scaling factor $\alpha/\sqrt{1-\alpha^2}$ in z_0 approaches infinity, and the factor $\Psi(z_0)$ in (10.109) can be replaced by the dashed line of Fig. 7.12; i.e., $\Psi(z_0)$ is replaced by $-z_0$ for $z_0\leqslant 0$ and by 0 for $z_0>0$. The upper limit of integration v_0 is determined from (10.110), and the integral can then be expressed as

$$v(q) = v_0^u\exp\left|-\frac{1}{2}v_0^2-\frac{1}{2}(q-Kv_0\mid v_0\mid)^2\right| \quad (10.112)$$

v_0^u is the zero-crossing frequency of $u(t)$. By solving the inequality $z_0(v_0)\leqslant 0$, the final asymptotic result is

$$v(q) \sim v_0^u\begin{cases}\exp\left(-\frac{1}{2}q^2\right), & \mid q\mid<\dfrac{1}{2K}\\[2mm]\exp\left|-\dfrac{1}{2K}\left(q-\dfrac{1}{4K}\right)\right|, & \mid q\mid\geqslant\dfrac{1}{2K}\end{cases} \quad (10.113)$$

In a narrow-band approximation of the force, the distribution of local maxima can be obtained by neglecting the positive minima and negative maxima. This gives the probability density function

$$f_m(q) \approx -\frac{1}{v(0)} \frac{d}{dq} v(q) \qquad (10.114)$$

and with the asymptotic result (10.113)

$$f_m(q) \approx \begin{vmatrix} q \exp\left(-\frac{1}{2} q^2\right), & 0 < q < \frac{1}{2K} \\ \frac{1}{2K} \exp\left| -\frac{1}{2K}\left(q - \frac{1}{4K}\right) \right|, & \frac{1}{2K} < q \end{vmatrix} \qquad (10.115)$$

This density was derived by direct consideration of trigonometric functions by Borgman (1965) and extended to include a moderate current by Moe and Crandall (1978). It is shown in normalized form in Fig. 10.5, and it consists of a Rayleigh distribution for $q < 1/2K$ and an exponential distribution for $q > 1/2K$. The transition point $q = 1/2K$ corresponds to the wave force $p = 1/2k_m^2 \omega_0^2 / k_d$. Also in this case a Gaussian approximation of $P(t)$ and the associated narrow-band Rayleigh distribution of P_{max} would underestimate the tails.

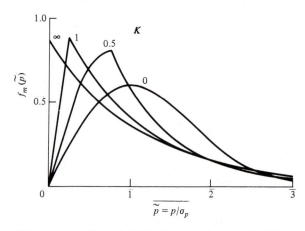

Figure 10.5   Asymptotic probability density of normalized local maxima.

Example 10.2

The nonlinearity of the drag term leads to increased importance of the peaks, and thereby to increased accumulation of fatigue damage as expressed by the expected sum of powers of the stress ranges, $E[\sum S_i^m]$ (Section 9.2.3). For quasistatic, highly correlated excitation the distribution of the stress corresponds to that of the wave force $q(t)$. With the narrow-band result (10.115) the expected sum for a normalized stress is

$$E[\textstyle\sum S_i^m] = v_0^u\, T\, E\left[\left(2\frac{q_m}{\sigma_q}\right)^m\right] = v_0^u T\left(\frac{2}{\sigma_q}\right)^m\left\{\int_0^{1/2K} q^{m+1}\exp\left(-\frac{1}{2}q^2\right)dq\right.$$

$$\left.+\frac{1}{2K}\int_{1/2K}^{\infty} q^m \exp\left[-\frac{1}{2K}\left(q-\frac{1}{4K}\right)\right]dq\right\}$$

By variable transformations the integral can be expressed in terms of the incomplete gamma function

$$\gamma(a,x) = \int_0^x t^{a-1} e^{-t}\, dt$$

In particular, $\gamma(a,\infty)=\Gamma(a)$ and for integer $a=n+1$,

$$\gamma(n+1,x) = n!\left(1 - e^{-x}\sum_{k=0}^{n}\frac{x^k}{k!}\right)$$

With the variance from (10.98),

$$\sigma_q^2 = 3K^2 + 1$$

the result is

$$E[\textstyle\sum S_i^m] = v_0^u T\left(\frac{4}{3K^2+1}\right)^{m/2}\left\{2^{m/2}\gamma\left(1+\frac{m}{2},\frac{1}{8K^2}\right)\right.$$

$$\left.+(2K)^m\, e^{1/(8K^2)}\left[\Gamma(m+1) - \gamma\left(m+1,\frac{1}{4K^2}\right)\right]\right\}$$

In the limit $K \to 0$, i.e., for dominating inertial forces

$$E[\textstyle\sum S_i^m] = v_0^u\, T\, (2\sqrt{2})^m\, \Gamma(1+m/2)$$

whereby the result (9.63) is recovered. In the other limit, $K \to \infty$, the non-linear drag forces dominate, giving

$$E[\textstyle\sum S_i^m] = v_0^u\, T\left(\frac{4}{\sqrt{3}}\right)^m\, \Gamma(1+m)$$

Drag force dominance gives the greatest damage accumulation rate, and the difference increases rapidly with m. This is illustrated by Fig. 10.6 giving the relative damage per stress cycle for selected values of m corresponding to curves adopted by DnV (1982) and AWS (1983). A more detailed discussion has been given by Brouwers and Verbeek (1983).

In addition to the statistical properties of the wave force at a particular cross section at one particular time, it is of interest to consider the correlation of wave forces at different positions and different times. Here the correlation of the wave forces p_1 and p_2 is derived in terms of the correlation between the corresponding particle velocities u_1 and u_2 and particle accelerations $\dot{u}_1$ and $\dot{u}_2$. The correlations between particle velocities and accelerations are determined by use of an appropriate wave theory as indicated briefly in the next section. In the present case $u_1, u_2, \dot{u}_1$, and $\dot{u}_2$ are

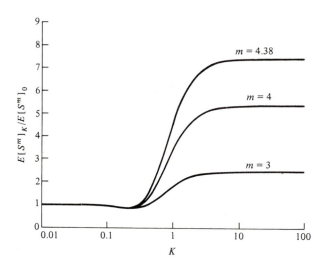

Figure 10.6 Relative damage per stress cycle normalized to unity for $K = 0$.

assumed to be zero-mean joint normal variables. A steady current resulting in nonzero mean values of u_1 and u_2 has been included by Borgman (1967).

The covariance of the wave forces p_1 and p_2 is

$$c_{pp} = E[p_1 p_2] = k_{d1} k_{d2} E[u_1 \mid u_1 \mid u_2 \mid u_2 \mid] + k_{m1} k_{m2} E[\dot{u}_1 \dot{u}_2] \qquad (10.116)$$
$$+ k_{d1} k_{m2} E[u_1 \mid u_1 \mid \dot{u}_2] + k_{m1} k_{d2} E[\dot{u}_1 u_2 \mid u_2 \mid]$$

The first term is conveniently treated by introducing normalized variables $v_1 = u_1 / \sigma_{u_1}$ and $v_2 = u_2 / \sigma_{u_2}$. The problem is to evaluate the function

$$G(\rho) = E[v_1 \mid v_1 \mid v_2 \mid v_2 \mid] \qquad (10.117)$$

where v_1 and v_2 are normalized joint variables with zero mean and correlation coefficient ρ. This is accomplished by repeated use of the identity (8.123). Differentiation with respect to ρ and integration by parts lead to

$$\frac{d^n G}{d\rho^n} = E\left[\left(\frac{\partial^2}{\partial v_1 \partial v_2} \right)^n v_1 \mid v_1 \mid v_2 \mid v_2 \mid \right] \qquad (10.118)$$

For $\rho = 0$ and $n = 0, 1, 2$ the expectation is easily evaluated:

$$G(0) = 0, \quad G'(0) = \frac{8}{\pi}, \quad G''(0) = 0 \qquad (10.119)$$

and for $n = 3$,

$$G'''(\rho) = \frac{8}{\pi \sqrt{1-\rho^2}} \qquad (10.120)$$

The function $G(\rho)$ then follows by integration of (10.80) with the boundary condition (10.79) (Borgman, 1967):

$$G(\rho) = \frac{2}{\pi}[(1 + 2\rho^2)\arcsin(\rho) + 3\rho\sqrt{1-\rho^2}] \qquad (10.121)$$

The power series expansion of $G(\rho)$ is closely related to an expansion of the normalized drag term $v|v|$ in terms of Hermite polynomials as explained in Section 8.4.2. It follows from (10.121) that the first terms are

$$G(\rho) = \frac{8}{\pi}\rho + \frac{4}{3\pi}\rho^3 + \cdots \qquad (10.122)$$

corresponding to

$$v|v| = \sqrt{8/\pi}\,\mathrm{He}_1(v) + \sqrt{2/(9\pi)}\,\mathrm{He}_3(v) + \cdots \qquad (10.123)$$

In the interval $-1 < \rho < 1$ the function $G(\rho)$ is nearly linear. Thus the linear term gives a maximum error of 15% at $\rho = \pm 1$, and inclusion of the cubic term limits the error to 1%.

As a consequence of the orthogonality properties of the Hermite polynomials, the covariance of a normal variable and a function, such as $v|v|$, of a joint normal variable is given by its first Hermite expansion term. With the first term given by (10.123), this gives the wave force covariance as

$$c_{pp} = k_{d1}\sigma_{u_1}^2 k_{d2}\sigma_{u_2}^2 G(\rho) + k_{m1}k_{m2}E[\dot{u}_1\dot{u}_2]$$

$$+ \sqrt{8/\pi}\,k_{d1}\sigma_{u_1}k_{m2}E[u_1\dot{u}_2] + \sqrt{8/\pi}\,k_{m1}k_{d2}\sigma_{u_2}E[\dot{u}_1 u_2] \qquad (10.124)$$

The necessary covariances for $u_1, u_2, \dot{u}_1$, and $\dot{u}_2$ are obtained in terms of the spectrum of the surface elevation as described in the next section.

10.2.4 Wave Forces on Structures

The results of the preceding section assumed knowledge of the stochastic processes describing particle velocities and accelerations. These processes were assumed joint normal with zero mean, and the covariance functions therefore give a complete description. Within the framework of the linearized wave theory of Section 10.2.2, the covariance functions can be obtained in terms of the spectrum of the surface elevation. The wave theory results in a spectral representation in which the wavelength, speed of propagation, and depth variation depend on the frequency ω. The high frequency components correspond to short wavelengths (large k) and therefore decrease more rapidly with depth. Thus increase in depth has a smoothing effect by diminishing the high-frequency part of the spectrum.

The variance of the particle velocity component $u(x,z,t)$ follows from its relation to the surface elevation $\eta(x,t)$. At any specific frequency, combination of (10.82) and (10.83) gives

$$u(x,z,t) = \omega \frac{\cosh k(z+d)}{\sinh(kd)} \eta(x,t) = \omega T(z,\omega)\eta(x,t) \qquad (10.125)$$

where $T(z,\omega)$ is a transfer function playing a role similar to the system transfer function $H(\omega)$ in Chapter 8. The spectral density of $u(x,z,t)$ follows by analogy with (8.13):

$$S_{uu}(\omega) = [\omega T(z,\omega)]^2 S_{\eta\eta}(\omega) \qquad (10.126)$$

$S_{uu}(\omega)$ and $S_{\eta\eta}(\omega)$ are symmetric, two-sided spectra, and for $\omega > 0$ they are related to the one-sided surface elevation spectrum by $S_{\eta\eta}(\omega) = 1/2 S_\eta(\omega)$. The variance of the horizontal velocity component $u(x,z,t)$ is

$$\sigma_u(z)^2 = \int_{-\infty}^{\infty} S_{uu}(\omega)\,d\omega = \int_0^{\infty} [\omega T(z,\omega)]^2 S_\eta(\omega)\,d\omega \qquad (10.127)$$

The variance of the acceleration $\dot{u}(x,z,t)$ follows immediately:

$$\sigma_{\dot{u}}(z)^2 = \int_{-\infty}^{\infty} \omega^2 S_{uu}(\omega)\,d\omega = \int_0^{\infty} [\omega^2 T(z,\omega)]^2 S_\eta(\omega)\,d\omega \qquad (10.128)$$

For large ω the dispersion relation (10.80) degenerates into the deep-water form (10.85), whereby $k \sim \omega^2/g$ and

$$T(z,\omega) \sim \exp\left(\frac{\omega^2 z}{g}\right) \qquad (10.129)$$

This factor provides a systematic filtering of the spectrum as a function of the depth $-z$.

It is more difficult to account for horizontal separation in a rigorous way, but the following simple argument illustrates the essence and provides the result. A more detailed presentation has been given by Sigbjørnsson (1979). Consider the wave velocity components $u_1 = u(x_1,z_1,t_1)$ and $u_2 = u(x_2,z_2,t_2)$. The covariance function is represented in the frequency domain as

$$E[u_1 u_2] = \int_{-\infty}^{\infty} S_{uu}(\mathbf{x}_1,\mathbf{x}_2,\omega) e^{i\omega\tau}\,d\omega \qquad (10.130)$$

For long crested waves any combination of time separation $\tau = t_2 - t_1$ and horizontal separation $\Delta x = x_2 - x_1$ that leaves $\omega\tau - k\Delta x$ unchanged will not affect the corresponding frequency contribution. As a consequence, Δx must appear exclusively in the exponential factor. The result then follows immediately from (10.125)

$$S_{u_1 u_2}(\omega) = \omega^2 T(z_1,\omega) T(z_2,\omega) \exp(-ik\Delta x) S_{\eta\eta}(\omega) \quad (10.131)$$

From this formula the necessary covariances can be calculated numerically, and in principle the calculation of forces and moments on groups of fixed vertical piles is straightforward (see Borgman, 1967). The procedure is indicated briefly in the following example with reference to a single pile.

Example 10.3

The total Morison wave force on a fixed vertical pile is given by

$$F = \int_{-d}^{0} (k_d u \mid u \mid + k_m \dot{u}) dz$$

In the absence of a current the mean value is zero and the covariance function follows from (10.124), where the contributions of the last two terms cancel upon integration.

$$c_{FF}(\tau) = E[F(t) F(t+\tau)] = \int_{-d}^{0}\int_{-d}^{0} \{k_d^2 \sigma_u(z_1)^2 \sigma_u(z_2)^2 G(\rho) + k_m^2 E[\dot{u}_1 \dot{u}_2]\} dz_1 dz_2$$

The variance $\sigma_u(z)^2$ is evaluated from (10.127) for several values of z. The correlation function is then evaluated from (10.131) in the form

$$\rho(z_1, z_2) = \frac{1}{\sigma_u(z_1)\sigma_u(z_2)} \int_0^\infty \omega^2 T(z_1,\omega) T(z_2,\omega) S_\eta(\omega) \cos(\omega\tau) d\omega$$

The covariance of the accelerations is

$$E[\dot{u}_1 \dot{u}_2] = \int_0^\infty \omega^4 T(z_1,\omega) T(z_2,\omega) S_\eta(\omega) \cos(\omega\tau) d\omega$$

In the last term the integration with respect to z_1 and z_2 can be carried out analytically, while a similar simplification is not possible in the first term, not even if $G(\rho)$ is linearized. However, the linearized form leads to a direct formula for the spectrum of $F(t)$ in terms of a one-dimensional integral with respect to z. In terms of one-sided spectra

$$S_F(\omega) = \left\{ \frac{8}{\pi} \left| \int_{-d}^{0} k_d \omega \sigma_u(z) T(z,\omega) dz \right|^2 + \left| k_m \frac{\omega^2}{k} \right|^2 \right\} S_\eta(\omega)$$

Note that the last term has the asymptotic form $(k_m g)^2$, and therefore does not amplify the high-frequency part of the spectrum $S_\eta(\omega)$.

The Morison formula for the wave force does not account for the finite dimension of the structure, and the particle velocity and acceleration are calculated from the incident wave without any interaction effect. In the case of a vertical circular cylinder extending from the bottom to the surface, the linear wave scattering problem including finite dimension and interaction but excluding flow separation, can be solved explicitly (MacCamy and Fuchs, 1954). For large structures, $D > 5\lambda$, this solution gives the main part of the wave load, and for small diameters it provides an analytical estimate of the coefficient C_m, which turns out to be a function of the relative wavelength.

Let the z-axis of the $\{x,y,z\}$ coordinate system be the axis of a circular cylinder of diameter D. The potential ψ_i of an incident long crested wave in the x-direction is given by (10.81). The total potential is the sum of ψ_i and the potential ψ_s of the scattered waves. It is convenient to introduce polar coordinates such that

$$x = r\cos\theta, \quad y = r\sin\theta \qquad (10.132)$$

The exponential $\exp(-ikx)$ in the potential ψ_i is now expanded in terms of Bessel functions (Abramowitz and Stegun, 1965):

$$\psi_i = \frac{\omega a}{k}T(z,\omega)\,\mathrm{Re}\left|e^{i\omega t}\,2\sum_{n=0}^{\infty}{}'\,i^{-n}J_n(kr)\cos(n\theta)\right| \qquad (10.133)$$

The first term in the summation should be multiplied by 1/2. The horizontal variation of the potential ψ_s must satisfy (10.79), which in polar coordinates is Bessel's equation. The scattered waves must also diverge from the cylinder, and the radial variation is therefore described in terms of the Hankel functions (Abramowitz and Stegun, 1965):

$$H_n^{(2)}(kr) = J_n(kr) - iY_n(kr) \qquad (10.134)$$

Thus the potential ψ_s is of the form

$$\psi_s = \frac{\omega a}{k}T(z,\omega)\,\mathrm{Re}\left|e^{i\omega t}\sum_{n=0}^{\infty}{}'\,B_n\,H_n^{(2)}(kr)\cos(n\theta)\right| \qquad (10.135)$$

At the cylinder boundary, $r = D/2$, there is no normal velocity component, i.e.,

$$\frac{\partial}{\partial r}(\psi_i + \psi_r) = 0 \qquad (10.136)$$

Substitution of (10.133) and (10.135) determines the coefficients

$$B_n = -2i^{-n}\frac{J_n'(kD/2)}{H_n^{(2)'}(kD/2)} \qquad (10.137)$$

At the cylinder boundary the solution can be simplified by use of the relation

$$J_n'(\zeta)H_n^{(2)}(\zeta) - J_n(\zeta)H_n^{(2)'}(\zeta) = \frac{2i}{\pi\zeta} \qquad (10.138)$$

leading to the potential

$$\psi = \frac{\omega a}{k}T(z,\omega)\mathrm{Re}\left|e^{i\omega t}\frac{4}{\pi}\sum_{n=0}^{\infty}{}'\,\frac{i^{-n-1}}{kD/2}\frac{\cos(n\theta)}{H_n^{(2)'}(kD/2)}\right| \qquad (10.139)$$

In the present context the wave force per unit height is of main interest. It is obtained by integrating the excess pressure around the circumference

$$p = -\frac{D}{2}\int_0^{2\pi} p_e \cos\theta \, d\theta = \frac{D}{2}\int_0^{2\pi} \rho\frac{\partial\psi}{\partial t}\cos\theta \, d\theta \qquad (10.140)$$

The result can be expressed in the form

$$p = \frac{\pi}{4}D^2\omega^2 aT(z,\omega)\text{Re}[C_m(kD)e^{i\omega t}] \qquad (10.141)$$

with the complex function

$$C_m(kD) = \frac{4}{\pi}[(kD/2)^2 iH_1^{(2)'}(kD/2)]^{-1} \qquad (10.142)$$

The modulus and phase of the function $C_m(kD)$ are shown in Fig. 10.7.

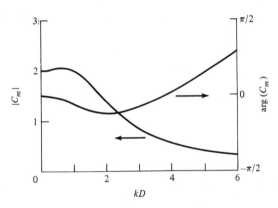

Figure 10.7 Modulus and phase of the coefficient $C_m(kD)$ from (10.142).

In the limit as $kD \to 0$ the coefficient $C_m(kD)$ becomes real and (10.141) reproduces the inertia term of the Morison formula with the theoretical value $C_m(0)=2$. The actual choice of coefficients C_d and C_m for design is subject to considerable uncertainty. A review of the literature may be found in Sarpkaya and Isaacson (1982, Section 5.3). C_d is typically of the order 0.6 to 1.2, while C_m is often found to be around 1.4 to 2.0. Ideally, a design should make use of coefficients determined by a simple theory from a comparable environment.

A final point to be considered is the flexibility of the structure. If the velocity of the structure is not negligible compared with the water particle velocities, the Morison equation should be modified to account for the relative velocity. The modified Morison equation is

$$p = \frac{1}{2}C_d\rho D(u-\dot{x})|u-\dot{x}| + (C_m-1)\rho A(\dot{u}-\ddot{x}) + \rho A\dot{u} \qquad (10.143)$$

D is the diameter of the structural member and A its cross-sectional area. u is the incident normal particle velocity and $\dot{x}$ the corresponding component of the velocity of the structure. The coefficient $(C_m - 1)$ indicates the additional mass contribution created by the presence of the structure. The form of the drag term suggests the introduction of the relative velocity $\dot{r} = u - \dot{x}$, and linearization of the drag term leads to

$$p \approx \frac{1}{2} C_d \rho D \sqrt{8/\pi}\, \sigma_{\dot{r}}(u - \dot{x}) + C_m \rho A \dot{u} - (C_m - 1)\rho A \ddot{x} \qquad (10.144)$$

This procedure was proposed by Malhotra and Penzien (1970). It identifies a hydrodynamic damping term and a hydrodynamic mass term — the terms containing $\dot{x}$ and $\ddot{x}$, respectively. However, it neglects excitation of frequencies around three times the dominating wave frequency created by the nonlinear form of the drag force term. This effect can be retained by a procedure by Taylor and Rajagopalan (1982). The nonlinear factor in the drag term is

$$(u - \dot{x})\,|\,u - \dot{x}\,| \;=\; \begin{vmatrix} u^2 - 2u\dot{x} + \dot{x}^2, & u > \dot{x} \\ -u^2 + 2u\dot{x} - \dot{x}^2, & u < \dot{x} \end{vmatrix} \qquad (10.145)$$

For $\dot{x} \ll \dot{u}$ this can be approximated by

$$(u - \dot{x})\,|\,u - \dot{x}\,| \;\approx\; u\,|\,u\,| - 2\,|\,u\,|\,\dot{x} \qquad (10.146)$$

The second term provides additional damping, and if the coefficient to $\dot{x}$ is replaced by its expectation, the following form is obtained:

$$(u - \dot{x})\,|\,u - \dot{x}\,| \;\approx\; u\,|\,u\,| - \sqrt{8/\pi}\,\sigma_u\,\dot{x} \qquad (10.147)$$

If this term is substituted into (10.143), the resulting formula identifies linear hydrodynamic damping and mass terms similar to (10.144), but now the nonlinearity of the drag term is retained in the form $u\,|\,u\,|$. It follows from the Hermite series expansion (10.123) and the discussion of Section 8.4.2 that the nonlinearity will give increased excitation of frequencies around three times the dominating wave frequency.

10.3 EARTHQUAKE LOADS

In comparison with wind and wave loads the loads from earthquakes are rare events. This leads to greater uncertainty in their description, and also makes it useful to consider earthquake loads conditioned on some seismic event, and then to attempt an independent assessment of the probability of the seismic event. A survey of the ingredients necessary for such a general analysis may be found in Lomnitz and Rosenblueth (1976).   The present discussion is limited to aspects of earthquake loads and nonlinear response effects directly related to the methods developed in Chapter 8.

10.3.1 The Kanai-Tajimi Spectrum and Markov Properties

In this section some features of earthquake excitation of simple structures are illustrated by reference to a linear single-degree-of-freedom model.   The stiffness and damping of the structure refer to the relative displacement u of the structure with respect to the ground, while the inertial forces are induced by the absolute acceleration $\ddot{u} + \ddot{u}_g$, where $\ddot{u}_g$ is the acceleration of the ground. The equation of motion then is

$$\ddot{u} + 2\zeta_0\omega_0\dot{u} + \omega_0^2 u = -\ddot{u}_g \tag{10.148}$$

If the ground acceleration $\ddot{u}_g(t)$ is modeled as a normal process a response analysis can be carried out as described in Chapter 8. However, a simple spectrum due to Kanai (1957) and Tajimi (1960) is often used, and it allows a simple, alternative formulation in terms of a Markov vector process.

The Kanai-Tajimi spectrum may be generated by a simple mechanical model, as illustrated in Fig. 10.8.

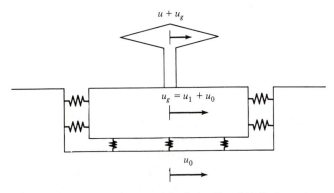

Figure 10.8   Mechanical model of the Kanai-Tajimi spectrum.

Let the ground immediately under the structure be considered as an elastically suspended mass with circular eigenfrequency ω_g and damping ratio ζ_g. The relative displacement u_1 of the local ground with respect to the incident disturbance u_0 then satisfies an equation analogous to (10.148),

$$\ddot{u}_1 + 2\zeta_g\omega_g\dot{u}_1 + \omega_g^2 u_1 = -\ddot{u}_0 \tag{10.149}$$

When the incident acceleration $\ddot{u}_0(t)$ has a white noise spectrum of intensity S_0, the ground acceleration $\ddot{u}_g = \ddot{u}_0 + \ddot{u}_1$ has the spectral density

$$S_g(\omega) = \frac{\omega_g^4 + 4\zeta_g^2\omega_g^2\omega^2}{(\omega_g^2 - \omega^2)^2 + 4\zeta_g^2\omega_g^2\omega^2} S_0 \tag{10.150}$$

This is the spectrum a of Table 8.1. Note that $S_g(0) = S_0 > 0$. The ground acceleration $\ddot{u}_g(t)$ therefore does not correspond to the derivative of a stationary velocity process $\dot{u}_g(t)$. The low-frequency part of the spectrum $S_g(\omega)$ can be removed by use of high-pass filter b of Table 8.1.

The generation of the response from white noise via ordinary differential equations of finite-order allows reformulation of the problem in terms of a Markov vector (see, e.g., Lin, 1976, Chapter 5). First the ground acceleration $\ddot{u}_g$ is eliminated from (10.148) by use of (10.149):

$$\ddot{u} + 2\zeta_0\omega_0\dot{u} + \omega_0^2 u - 2\zeta_g\omega_g\dot{u}_1 - \omega_g^2 u_1 = 0 \tag{10.151}$$

When the vector $\mathbf{Y} = (u, \dot{u}, u_1, \dot{u}_1)$ is introduced, (10.151) and (10.149) can be written in matrix form as

$$\frac{d}{dt}\mathbf{Y} = \mathbf{g}\mathbf{Y} + \mathbf{Q} \tag{10.152}$$

where $\mathbf{g}$ is a constant matrix

$$\mathbf{g} = \begin{vmatrix} 0 & 1 & 0 & 0 \\ -\omega_0^2 & -2\zeta_0\omega_0 & \omega_g^2 & 2\zeta_g\omega_g \\ 0 & 0 & 0 & 1 \\ 0 & 0 & -\omega_g^2 & -2\zeta_g\omega_g \end{vmatrix} \tag{10.153}$$

and $\mathbf{Q}^T = (0,0,0,-\ddot{u}_0)$ is a vector containing the excitation process. The vector $\mathbf{Y}(t)$ describes the processes $u(t)$ and $u_1(t)$ in the phase plane, and as the increments depend only on the current position and the white noise process $\mathbf{Q}(t)$, $\mathbf{Y}(t)$ is a Markov vector process.

In the particular case of a white noise process $\mathbf{Q}(t)$, a differential equation for the covariance matrix $\mathbf{C} = \mathrm{E}[\mathbf{Y}\mathbf{Y}^T]$ can be obtained directly from (10.152). Multiplying (10.152) with $\mathbf{Y}^T$ from the right, adding the transposed equation, and taking the expectation yields the equation

$$\frac{d}{dt}\mathbf{C} = \mathbf{g}\mathbf{C} + \mathbf{C}\mathbf{g}^T + \mathrm{E}[\mathbf{Q}\mathbf{Y}^T + \mathbf{Y}\mathbf{Q}^T] \tag{10.154}$$

The last term is evaluated by integration of (10.152) from an immediately preceding time $t - \tau$.

$$\mathbf{Y}(t) = \int_{t-\tau}^{t} \mathbf{g}\,\mathbf{Y}\,d\tau + \int_{t-\tau}^{t} \mathbf{Q}\,d\tau + \mathbf{Y}(t-\tau) \qquad (10.155)$$

$\mathbf{Q}(t)$ and $\mathbf{Y}(t-\tau)$ are uncorrelated and after multiplication with $\mathbf{Q}(t)^T$ the expectation is

$$E\,[\mathbf{Y}(t)\mathbf{Q}(t)^T] = \lim_{\tau \to 0} E\left[\int_{t-\tau}^{t}\mathbf{g}\mathbf{Y}d\tau\mathbf{Q}(t)^T\right] \qquad (10.156)$$

$$+ \lim_{\tau \to 0} E\left[\int_{t-\tau}^{t}\mathbf{Q}d\tau\mathbf{Q}(t)^T\right] = \frac{1}{2}\int_{-\infty}^{\infty} C^Q(s)ds$$

The first term is of order $O(\tau)$ and vanishes, while the second contributes half of the full integral of the autocorrelation function of $\mathbf{Q}(t)$. As a result, the differential equation is

$$\frac{d}{dt}\mathbf{C} = \mathbf{g}\,\mathbf{C} + \mathbf{C}\mathbf{g}^T + \mathbf{B}(t) \qquad (10.157)$$

The matrix $\mathbf{B}(t)$ contains the intensities of the white noise excitations. In the present case the only nonvanishing component is B_{44}. If the incident acceleration $\ddot{u}_0(t)$ is represented as a time-modulated white noise process

$$\ddot{u}_0(t) = \psi(t)\,n(t) \qquad (10.158)$$

with modulation function $\psi(t)$ and a white noise process $n(t)$ with spectral density S_0, then $B_{44} = 2\pi\psi(t)^2 S_0$.

In the case of stationary excitation $d\mathbf{C}/dt = \mathbf{0}$, and (10.157) is a system of linear equations for the covariance matrix $\mathbf{C}$. Similar results are easily obtained by the technique of Chapter 8. However, in earthquake applications the response is nonstationary, and (10.157) may be integrated directly numerically.

A number of modulation functions $\psi(t)$ have been proposed. Thus Shinozuka and Sato (1967) proposed a combination of exponentials

$$\psi(t) = e^{-\alpha t} - e^{-\beta t} \qquad (10.159)$$

while Amin and Ang (1969) used a representation with an initial phase, a constant level, and a final phase:

$$\psi(t) = \begin{vmatrix} t/t_1, & 0 \leqslant t \leqslant t_1 \\ 1, & t_1 < t < t_2 \\ e^{-\alpha(t-t_2)}, & t_2 \leqslant t \end{vmatrix} \qquad (10.160)$$

Other forms have been proposed by Iyengar and Iyengar (1969) and Ruiz and Penzien (1971), but differences between individual earthquakes makes it difficult to select a generally most appropriate form of the modulation function. Also, the parameters ω_g and ζ_g show some variation, depending on the particular earthquake and the soil properties of the point of observation. Thus Tajimi (1960) gives $\omega_g/2\pi = 1$ to 5 Hz, $\zeta_g = 0.3$, while Housner and Jennings (1964), Amin and Ang (1969) and Ruiz and Penzien (1971) obtain maximum ground acceleration frequencies in the range $\omega_g/2\pi = 2.5$ to 5 Hz and a somewhat higher equivalent ground damping ratio $\zeta_g = 0.5$ to 0.65.

10.3.2 Hysteretic Response

Many structures exhibit highly nonlinear and hysteretic response when exposed to severe earthquake loading, and it is important to include this in the response analysis. A characteristic force-displacement history is shown in Fig. 10.9.

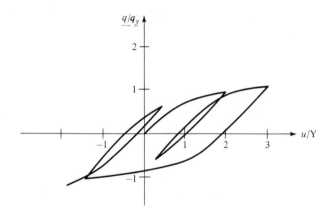

Figure 10.9 Typical hysteretic response.

Following Wen (1976, 1980), the restoring force can be represented in the form

$$f = g(u, \dot{u}) + z \qquad (10.161)$$

where $g(u, \dot{u})$ is an explicit function of u and $\dot{u}$ constituting the nonhysteretic part, while the hysteretic part z depends on u and $\dot{u}$ by means of an incremental relation. A smooth, hysteretic dependence is provided by the relation

$$\dot{z} = \dot{u}[A - (\beta + \gamma \operatorname{sgn}(\dot{u}z)) \mid z \mid^n] \qquad (10.162)$$

where γ, β, n, and A are parameters. When both z and $\dot{u}$ are positive, z^n has the limiting value $A/(\gamma + \beta)$.

In the following the equivalent linearization technique described in Section 8.4.1 is illustrated for linear damping and restoring force $g(u,\dot{u})$ and $n = 1$. The equation of motion can then be written in the form

$$\ddot{u} + 2\zeta_0\omega_0\,\dot{u} + \alpha\omega_0^2 u + (1-\alpha)\omega_0^2 z = f(t) \qquad (10.163)$$

For $A = 1$, ω_0 is the eigenfrequency for small values of z, and α represents the fraction of stiffness left for large u. The equilibrium equation (10.163) is already in linear form, and the constitutive relation (10.162) is now replaced by the linearized form

$$\dot{z} + C\dot{u} + Kz = 0 \qquad (10.164)$$

where, according to (8.114) and (8.115)

$$C = E[-\frac{\partial \dot{z}}{\partial \dot{u}}] = \beta E[\,|z\,|\,] + \gamma E[z\,\text{sgn}\,\dot{u}] - A \qquad (10.165)$$

$$K = E[-\frac{\partial \dot{z}}{\partial z}] = \beta E[\dot{u}\,\text{sgn}\,z] + \gamma E[\,|\dot{u}\,|\,] \qquad (10.166)$$

With the assumption that $\dot{u}$ and z are joint-normal, zero-mean variables, the expectations can be calculated — e.g., by use of conditional expectations — resulting in

$$C = \sqrt{2/\pi}\,\sigma_z\,(\beta + \gamma\rho_{\dot{u}z}) - A \qquad (10.167)$$

$$K = \sqrt{2/\pi}\,\sigma_{\dot{u}}\,(\beta\rho_{\dot{u}z} + \gamma) \qquad (10.168)$$

where σ_z and $\sigma_{\dot{u}}$ are the standard deviations and $\rho_{\dot{u}z}$ the correlation coefficient. It should be noted that in the normalized form (10.163) of the equations of motion, the expected rate of hysteretic energy dissipation is

$$\dot{\epsilon} = (1-\alpha)\omega_0^2\,E[\dot{u}z] = (1-\alpha)\omega_0^2\sigma_{\dot{u}}\sigma_z\rho_{\dot{u}z} \qquad (10.169)$$

If $f(t) = -\ddot{u}_g(t)$ is the time-modulated white noise process (10.158), the linearized system of differential equations (10.163) and (10.164) can be rearranged in the Markov vector form (10.152) with $\mathbf{Y}^T = (u,\dot{u},z)$, $\mathbf{Q}^T = (0,-\ddot{u}_g,0)$ and the matrix

$$\mathbf{g} = \begin{vmatrix} 0 & 1 & 0 \\ -\alpha\omega_0^2 & -2\zeta_0\omega_0 & -(1-\alpha)\omega_0^2 \\ 0 & -C & -K \end{vmatrix} \qquad (10.170)$$

The only nonvanishing element of $\mathbf{B}$ is $B_{22} = 2\pi\psi(t)^2 S_0$. Thus the covariance matrix $\mathbf{C}$ is determined by the differential equation (10.157). In the present case the matrix $\mathbf{g}$ depends on the elements of $\mathbf{C}$ via the linearization parameters C and K. The differential equation is therefore nonlinear and must be solved iteratively. The procedure is easily modified to ground accelerations with the Kanai-Tajimi spectrum by including u_1 and $\dot{u}_1$ in the Markov vector $\mathbf{Y}$.

Examples covering a wide parameter range have been calculated by Wen (1980), and the method has been extended to degrading systems, in which the hysteretic limiting parameter A decreases with the hysteretic energy dissipation ε (Baber and Wen, 1981). All components of the covariance matrix $\mathbf{C}$ were found to agree well with numerical simulation as well as the analog simulation by Iwan and Lutes (1968) of a nearly bilinear hysteretic system. The reason for the success of the linearization method in this highly nonlinear problem is illustrated in Fig. 10.10, showing the frequency content of the stationary response $u(t)$ to white noise excitation.

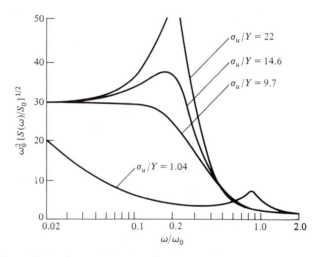

Figure 10.10 Spectral density of a hysteretic structure excited by white noise, from Baber and Wen (1981).

The system has no viscous damping, $\zeta_0 = 0$, $\beta = \gamma = 0.5$, $A = 1$, and $\alpha = 1/21$. Y is the equivalent elastic displacement for a force corresponding to yield. The figure shows a transition with increasing excitation intensity from a spectrum with a peak around $\omega = \omega_0$, via a wide-band spectrum, to a limiting narrow-band spectrum with a peak at $\omega = \alpha^{1/2}\omega_0 = 0.22\omega_0$. This transition agrees with results from analog simulation by Iwan and Lutes (1968) and numerical solution of the associated Fokker-Planck equation (Wen, 1976). In particular, the presence of a large low-frequency contribution for the technically important level of excitation somewhat beyond yield, $\sigma_u/Y \approx 1-2$, should be noted.

10.4 TRAFFIC LOAD MODELING

This section presents a simple model for traffic loading due to Ditlevsen (1971). The model is particularly relevant for loads on long-span bridges. Only the loading in a single lane is modeled, and the load from a vehicle is assumed constant in time. More comprehensive load models are proposed by, e.g., Harman and Davenport (1979) and Larrabee (1979).

10.4.1 Load Model

The model is based on the assumption that all vehicles travel at the same speed v. The load pattern as shown in Fig. 10.11 does therefore not change but is translated with speed v. This assumption is similar to the assumption of frozen turbulence in Section 10.1. The load from a vehicle is described as a uniformly distributed load P_i over the length of the vehicle L_i. The distance between vehicles i and $i+1$ is denoted D_i. It is assumed that triples (P_i, L_i, D_i) are mutually independent and identically distributed. For each vehicle it is further assumed that D_i is independent of (P_i, L_i). The load intensity $Q(x,t)$ is assumed to form a homogeneous and ergodic process. The computations of the mean value and covariance function are demonstrated and it is shown how these statistics, together with the extreme-value theory for Gaussian processes in Chapter 7, can be used to determine extreme values of linear load effects.

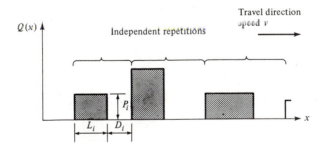

Figure 10.11   Traffic load pattern.

Since the traffic travels at constant speed, the load intensity depends on x and t solely in the combination $x - vt$. The mean value can therefore be expressed as

$$E[Q(x,t)] = E[Q(x-vt)] = \mu_Q(x-vt) = \mu_Q \qquad (10.171)$$

and the covariance function as

$$\text{Cov}\,[Q(x,t),Q(x+\xi,t+\tau)] = \text{Cov}\,[Q(x-vt),Q(x+\xi-v(t+\tau))] \quad (10.172)$$

$$= C_{QQ}(\xi-v\tau)$$

Besides the expected value μ_Q, it is thus sufficient to determine the one-dimensional covariance function, $C_{QQ}(x)$, where the position coordinate is chosen for simplicity. This corresponds to analysis of the traffic stream at $t=0$.

10.4.2 Mean Value and Covariance Function

The load intensity process is ergodic and to determine the mean value and covariance function a random point x must first be selected. The probability that the load intensity is nonzero at x is

$$P(Q(x)>0) = \frac{\text{E}[L]}{\text{E}[L]+\text{E}[D]} \quad (10.173)$$

This follows from the ergodicity of the load process since the right-hand side is the fraction of the x-axis occupied by vehicles. If the load intensity is nonzero at x, an interval covered by a vehicle has been selected. The length of this interval is denoted by L_1, as illustrated in Fig. 10.12.

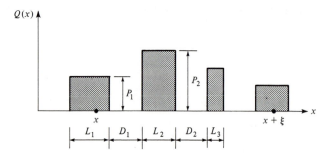

Figure 10.12 Traffic load pattern with vehicle at randomly selected point x.

Since x is selected at random, the distribution of L_1 is not the same as the distribution of the length of a random load interval L. Rather, it follows from an extension of (7.38) that the joint distribution of (P_1,L_1) is given by

$$f_{P_1 L_1}(p,l) = \frac{l\,f_{PL}(p,l)}{\text{E}[L]} \quad (10.174)$$

where $f_{PL}(p,l)$ is the joint distribution of (P,L) for an arbitrary interval. The expected value of P_1 follows from (10.174) as

$$E[P_1] = \int_0^\infty p\, f_{P_1}(p)dp = \int_0^\infty \int_0^\infty \frac{p\, l\, f_{PL}(p,l)}{E[L]} dp\, dl \qquad (10.175)$$

$$= \frac{E[PL]}{E[L]}$$

From this the expected value of $Q(x)$ follows directly as

$$E[Q(x)] = E[Q(x) \mid Q(x)>0]\, P(Q(x)>0) \qquad (10.176)$$

$$= E[P_1]\, P(Q(x)>0) = \frac{E[PL]}{E[L]+E[D]}$$

The expected value of $Q(x)^2$ is similarly

$$E[Q(x)^2] = \frac{E[P^2 L]}{E[L]+E[D]} \qquad (10.177)$$

leading to the variance of $Q(x)$:

$$\text{Var}[Q(x)] = \frac{E[P^2 L]}{E[L]+E[D]} - \left(\frac{E[PL]}{E[L]+E[D]}\right)^2 \qquad (10.178)$$

Computation of the covariance function is somewhat more complicated. Conditioned that the load intensity is nonzero at x, $Q(x)Q(x+\xi)$ follows as (see Fig. 10.12),

$$Q(x)Q(x+\xi) \mid Q(x)>0 = \qquad (10.179)$$

$$
\begin{cases}
P_1^2 & 0 \leq \xi < L_1' \\
0 & L_1' \leq \xi < L_1'+D_1 \\
P_1 P_2 & L_1'+D_1 \leq \xi < L_1'+L_2+D_1 \\
0 & L_1'+L_2+D_1 \leq \xi < L_1'+L_2+D_1+D_2 \\
\vdots & \vdots \\
P_1 P_{i+1} & L_1'+\sum_{j=2}^{i}(L_j+D_j) \leq \xi < L_1'+\sum_{j=2}^{i}(L_j+D_j)+L_{i+1} \\
0 & L_1'+\sum_{j=2}^{i+1}(L_j+D_j)+L_{i+1} \leq \xi < L_1'+\sum_{j=2}^{i+1}(L_j+D_j) \\
\vdots & \vdots
\end{cases}
$$

Only $\xi>0$ is considered here, since the covariance function is an even function. The probability of $L_1'>\xi$ follows from the waiting-time problem of a renewal process, and by integration of (7.41),

$$P(L_1' > \xi \mid Q(x) > 0) = \int_\xi^\infty \frac{1 - F_L(z)}{E[L]} \, dz \qquad (10.180)$$

Random variables V_i and W_i are defined to simplify (10.179):

$$V_i = L_1' + \sum_{j=2}^i L_1 \qquad (10.181)$$

$$W_i = \sum_{j=1}^i D_i \qquad (10.182)$$

Under the additional assumption that in each pair (P_i, L_i) the two random variables are independent, the expected value of $Q(x)Q(x+\xi)$ now follows directly from (10.179) as

$$E[Q(x)Q(x+\xi)] = E[Q(x)Q(x+\xi) \mid Q(x) > 0] \, P(Q(x) > 0) \quad (10.183)$$

$$= \left| E[P_1^2] \, P(L_1' > \xi) + \sum_{i=1}^\infty E[P_1 P_{i+1}] \right.$$

$$\left. \times P(V_i + W_i \leqslant \xi < V_{i+1} + W_i) \right| P(Q(x) > 0) \qquad (10.183)$$

$$= \left| E[P^2] \, P(L_1' > \xi) + E[P]^2 \right.$$

$$\left. \times \sum_{i=1}^\infty P(V_i + W_i \leqslant \xi < V_{i+1} + W_i) \right| \frac{E[L]}{E[L] + E[D]}$$

since all P_i including P_1 are now identically distributed. The covariance function of $Q(x,t)$ is

$$C_{QQ}(\xi) = E[Q(x)Q(x+\xi)] - E[Q(x)]^2 \qquad (10.184)$$

$$= \left| E[P^2] \, P(L_1' > \xi) + E[P]^2 \sum_{i=1}^\infty g_i(\xi) \right| \frac{E[L]}{E[L] + E[D]}$$

$$- \left(\frac{E[P]E[L]}{E[L] + E[D]} \right)^2$$

where $g_i(\xi)$ has been introduced for brevity as

$$g_i(\xi) = P(V_i + W_i \leqslant \xi < V_{i+1} + W_i) \qquad (10.185)$$

The covariance function is computed for three different cases of traffic pattern in Example 10.4.

Example 10.4

As the first case, assume that the length of a vehicle is a constant a and the distance between two vehicles is a constant d. From (10.180) follows

$$P(L_1' > \xi) = \int_{\xi}^{\infty} \frac{1 - H(z - l)}{l} dz = \frac{l - \xi}{l}$$

where $H(\ )$ is Heaviside's function. $g_i(\xi)$ becomes

$$g_i(\xi) = \begin{vmatrix} \dfrac{\xi + l - i(l + d)}{l}, & i(l + d) - l \leqslant \xi < i(l + d) \\ \dfrac{l - \xi + i(l + d)}{l}, & i(l + d) \leqslant \xi < i(l + d) + l \\ 0, & \text{otherwise} \end{vmatrix}$$

Figure 10.13 shows the autocorrelation function for $d/l = 3$ for various ratios σ_P/μ_P.

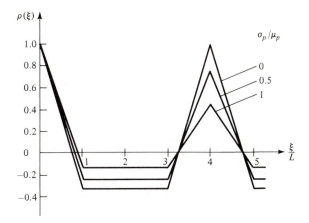

Figure 10.13 Autocorrelation function for traffic load pattern with fixed vehicle length and distance

In the second case the length of a vehicle is assumed to be exponentially distributed and the distance between two vehicles is a constant d. In this case L_1' and an arbitrary L_i are identically distributed, as discussed in Section 7.3.2. V_i is the sum of i independent and identical exponentially distributed random variables. The result is a gamma-distributed random variable, and with appropriate change in notation, it follows from (7.21) that

$$f_{V_i}(v) = \frac{1}{\mu_L(i - 1)!} \left(\frac{v}{\mu_L}\right)^{i-1} \exp\left(-\frac{v}{\mu_L}\right), \quad v > 0$$

$g_i(\xi)$ is computed as

$$g_i(\xi) = \begin{vmatrix} 0, & \xi < id \\ \dfrac{1}{i!} \left(\dfrac{\xi - id}{\mu_L}\right)^i \exp\left(-\dfrac{\xi - id}{\mu_L}\right), & \xi \geqslant id \end{vmatrix}$$

Figure 10.14 shows the autocorrelation function for $d/\mu_L = 3$ for various ratios σ_P/μ_P.

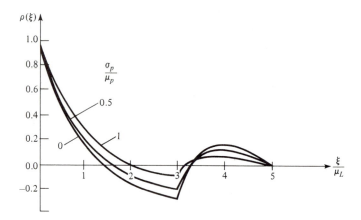

Figure 10.14 Autocorrelation function for traffic load pattern with random vehicle length and fixed distance.

In the third case the length of a vehicle is assumed to be exponentially distributed and the distance between two vehicles is also assumed to be exponentially distributed. V_i is then distributed as above and the distribution of W_i is similarly

$$f_{W_i}(w) = \frac{1}{\mu_D(i-1)!} \left(\frac{w}{\mu_D}\right)^{i-1} \exp\left(-\frac{w}{\mu_D}\right), \quad w > 0$$

Straightforward although rather lengthy calculations lead to the result for $g_i(\xi)$ (Madsen, 1979):

$$g_i(\xi) = \begin{cases} \dfrac{1}{(2i)!} \left(\dfrac{\xi}{\mu_L}\right)^{2i} \exp\left(-\dfrac{\xi}{\mu_L}\right), & \mu_L = \mu_D \\[2em] \exp\left(-\dfrac{\xi}{\mu_L}\right) \displaystyle\sum_{k=0}^{i} \dfrac{(\mu_L \mu_D)^k}{(\mu_L - \mu_D)^{i+k}} \xi^{i-k} (-1)^k \dfrac{(i+k-1)!}{k!(i-k)!(i-1)!} \\ \quad \times \left[1 - \exp\left(-\xi\dfrac{\mu_L - \mu_D}{\mu_L \mu_D}\right) \displaystyle\sum_{j=0}^{i+k-1} \dfrac{\xi^j}{j!} \left(\dfrac{\mu_L - \mu_D}{\mu_L \mu_D}\right)^j\right], & \mu_L \neq \mu_D \end{cases}$$

The autocorrelation function for this case with $\mu_D/\mu_L = 3$ is shown in Fig. 10.15 for various ratios σ_P/μ_P.

It is difficult, and unnecessary, to obtain a general formula for $g_i(\xi)$. What is needed is a formula for the sum $\sum_{i=1}^{\infty} g_i(\xi)$, and this can be obtained by separation of the two limits in (10.185) and use of a recurrence formula. Introduce the distribution function

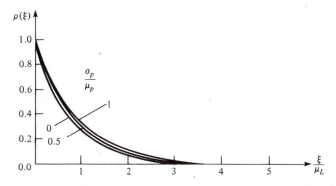

Figure 10.15 Autocorrelation function for traffic load pattern with random vehicle length and distance.

$$\Psi_i(\xi) \;=\; P(V_i + W_i \leq \xi) \tag{10.186}$$

whereby the definition (10.185) of $g_i(\xi)$ takes the form

$$g_i(\xi) \;=\; \Psi_i(\xi) - \int_{F_L=0}^{1}\Psi_i(\xi-l)dF_L(l) \tag{10.187}$$

It then follows by summation that

$$\sum_{i=1}^{\infty} g_i(\xi) \;=\; \Psi(\xi) - \int_{l=0}^{\infty}\Psi(\xi-l)dF_L(l) \tag{10.188}$$

where $\Psi(\xi)$ is the sum $\sum_{i=1}^{\infty}\Psi_i(\xi)$. This sum is evaluated by recurrence.

$$\Psi(\xi) \;=\; \sum_{i=1}^{\infty}\Psi_i(\xi) \;=\; \Psi_1(\xi) + \sum_{i=2}^{\infty} P(V_{i-1} + W_{i-1} \leq \xi - L_i - D_i) \tag{10.189}$$

$$=\; \Psi_1(\xi) + \int_{l=0}^{\xi}\int_{d=0}^{\xi-l}\Psi(\xi-l-d)dF_L(l)dF_D(d)$$

It is easily seen that the values of $\Psi(\xi)$ necessary to evaluate the integral all correspond to arguments less than or equal to ξ, and thus $\Psi(\xi)$ can be evaluated by a recursive numerical scheme from (10.189), provided that $\Psi_1(\xi)$ is known. It follows from (10.186) that $\Psi_1(\xi)$ is the distribution function for the sum $L_1' + D_1$, and therefore

$$\Psi_1(\xi) \;=\; \frac{1}{E[L]+E[D]} \int_{l=0}^{\xi} F_D(\xi-l)(1-F_L(l))dl \tag{10.190}$$

The function $\Psi(\xi)$ satisfies $\Psi(\xi)=0$ for $\xi<0$, and the equations (10.188) to (10.190) can therefore also be reformulated in terms of Laplace transforms.

10.4.3 Linear Load Effects

The theory of extremes of stationary Gaussian processes described in Chapter 7 can be applied to compute the extreme-value distribution of various load effects in bridges. The assumption of a Gaussian process can be reasonable for a long-span bridge where numerous loads are superimposed but is most likely not for a short-span bridge. A linear load effect $S(t)$ is expressed as

$$S(t) = \int_0^a i_S(x)Q(x,t)dx \qquad (10.191)$$

where a is the length of the bridge and $i_S(x)$ is the influence function. The approximation to the extreme-value distribution of $S(t)$ [(7.131)], requires knowledge of the expected value and variance of $S(t)$ and the variance of the time derivative $\dot{S}(t)$. The expected value and covariance function of $Q(x,t)$ are

$$E[Q(x,t)] = \mu_Q \qquad (10.192)$$

$$\text{Cov}[Q(x,t),Q(y,s)] = C_{QQ}(y - x - v(s - t)) \qquad (10.193)$$

The expected value of $S(t)$ is therefore

$$E[S(t)] = E\left[\int_0^a i_S(x)Q(x,t)dx\right] = \int_0^a i_S(x)E[Q(x,t)]dx \qquad (10.194)$$

$$= \mu_Q \int_0^a i_S(x)dx$$

and the covariance function is similarly

$$C_{SS}(\tau) = \text{Cov}[S(t),S(t+\tau)] \qquad (10.195)$$

$$= \text{Cov}\left[\int_0^a i_S(x)Q(x,t)dx, \int_0^a i_S(y)Q(y,t+\tau)dy\right]$$

$$= \int_0^a \int_0^a i_S(x)i_S(y)\text{Cov}[Q(x,t),Q(y,t+\tau)]dxdy$$

$$= \int_0^a \int_0^a i_S(x)i_S(y)C_{QQ}(y - x - v\tau)dxdy$$

From this expression the variance of $S(t)$ is found by setting $\tau=0$. The variance of the time derivative process $\dot{S}(t)$ is

$$\text{Var}[\dot{S}(t)] = C_{\dot{S}\dot{S}}(0) = -\frac{d^2}{d\tau^2}C_{SS}(\tau)\,|_{\tau=0} \qquad (10.196)$$

If the second derivative $C_{QQ}''(x)$ is continuous at $x=0$ it is continuous for all x (Miller, 1974). The variance of $\dot{S}(t)$ can then be expressed as

$$\text{Var}[\dot{S}(t)] = -v^2 \int_0^a \int_0^a i_S(x)\, i_S(y)\, C_{QQ}''(y - x)\, dxdy \qquad (10.197)$$

Examples of $C_{QQ}(x)$ shown in Figs. 10.13 to 10.15 do not have a second-order derivative at $x = 0$. In this case the variance of $\dot{S}(t)$ can still be computed if $i_S'(x)$ is piecewise continuous. The first step in the calculation is to write

$$\text{Var}[\dot{S}(t)] = -\frac{d^2}{d\tau^2}\left|\int_0^a\int_0^a i_S(x)i_S(y)C_{QQ}(y-x-v\tau)dxdy\right| \tag{10.198}$$

$$= -\frac{d^2}{d\tau^2}\left|\int_0^a\int_{v\tau}^{a+v\tau} i_S(y)i_S(x-v\tau)C_{QQ}(y-x)dxdy\right|$$

$$= -\frac{d}{d\tau}\left\{\begin{array}{l} -v\int_0^a\int_{v\tau}^{a+v\tau} i_S(y)i_S'(x-v\tau)C_{QQ}(y-x)dxdy \\ +v\int_0^a i_S(y)i_S(a)C_{QQ}(y-a-v\tau)dy \\ -v\int_0^a i_S(y)i_S(0)C_{QQ}(y-v\tau)dy \end{array}\right\}$$

$$= -\frac{d}{d\tau}\left\{\begin{array}{l} -v\int_0^a\int_0^a i_S(y)i_S'(x)C_{QQ}(y-x-v\tau)dxdy \\ +vi_S(a)\int_0^a i_S(y)C_{QQ}(y-a-v\tau)dy \\ -vi_S(0)\int_0^a i_S(y)C_{QQ}(y-v\tau)dy \end{array}\right\}$$

A similar change of the y-variable in the integration leads to the final result:

$$\text{Var}[\dot{S}(t)] = v^2\left|\int_0^a\int_0^a i_S'(x)i_S'(y)C_{QQ}(y-x)dxdy\right. \tag{10.199}$$

$$- 2i_S(a)\int_0^a i_S'(x)C_{QQ}(a-x)dx + 2i_S(0)\int_0^a i_S'(x)C_{QQ}(x)dx$$

$$\left. + (i_S(0)^2 + i_S(a)^2)C_{QQ}(0) - 2i_S(0)i_S(a)C_{QQ}(a)\right|$$

An approximation to the extreme-value distribution of $S(t)$ is then given from (7.66) and (7.131) as

$$P\left(\max_{0\leq t\leq T} S(t)\leq \xi\right) \approx F_S(\xi)\exp(-v_\xi^S T) \tag{10.200}$$

$$= \Phi\left(\frac{\xi-E[S(t)]}{\sqrt{\text{Var}[S(t)]}}\right)\exp\left|-\frac{1}{2\pi}\left(\frac{\text{Var}[\dot{S}(t)]}{\text{Var}[S(t)]}\right)^{1/2}T\exp\left(-\frac{(\xi-E[S(t)])^2}{2\text{Var}[S(t)]}\right)\right|$$

10.5 LIVE-LOAD MODELING

Stochastic models for the variation of live loads in buildings have gained wide interest in connection with reformulation of building codes. In these codes live loads are usually prescribed as uniformly distributed loads with intensity depending on, e.g., the use of the building, the number of floors, and the area of the rooms. The real load pattern is much more complicated, but the uniform load is chosen for simplicity. The stochastic load models presented here are rather crude but are believed to capture the essential features of live load variation in both time and space.

Live loads in buildings include those loads imposed on the structure by the weight of furniture, light machines, and other equipment, and the load from the people occupying the building. The load from heavy machinery is not included, but in a design situation these loads are usually known in advance and accounted for separately. The live load is modeled as the sum of two load processes, designated the *sustained load* (or ordinary load) and the *transient load* (or extraordinary load). The sustained load is always acting and includes the weight of furniture and normal occupants. It does not vary greatly except at each major change of occupancy. The transient load represents rare loads of short duration, e.g., during crowding at parties or stacking of furniture for painting and remodeling.

Two spatial load models are presented, of which one is applied to the sustained load and the other to the transient load. Models for the time variation of the two loads are then presented and the load combination problem is addressed. Many surveys have been conducted to obtain estimates of the model parameters. Most surveys are for offices and residences, but surveys have also been performed for schools, hotels, hospitals, and warehouses. No results from these surveys are given here but can be found in, e.g., Sentler (1975), Corotis and Doshi (1977), and Madsen and Turkstra (1979).

10.5.1 Spatial Load Field Modeling

Analogous to the two usual types of load representation, distributed loads and single force loads, there are two random field representations. Corresponding to the concept of a deterministic, distributed load, the following model for the load intensity $Q(x,y)$ of the sustained load at a fixed point in time has been suggested by Peir and Cornell (1973). The load intensity in a particular room on a particular floor in a particular building is represented by

$$Q(x,y) = \mu_Q + L_B + L_F + L_R + D(x,y) \qquad (10.201)$$

where μ_Q is the overall mean load intensity for the class of buildings, L_B a mean-zero random variable representing the deviation of the building mean load intensity from μ_Q, L_F a mean-zero random variable representing the deviation of the floor mean load intensity from $\mu_Q + L_B$, L_R a mean-zero random variable representing the deviation of the mean room load intensity from $\mu_Q + L_B + L_F$, and

$D(x,y)$ a mean-zero random field which represents the deviations of the load intensity from $\mu_Q + L_B + L_F + L_R$. The random variables L_B, L_F, and L_R thus represent systematic variations for the particular house, floor, and room. These random variables are assumed mutually independent and independent of $D(x,y)$.

A linear load effect, S, is obtained by integrating the load intensity over the influence surface:

$$S = \int_A i_S(x,y)\, Q(x,y)dA \qquad (10.202)$$

where A is the influence area and $i_S(x,y)$ is the influence function. Figure 10.16 shows the influence area and influence function for three load effects in a beam-column supported floor.

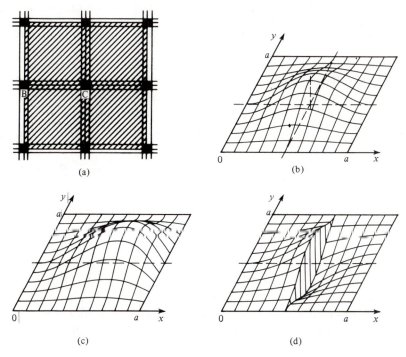

Figure 10.16 Beam column supported floor with influence functions: (a) beam column system; (b) influence surface for axial column force; (c) influence surface for beam end moment; (d) influence surface for shear at beam center, from Sentler (1975).

The expected value of S is

$$E[S] = E\left[\int_A i_S(x,y)Q(x,y)dA\right] = \int_A i_S(x,y)E[Q(x,y)]dA \qquad (10.203)$$

$$= \mu_Q \int_A i_S(x,y)dA$$

Assuming that the influence area for S is within one room, the variance is, similarly,

$$Var[S] = Cov \left[\int_A i_S(x_1,y_1)Q(x_1,y_1)dA_1, \int_A i_S(x_2,y_2)Q(x_2,y_2)dA_2 \right] \quad (10.204)$$

$$= \int_A \int_A i_S(x_1,y_1)i_S(x_2,y_2)Cov[Q(x_1,y_1),Q(x_2,y_2)]dA_1dA_2$$

$$= \int_A \int_A i_S(x_1,y_1)i_S(x_2,y_2)$$

$$\times (\sigma_B^2 + \sigma_F^2 + \sigma_R^2 + Cov[D(x_1,y_1),D(x_2,y_2)])dA_1dA_2$$

$$= (\sigma_B^2 + \sigma_F^2 + \sigma_R^2)\left(\int_A i_S(x,y)dA \right)^2$$

$$+ \int_A \int_A i_S(x_1,y_1)i_S(x_2,y_2)Cov[D(x_1,y_1),D(x_2,y_2)]dA_1dA_2$$

were σ_B^2, σ_F^2, and σ_R^2 are the variances of L_B, L_F, and L_R, respectively.

The random field $D(x,y)$ is assumed homogeneous, thus neglecting effects such as a higher concentration of furniture near walls. No specific choice of covariance function for $D(x,y)$ is theoretically well founded, but common choices are

$$Cov[D(x,y),D(x+u,y+v)] \quad (10.205)$$

$$= C_D(u,v) = \begin{vmatrix} \sigma_D^2\delta(u)\delta(v), & R = \dfrac{1}{\sqrt{\pi}} \\[2mm] \sigma_D^2\exp\left(-\dfrac{\sqrt{u^2+v^2}}{r}\right), & R = r\sqrt{2} \\[2mm] \sigma_D^2\exp\left(-\dfrac{u^2+v^2}{r^2}\right), & R = r \\[2mm] \sigma_D^2\exp\left(-\dfrac{|u|}{r}\right)\exp\left(-\dfrac{|v|}{r}\right), & R = \dfrac{2}{\sqrt{\pi}}r \end{vmatrix}$$

R is the *correlation radius* or *scale of fluctuation*, which for a homogeneous random field is defined from the equation (Vanmarcke, 1983)

$$\pi R^2 = \int_{-\infty}^{\infty}\int_{-\infty}^{\infty} \frac{Cov[D(x,y),D(x+u,y+v)]}{Cov[D(x,y),D(x,y)]} dudv \quad (10.206)$$

The first choice corresponds to a white noise random field, the second and third choices result in isotropic random fields, and the fourth choice gives an anisotropic random field.

The load effect S_L from a uniformly distributed random load with intensity L is

$$S_L = \int_A i_S(x,y)LdA = L\int_A i_S(x,y)dA \quad (10.207)$$

The mean value and variance of S_L are

$$E[S_L] = \mu_L \int_A i_S(x,y)dA \qquad (10.208)$$

$$Var[S_L] = \sigma_L^2 \left(\int_A i_S(x,y)dA \right)^2 \qquad (10.209)$$

where μ_L and σ_L^2 are the mean and variance of L, respectively.

The equivalent uniformly distributed load $EUDL_S$ is defined as the intensity of the uniformly distributed load that produces the same load effect as the actual load process. The mean value and variance of $EUDL_S$ for a load effect S with influence area within one room follow from (10.203), (10.204), (10.208) and (10.209):

$$E[EUDL_S] = \mu_Q \qquad (10.210)$$

$$Var[EUDL_S] = \sigma_B^2 + \sigma_F^2 + \sigma_R^2 + k_S\sigma_D^2 \qquad (10.211)$$

where the variance reduction factor k_S is defined as

$$k_S = \min\left\{ 1, \frac{\int_A\int_A i_S(x_1,y_1)i_S(x_2,y_2)C_D(x_2-x_1,y_2-y_1)dA_1dA_2}{\sigma_D^2(\int_A i_S(x,y)dA)^2} \right\} \qquad (10.212)$$

k_S is less than or equal to 1, since $C_D(,)$ is less than the variance of the process. This has effect for small areas for the white noise random field in (10.205). Figure 10.17 shows an example of the variance reduction factor for a square floor (Madsen and Turkstra, 1979). Plotted on the abscissa is the square root of the area normalized with respect to the correlation radius. A comparison in terms of standard deviations, i.e., the square root of the ordinate, shows that the reduction is rather insensitive to the detail of the correlation structure but is well determined simply by the correlation radius. The reduction is also seen to be approximately the same for all three load effects considered.

The model implications described above have been justified from load survey data and are included in building codes. In such codes the $EUDL$ generally also decreases with the number of floors contributing to a load effect. To demonstrate this, the axial force in a column supporting n floors is considered. Assuming that the column receives load from one room on each floor, the axial load S can be written as

$$S = \left(n\mu_Q + nL_B + \sum_{i=1}^{n} L_{Fi} + L_{Ri} \right)\int_A i_S(x,y)dA \qquad (10.213)$$

$$+ \sum_{i=1}^{n} \int_A i_S(x,y)D_i(x,y)dA$$

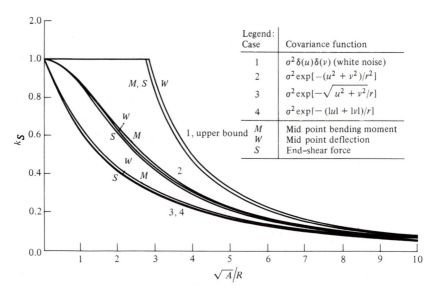

Figure 10.17 Example of variance reduction factor for beam in floor support system.

The random variables L_{Fi} and L_{Ri} and the random processes $D_i(x,y)$ are assumed mutually independent. For the equivalent uniformly distributed load within each room, $EUDL_S$ is then found:

$$E[EUDL_S] = \mu_Q \qquad (10.214)$$

$$Var[EUDL_S] = \sigma_B^2 + \sigma_F^2/n + \sigma_R^2/n + k_S\sigma_D^2/n \qquad (10.215)$$

Code-specified values of $EUDL$ correspond to high fractiles in the distribution. For a distribution of $EUDL$ with mean value and variance from (10.214) and (10.215), such fractiles decrease with the number of floors.

Corresponding to the concept of a deterministic single force load model, Peir and Cornell (1973) and McGuire and Cornell (1974) have proposed a similar probabilistic load model for transient floor loads. As illustrated in Fig. 10.18, the loading during a load event is modeled as the sum of concentrated loads which are placed in load cells over the influence area. Each load corresponds to a person or some item of furniture of a transient nature. The number of load cells is a random variable M. The number of loads within load cell i is a random variable denoted N_i. The magnitude of a load is a random variable denoted P_{ij} for load j in load cell i. The load from load cell i is idealized as a single force Q_i:

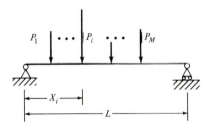

Figure 10.18   Random field model for concentrated loads.

$$Q_i = \sum_{j=1}^{N_i} P_{ij} \qquad (10.216)$$

The contribution R_i to a load effect S from load cell i is

$$R_i = Q_i I_{Si} \qquad (10.217)$$

where I_{Si} is the influence coefficient of the randomly located load group i. The difference in influence coefficient for loads within each load cell is thus neglected. The mean value of I_{Si} is expressed in terms of the probability density function for the random location (X, Y) as

$$E[I_{Si}] = \int_A i_S(x,y) f_{X,Y}(x,y) dA \qquad (10.218)$$

The variance is computed in a similar way and the distribution function is

$$F_{I_{Si}}(z) = \int_{i_S(x,y) \leq z} f_{X,Y}(x,y) dA \qquad (10.219)$$

The total load effect is

$$S = \sum_{i=1}^{M} R_i \qquad (10.220)$$

The random variables M, N_i, I_{Si}, and Q_{ij} are modeled as being mutually independent with mean values μ_M, μ_N, μ_I, and μ_Q and coefficients of variation v_M, v_N, v_I, and v_Q, respectively. To compute the mean value and coefficient of variation of S, two general results for random variables are used (see, e.g., Ditlevsen, 1981, pp. 70 and 162). The first result concerns the sum Y of a random number N of independent random variables $X_1, X_2, \ldots, X_N$ with common mean value μ_X and coefficient of variation v_X.

$$Y = \sum_{i=1}^{N} X_i \tag{10.221}$$

The mean value and coefficient of variation are

$$\mu_Y = \mu_N \mu_X \tag{10.222}$$

$$v_Y^2 = \frac{v_X^2}{\mu_N} + v_N^2 \tag{10.223}$$

where μ_N and v_N are the mean value and coefficient of variation of N. The second result concerns the product Z of two independent random variables X and Y.

$$Z = XY \tag{10.224}$$

The mean value and coefficient of variation are

$$\mu_Z = \mu_X \mu_Y \tag{10.225}$$

$$v_Z^2 = (1 + v_X^2)(1 + v_Y^2) - 1 \tag{10.226}$$

Application of these formulas gives

$$\mu_Q = \mu_N \mu_P \tag{10.227}$$

$$v_Q^2 = \frac{v_P^2}{\mu_N} + v_N^2 \tag{10.228}$$

$$\mu_R = \mu_Q \mu_I = \mu_N \mu_P \mu_I \tag{10.229}$$

$$v_R^2 = (1 + v_Q^2)(1 + v_I^2) - 1 = \left(1 + \frac{v_P^2}{\mu_N} + v_N^2\right)(1 + v_I^2) - 1 \tag{10.230}$$

$$\mu_S = \mu_M \mu_R = \mu_M \mu_N \mu_P \mu_I \tag{10.231}$$

$$v_S^2 = \frac{v_R^2}{\mu_M} + v_M^2 = \frac{1}{\mu_M}\left[\left(1 + \frac{v_P^2}{\mu_N} + v_N^2\right)(1 + v_I^2) - 1\right] + v_M^2 \tag{10.232}$$

The mean value and variance of the equivalent uniformly distributed load, $EUDL_S$, are in this case

$$E[EUDL_S] = \frac{\mu_S}{A \mu_I} = \frac{\mu_M \mu_N \mu_P}{A} \tag{10.233}$$

$$Var[EUDL_S] = \left(\frac{\mu_S v_S}{A \mu_I}\right)^2 = \left(\frac{\mu_M \mu_N \mu_P v_S}{A}\right)^2 \tag{10.234}$$

A model without load cells has $\mu_N = 1$ and $v_N = 0$. It is thus noted that the assumption about clustering of loads in load cells leads to a larger coefficient of variation of the $EUDL_S$. The distribution type of the $EUDL$ for the transient load is not determined, but a gamma distribution is often chosen, although somewhat arbitrarily. Parameters in the gamma distribution are determined from (10.214) and (10.215). The transient load model is illustrated for a simple beam example.

Example 10.5

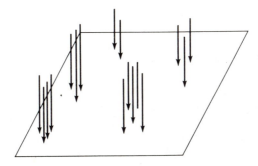

Figure 10.19   Random line load model on beam structure.

Figure 10.19 shows a simply supported beam of length L loaded by a random number M of single forces of random magnitudes P_i. The bending moment at midspan is

$$S = \sum_{i=1}^{M} P_i I_i$$

where $I_i = i(X_i)$ is the influence number for the ith force. $i(x)$ is the influence function and X_i is the random position of the ith force. The influence function is

$$i(x) = \begin{cases} \dfrac{1}{2}x, & 0 \leqslant x \leqslant \dfrac{L}{2} \\ \dfrac{1}{2}(L - x), & \dfrac{L}{2} < x < L \end{cases}$$

If each force has a uniform distribution over the length of the beam, the probability density function for X_i is

$$f_X(x) = \frac{1}{L}, \quad 0 \leqslant x \leqslant L$$

The expected value and variance of I_i are

$$E[I_i] = E[i(X_i)] = \int_0^L i(x) f_X(x) dx = \frac{1}{8}L$$

$$\text{Var}[I_i] = \text{Var}[i(X_i)] = \frac{1}{192}L^2$$

giving a coefficient of variation of $1/\sqrt{3}$ for I_i.

Let the random variables M, P_i, and I_i be mutually independent and

$$\mu_M = 5, \qquad \nu_M = 0.2$$
$$\mu_P = 80 \text{ kN}, \qquad \nu_P = 0.3$$

For $L = 10$ m the expected value and coefficient of variation of the bending moment are given by (10.231) and (10.132) with $\mu_N = 1$ and $v_N = 0$.

$$\mu_S = \mu_M \mu_P \mu_I = 500 \text{ kNm}$$

$$v_S = \left| \frac{1}{\mu_M}[(1 + v_P^2)(1 + v_I^2) - 1] + v_M^2 \right|^{1/2} = 0.36$$

10.5.2 Maximum Live Load

Time variation of the sustained load is generally taken as a Poisson square-wave process with intensity λ. The time variation of the transient load is similarly taken as a Poisson "spike" process with intensity μ. The time variation is illustrated in Fig. 10.20 and definitions of these filtered Poisson processes are given in Section 7.3.4.

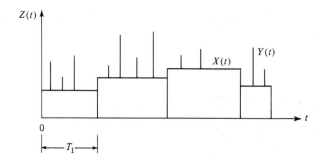

Figure 10.20 Combination of sustained live load and transient live load processes.

The load processes can represent uniformly distributed loads and are here denoted $X(t)$ for the sustained load and $Y(t)$ for the transient load. The two processes are assumed independent and extreme-value distributions for the individual processes are given in Section 7.3.4.

The total load is denoted $Z(t)$:

$$Z(t) = X(t) + Y(t) \tag{10.235}$$

The random variable $M(T)$ is defined as the maximum value of $Z(t)$ in the time interval $[0,T]$.

$$M(T) = \max_{0 \leqslant t \leqslant T} Z(t) \tag{10.236}$$

The distribution of $M(T)$ has been determined by Hasofer (1974) and an alternative derivation by Gaver and Jacobs (1981) is given here. T_1 denotes the time to the first change in the sustained-load process. T_1 is exponentially distributed [(7.16)] with mean value $1/\lambda$. An integral equation for the distribution function of $M(T)$ can now be formulated by a procedure similar to that leading to (7.124):

$$F_M(z,T) = P(M(T) \leqslant z, T_1 > T) + P(M(T) \leqslant z, T_1 \leqslant T) \tag{10.237}$$

$$= e^{-\lambda T} \int_0^z f_X(x) \exp[-\mu T(1 - F_Y(z-x))] dx$$

$$+ \int_0^T \lambda e^{-\lambda t} F_M(z, T-t) \left\{ \int_0^z f_X(x) \exp[-\mu t(1 - F_Y(z-x))] dx \right\} dt$$

In the last term, t is the latest change in the sustained load process before T. The integral is of convolution type and it is therefore convenient to introduce $l_M(z,\xi)$ as the Laplace transform of $F_M(z,T)$ with respect to T:

$$l_M(z,\xi) = \int_0^\infty e^{-\xi T} F_M(z,T) dT \tag{10.238}$$

Taking Laplace transforms with respect to T in (10.237) yields

$$l_M(z,\xi) = \Theta(z,\xi) + \lambda\Theta(z,\xi)l_M(z,\xi) \tag{10.239}$$

where the function $\Theta(z,\xi)$ is determined as

$$\Theta(z,\xi) = \int_0^\infty e^{-\xi T} e^{-\lambda T} \int_0^z f_X(x) \exp[-\mu T(1 - F_Y(z-x))] dx\, dT \tag{10.240}$$

$$= \int_0^z \frac{f_X(x)}{\xi + \lambda + \mu(1 - F_Y(z-x))} dx$$

The Laplace transform of the distribution function for the maximum value is thus

$$l_M(z,\xi) = \frac{\Theta(z,\xi)}{1 - \lambda\Theta(z,\xi)} \tag{10.241}$$

It is difficult to invert this equation and obtain $F_M(z,T)$ explicitly. Useful information of asymptotic properties of $M(T)$ and the first-passage time have been obtained by Gaver and Jacobs (1981). The expected value of the time $T_M(z)$ to the first passage of level z is thus

$$E[T_M(z)] = l_M(z,0) = \frac{\Theta(z,0)}{1 - \lambda\Theta(z,0)} \tag{10.242}$$

The extreme-value distribution for the total load can thus be found for a situation in which the full sustained load changes at the same time. With reference to (10.201), the procedure therefore applies directly to all load effects if the use of the building changes at each change of the sustained load. If, on the other hand, load changes occur independently from floor to floor, the procedure applies only to load effects from loads on a single floor. Instead of (10.235) the total *EUDL* must then be written as

$$Z(t) = L_B + X(t) + Y(t) \qquad (10.243)$$

where the *EUDL* from the sustained load, $X(t)$, does not include loads from L_B. The maximum of $X(t)+Y(t)$ is first determined by the method described above and the random variable L_B is added. A load effect receiving load from several floors, such as the axial load (10.213), must be analyzed in a different way. $Z(t)$ is written as

$$Z(t) = L_B + \frac{1}{n}\sum_{i=1}^{n} X_i(t) + Y(t) = L_B + \bar{X}(t) + Y(t) \qquad (10.244)$$

where $X_i(t)$ is the *EUDL* for the ith floor. The difference between $X(t)$ in (10.243) and $\bar{X}(t)$ in (10.244) is illustrated in Fig. 10.21.

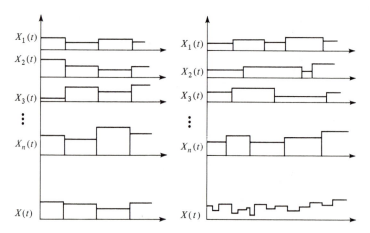

Figure 10.21   Difference between simultaneous and independent load changes on each floor of a multistory house.

$\bar{X}(t)$ is a Poisson square-wave process but the wave heights are not mutually independent. The intensity $\tilde{\lambda}$ of $\bar{X}(t)$ is

$$\tilde{\lambda} = n\lambda \qquad (10.245)$$

Madsen and Ditlevsen (1981) have presented two methods by which good approximations to the extreme-value distribution of $\bar{X}(t)$ can be obtained. An approximation to the combination of $\bar{X}(t)$ and $Y(t)$ can then be obtained by Turkstra's rule (see Section 10.6.3).

10.6 THEORY OF STOCHASTIC LOAD COMBINATION

In Chapter 7 and in this chapter various scalar models for time-varying loads are considered and the extreme-value distribution is derived. The extreme-value distribution contains sufficient information on the loading when only one time-varying load acts on the structure and failure is defined as the crossing of some level by the load process. The theory of stochastic load combination is applied in situations where a structure is subjected to two or more time-varying scalar loads acting simultaneously. The scalar loads can be components of the same load process or be components of different load processes. To evaluate the reliability of the structure, each load cannot be characterized by its extreme-value distribution alone; a more detailed characterization of the stochastic process is necessary. The reason is that the loads in general do not obtain their extreme values at the same time.

A structure subjected to loads defining a vector-valued load process $\mathbf{Q}(t)$ is considered. Failure of the structure is assumed to occur at the time of the first exceedence of the deterministic function $\xi(t)$ by the random function $b(\mathbf{Q}(t))$. Here $\xi(t)$ represents a strength threshold and the b-function combines the load processes to the load effect process corresponding to this strength. A *linear load combination* is said to exist when the b-function is linear. Otherwise, the load combination is nonlinear. The failure event is illustrated geometrically in Fig. 10.22 for a combination of two loads and a constant threshold $\xi(t)$. The figure shows that failure can be thought of as either the first upcrossing of $\xi(t)$ by the process $b(\mathbf{Q}(t))$ or as the first outcrossing of the set $B(t) = \{\mathbf{q} \mid b(\mathbf{q}) \leqslant \xi(t)\}$ by the vector process $\mathbf{Q}(t)$. In both cases this is true under the condition that failure does not occur at time zero.

10.6.1 Bounds on the Extreme-Value Distribution

A simple upper bound on the probability of failure P_F in the time interval $[0,T]$ is derived. Let $N_\xi(T)$ denote the number of upcrossings in $[0,T]$ of $\xi(t)$ by $b(\mathbf{Q}(t))$ or, equivalently, the number of outcrossings in $[0,T]$ of the set $B(t)$ by $\mathbf{Q}(t)$. The failure probability is expressed as

$$P_F = P(\text{failure at } t=0) + P(N_\xi(T) \geqslant 1) \tag{10.246}$$

$$- P(\text{failure at } t=0 \text{ and } N_\xi(T) \geqslant 1)$$

The negative term is numerically smaller than the smallest positive term. An upper bound on P_F is

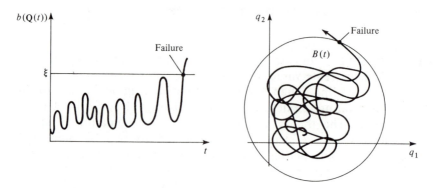

Figure 10.22 Geometrical illustration of the failure event from combined loading.

$$P_F \leqslant P_0 + P(N_\xi(T) \geqslant 1) \tag{10.247}$$

where $P_0 = P(\text{failure at } t=0)$. This upper bound is a good approximation to P_F, at least if one of the terms on the right-hand side is much larger than the other term. The upper bound is further developed as

$$P_F \leqslant P_0 + \sum_{n=1}^{\infty} P(N_\xi(T)=n) \tag{10.248}$$

$$\leqslant P_0 + \sum_{n=1}^{\infty} n\, P(N_\xi(T)=n) = P_0 + \mathrm{E}[N_\xi(T)]$$

The right-hand side is a good approximation to P_F if (10.247) is a good approximation and if, further,

$$P(N_\xi(T)=1) \;>>\; \sum_{n=2}^{\infty} n\, P(N_\xi(T)=n) \tag{10.249}$$

Equation (10.249) will often be valid in practical situations with high-reliability structures when clustering of crossings can be neglected. Similar to (7.131), the failure probability can also be approximated by

$$P_F \approx 1 - (1-P_0)\exp\left| -\frac{\mathrm{E}[N_\xi(T)]}{1-P_0} \right| \tag{10.250}$$

When the probability distribution of $\mathbf{Q}(0)$ is known, the probability of failure at $t=0$ can be calculated by the methods described in Chapter 5. The second term in (10.248) is written as

$$\mathrm{E}[N_\xi(T)] = \int_{t=0}^{T} \nu(\xi,t)\,dt \tag{10.251}$$

Here $\nu(\xi,t)$ is the mean-upcrossing rate of $\xi(t)$ or mean outcrossing rate of $B(t)$ at time t. In both situations $v(\xi,t)$ can be calculated by Rice's formula (7.59) or a generalization of it. If $v(\xi,t)$ is interpreted as the mean-upcrossing rate of $\xi(t)$, it follows directly from (7.59) that

$$v(\xi,t) = \int_{\dot{s}=\xi}^{\infty} (\dot{s} - \dot{\xi}) f_{S\dot{S}}(\xi,\dot{s},t)\, d\dot{s} \qquad (10.252)$$

where the stochastic process $b(\mathbf{Q}(t))$ is denoted by $S(t)$. If $v(\xi,t)$ is thought of as the mean-outcrossing rate of $B(t)$, it follows from a generalization of the arguments leading to Rice's formula that

$$v(\xi,t) = \int_{\partial B} \int_{\dot{q}_N=0}^{\infty} \dot{q}_N\, f_{Q\dot{Q}_N}(\mathbf{q},\dot{q}_N,t)\, d\dot{q}_N\, d(\partial B) \qquad (10.253)$$

It is here assumed that the set $B(t)$ is constant in time. $\dot{Q}_N$ denotes the projection of $\dot{\mathbf{Q}}(t)$ on the outward normal to B at a point on the boundary ∂B. The inner integral in (10.253) can be viewed as a local outcrossing rate. A generalization of (10.253) to a time-varying set $B(t)$ is conceptually straightforward.

Very few closed-form results for $v(\xi,t)$ exist for general processes and general safe regions. Among these results can be mentioned results for normal processes and different safe regions found by Veneziano et al. (1977), Ditlevsen (1983), and Fuller (1982), and results for combinations of rectangular filtered Poisson processes found by Waugh (1977) and Breitung and Rackwitz (1982). Example 10.6 shows the result for a combination of three stationary normal processes.

Example 10.6

Stresses in a linear system under excitation by time-varying loading are considered as stationary normal processes. In this example the nonlinear combination of stress components to the von Mieses stress arising from the von Mieses yield criterion is addressed. According to this criterion, yielding occurs if

$$(\sigma_x - \sigma_y)^2 + (\sigma_y - \sigma_z)^2 + (\sigma_x - \sigma_z)^2 + 6(\tau_{xy}^2 + \tau_{xz}^2 + \tau_{yz}^2) \geq 2\sigma_F^2$$

where σ_F is the yield stress. In many cases in practice several stress components are zero. The analysis is therefore carried out only for the special case where $\sigma_z(t) = \tau_{xz}(t) = \tau_{yz}(t) = 0$. Yielding then occurs if

$$\sigma_x^2 + \sigma_y^2 - \sigma_x \sigma_y + 3\tau_{xy}^2 \geq \sigma_F^2$$

Other stress combinations can be analyzed in a similar way.

The stress vector $\tau(t)$ is introduced as

$$\tau(t) = (\tau_1(t),\tau_2(t),\tau_3(t)) = (\sigma_x(t),\sigma_y(t),\tau_{xy}(t))$$

The von Mieses stress $\sigma_{vM}(t)$ is then defined as

$$\begin{aligned}
\sigma_{vM}(t) &= \sqrt{\sigma_x(t)^2 + \sigma_y(t)^2 - \sigma_x(t)\sigma_y(t) + 3\tau_{xy}(t)^2} \\
&= \sqrt{\tau_1(t)^2 + \tau_2(t)^2 - \tau_1(t)\tau_2(t) + 3\tau_3(t)^2}
\end{aligned}$$

The cross-spectral matrix of the stress components is $\{S_{ij}(\omega)\}$. It is determined as (see Chapter 8)

$$S_{ij}(\omega) = S^Q(\omega)H_i(\omega)\overline{H_j(\omega)}, \quad i = 1,2,3; \quad j = 1,2,3$$

where $S^Q(\omega)$ is the spectral density function of the excitation process $Q(t)$, and where an overbar denotes a complex conjugate.

A normal vector $(\mathbf{T}(t), \dot{\mathbf{T}}(t)) = (T_1(t), T_2(t), T_3(t), \dot{T}_1(t), \dot{T}_2(t), \dot{T}_3(t))$ is introduced next. Here a dot denotes a time derivative. The stress process is stationary, thus implying zero mean values for $\dot{T}_1(t)$, $\dot{T}_2(t)$ and $\dot{T}_3(t)$. The covariance matrix for $(\mathbf{T}(t), \dot{\mathbf{T}}(t))$ is

$$\mathbf{c} = \begin{vmatrix} \mathbf{C}_{TT} & \mathbf{C}_{T\dot{T}} \\ \mathbf{C}_{\dot{T}T} & \mathbf{C}_{\dot{T}\dot{T}} \end{vmatrix}$$

The variances and covariances in these submatrices are obtained from the cross-spectral density matrix as (see Chapter 8).

$$\text{Cov}[T_i(t), T_j(t)] = \int_{-\infty}^{\infty} S_{ij}(\omega)d\omega = 2\int_0^{\infty} \text{Re}[S_{ij}(\omega)]d\omega$$

$$\text{Cov}[T_i(t), \dot{T}_j(t)] = \int_{-\infty}^{\infty} (i\omega) S_{ij}(\omega)d\omega = -2\int_0^{\infty} \omega \, \text{Im}[S_{ij}(\omega)]d\omega$$

$$\text{Cov}[\dot{T}_i(t), \dot{T}_j(t)] = \int_{-\infty}^{\infty} \omega^2 S_{ij}(\omega)d\omega = 2\int_0^{\infty} \omega^2 \text{Re}[S_{ij}(\omega)]d\omega$$

where Re[ ] denotes the real part and Im[ ] denotes the imaginary part.

For white noise input

$$S^Q(\omega) = S_0, \quad -\infty < \omega < \infty$$

and transfer functions for a single-degree-of-freedom oscillator

$$H_j(\omega) = \frac{G_j}{\omega_j^2 - \omega^2 + 2i\zeta_j\omega\omega_j}, \quad j = 1,2,3$$

these covariances are, according to (8.78):

$$\text{Cov}[T_i(t), T_j(t)] = S_0 G_i G_j \alpha_{ij}$$

$$\text{Cov}[T_i(t), \dot{T}_j(t)] = S_0 G_i G_j \beta_{ij}$$

$$\text{Cov}[\dot{T}_i(t), \dot{T}_j(t)] = S_0 G_i G_j \gamma_{ij}$$

where the functions α_{ij}, β_{ij}, and γ_{ij} are defined in (8.79) to (8.81).

The von Mises stress $\sigma_{vM}(t)$ exceeds σ_F in the time interval $[0,D]$ if the process $\mathbf{T}(t)$ is outside the ellipsoid with equation $\tau_1^2 + \tau_2^2 - \tau_1\tau_2 + 3\tau_3^2 = \sigma_F^2$ at any point in time during the interval. The equation for this ellipsoid can be written in the form

$$\left| \frac{\tau_1 + \tau_2}{2\sigma_F} \right|^2 + \left| \frac{\tau_1 - \tau_2}{\frac{2}{\sqrt{3}}\sigma_F} \right|^2 + \left| \frac{\tau_3}{\frac{1}{\sqrt{3}}\sigma_F} \right|^2 = 1$$

The ellipsoid has principal axes in the (τ_1, τ_2) plane at angles $\pi/4$ with the τ_1- and τ_2-axes. The third principal axis is the τ_3-axis. The semiaxes of the ellipsoid are $2\sigma_F$, $2\sigma_F/\sqrt{3}$, and $\sigma_F/\sqrt{3}$, respectively. This is illustrated further in Fig. 10.23.

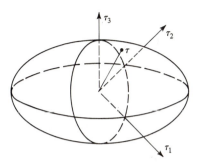

Figure 10.23 Yield surface.

The mean-outcrossing rate is computed by Rice's formula (10.253), which can be rewritten as

$$\nu^+(\sigma_F) = \int_{\partial B} f_{\mathbf{T}}(\boldsymbol{\tau}) \int_{\dot{\tau}_N=0}^{\infty} \dot{\tau}_N f_{\dot{T}_N \mid \mathbf{T}=\boldsymbol{\tau}}(\dot{\tau}_N \mid \boldsymbol{\tau}) d\dot{\tau}_N \, d\partial B$$

The inner integral can be viewed as a local mean-outcrossing rate. Integration variables θ and φ are introduced on the ellipsoid surface as

$$\tau_1 = \sigma_F \left(\cos\theta\sin\varphi + \frac{1}{\sqrt{3}}\sin\theta\sin\varphi \right), \quad 0 \le \theta \le 2\pi$$

$$\tau_2 = \sigma_F \left(\cos\theta\sin\varphi - \frac{1}{\sqrt{3}}\sin\theta\sin\varphi \right), \quad 0 \le \varphi \le \pi$$

$$\tau_3 = \sigma_F \frac{1}{\sqrt{3}} \cos\varphi$$

The outward unit normal vector is

$$\mathbf{n} = \frac{(\cos\theta\sin\varphi + \sqrt{3}\sin\theta\sin\varphi, \cos\theta\sin\varphi - \sqrt{3}\sin\theta\sin\varphi, 2\sqrt{3}\cos\varphi)}{\sqrt{2\sin^2\varphi(\cos^2\theta + 3\sin^2\theta) + 12\cos^2\varphi}}$$

and the time derivative $\dot{\tau}_N$ is

$$\dot{\tau}_N = \mathbf{n}^T \dot{\boldsymbol{\tau}}$$

where a superscript T denotes the transpose. The distribution of $\dot{\mathbf{T}}$ conditioned on $\mathbf{T}=\boldsymbol{\tau}$ is a normal distribution with mean value and covariance matrix [see (7.137) to (7.139)]

$$E[\dot{\mathbf{T}} \mid \mathbf{T}=\boldsymbol{\tau}] = \mathbf{C}_{\dot{\mathbf{T}}\mathbf{T}} \mathbf{C}_{\mathbf{T}\mathbf{T}}^{-1}(\boldsymbol{\tau} - E[\mathbf{T}])$$

$$\mathbf{C}_{\dot{\mathbf{T}} \mid \mathbf{T}=\boldsymbol{\tau}} = \mathbf{C}_{\dot{\mathbf{T}}\dot{\mathbf{T}}} - \mathbf{C}_{\dot{\mathbf{T}}\mathbf{T}} \mathbf{C}_{\mathbf{T}\mathbf{T}}^{-1} \mathbf{C}_{\mathbf{T}\dot{\mathbf{T}}}$$

$\dot{T}_N \mid \mathbf{T}=\boldsymbol{\tau}$ is then normally distributed, with

$$E[\dot{T}_N \mid \mathbf{T}=\boldsymbol{\tau}] = \mu_N(\boldsymbol{\tau}) = \mathbf{n}^T E[\dot{\mathbf{T}} \mid \mathbf{T}=\boldsymbol{\tau}] = \mathbf{n}^T \mathbf{C}_{\dot{\mathbf{T}}\mathbf{T}} \mathbf{C}_{\mathbf{T}\mathbf{T}}^{-1}(\boldsymbol{\tau} - E[\mathbf{T}])$$

$$\mathrm{Var}[\dot{T}_N \mid \mathbf{T}=\boldsymbol{\tau}] = \sigma_N(\boldsymbol{\tau})^2 = \mathbf{n}^T \mathbf{C}_{\dot{\mathbf{T}} \mid \mathbf{T}=\boldsymbol{\tau}} \mathbf{n} = \mathbf{n}^T (\mathbf{C}_{\dot{\mathbf{T}}\dot{\mathbf{T}}} - \mathbf{C}_{\dot{\mathbf{T}}\mathbf{T}} \mathbf{C}_{\mathbf{T}\mathbf{T}}^{-1} \mathbf{C}_{\mathbf{T}\dot{\mathbf{T}}}) \mathbf{n}$$

The inner integral can now be computed directly:

$$\int_{\dot\tau_N=0}^{\infty} \dot\tau_N f_{\dot T_N | T=\tau}(\dot\tau_N | \tau)\, d\dot\tau_N = \int_{\dot\tau_N=0}^{\infty} \dot\tau_N \frac{1}{\sqrt{2\pi}\sigma_N(\tau)} \exp\left[-\frac{1}{2}\left(\frac{\dot\tau_N - \mu_N(\tau)}{\sigma_N(\tau)}\right)^2\right] d\dot\tau_N$$

$$= \sigma_N(\tau)\,\varphi\!\left(\frac{\mu_N(\tau)}{\sigma_N(\tau)}\right) + \mu_N(\tau)\,\Phi\!\left(\frac{\mu_N(\tau)}{\sigma_N(\tau)}\right)$$

$$= \sigma_N(\tau)\,\Psi\!\left(-\frac{\mu_N(\tau)}{\sigma_N(\tau)}\right)$$

with $\Psi(\ )$ defined in (7.67). The mean-outcrossing rate is thus

$$v^+(\sigma_F) = \int_{\theta=0}^{2\pi}\int_{\varphi=0}^{\pi} \frac{1}{(2\pi)^{3/2}\,|\det C_{TT}|^{1/2}} \exp\left[-\frac{1}{2}(\tau - E[T])^T C_{TT}^{-1}(\tau - E[T])\right]$$

$$\times \sigma_N(\tau)\Psi\!\left(-\frac{\mu_N(\tau)}{\sigma_N(\tau)}\right)\frac{1}{3}\sqrt{2(\cos^2\theta + 3\sin^2\theta)\sin^2\varphi + 12\cos^2\varphi}\,\sigma_F^2 \sin\varphi\, d\varphi d\theta$$

where $1/3\,\sqrt{2(\cos^2\theta + 3\sin^2\theta)\sin^2\varphi + 12\cos^2\varphi}\,\sigma_F^2\sin\varphi\,d\varphi d\theta$ is the surface area element.

10.6.2 Linear Load Combinations

In a general study for code purposes the case of stationary and mutually independent processes combined linearly is very important. The sum of two processes is first considered, so let

$$Q(t) = Q_1(t) + Q_2(t) \tag{10.254}$$

The expected rate of upcrossings of the constant level ξ, $v_Q(\xi)$, is calculated by Rice's formula (7.59):

$$v_Q(\xi) = \int_{\dot q=0}^{\infty} \dot q f_{Q\dot Q}(\xi,\dot q)\, d\dot q \tag{10.255}$$

The joint probability density function $f_{Q\dot Q}(\ ,\ )$ is expressed in terms of the density functions $f_{Q_1\dot Q_1}(\ ,\ )$ and $f_{Q_2\dot Q_2}(\ ,\ )$ by the convolution integral

$$f_{Q\dot Q}(q,\dot q) = \int_{q_1=-\infty}^{\infty}\int_{\dot q_1=-\infty}^{\infty} f_{Q_1\dot Q_1}(q_1,\dot q_1) \tag{10.256}$$

$$\times f_{Q_2\dot Q_2}(q-q_1,\dot q - \dot q_1)\,d\dot q_1 dq_1$$

Inserting this result in (10.255) and substituting $\dot q = \dot q_1 + \dot q_2$ gives

$$v_Q(\xi) = \int_{q=-\infty}^{\infty}\int_{\dot q_1=-\infty}^{\infty}\int_{\dot q_2=-\dot q_1}^{\infty} \dot q_1 f_{Q_1\dot Q_1}(q,\dot q_1) \tag{10.257}$$

$$\times f_{Q_2\dot Q_2}(\xi-q,\dot q_2)\,d\dot q_2 d\dot q_1 dq$$

$$+ \int_{q=-\infty}^{\infty}\int_{\dot q_1=-\infty}^{\infty}\int_{\dot q_2=-\dot q_1}^{\infty} \dot q_2 f_{Q_1\dot Q_1}(q,\dot q_1)$$

$$\times f_{Q_2\dot Q_2}(\xi-q,\dot q_2)\,d\dot q_2 d\dot q_1 dq$$

The two triple integrals can be evaluated analytically only in special cases. Simple upper and lower bounds on the mean-upcrossing rate can, however, be found by changing the area of integration in the $(\dot{q}_1, \dot{q}_2)$-plane for the two integrals in (10.257). Figure 10.24a shows the common area of integration of the integrals. The vertical and horizontal hatching illustrate the integrations with the integrands of the first and second integrals, respectively. Figure 10.24b correspondingly illustrates the integrations in an upper bound and Fig. 10.24c the integrations in a lower bound.

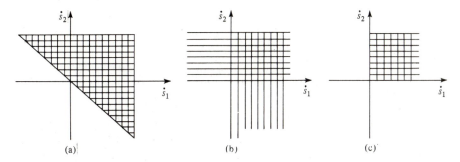

Figure 10.24   Areas of integration.

The upper bound is

$$v_Q(\xi) \leqslant \int_{q=-\infty}^{\infty}\int_{\dot{q}_1=0}^{\infty}\int_{\dot{q}_2=-\infty}^{\infty} \dot{q}_1 f_{Q_1\dot{Q}_1}(q,\dot{q}_1) \tag{10.258}$$

$$\times f_{Q_2\dot{Q}_2}(\xi-q,\dot{q}_2)d\dot{q}_2 d\dot{q}_1 dq$$

$$|\ \int_{q=-\infty}^{\infty}\int_{\dot{q}_1=-\infty}^{\infty}\int_{\dot{q}_2=0}^{\infty} \dot{q}_2 f_{Q_1\dot{Q}_1}(q,\dot{q}_1)$$

$$\times f_{Q_2\dot{Q}_2}(\xi-q,\dot{q}_2)d\dot{q}_2 d\dot{q}_1 dq$$

$$= \int_{q=-\infty}^{\infty} v_{Q_1}(q)f_{Q_2}(\xi-q)dq + \int_{q=-\infty}^{\infty} v_{Q_2}(q)f_{Q_1}(\xi-q)dq$$

$$= v_{Q_1}*f_{Q_2} + v_{Q_2}*f_{Q_1}$$

where * means convolution. The terms in the upper bound are generally called the *point crossing terms* (Larrabee and Cornell, 1981).

The lower bound is similarly

$$v_Q(\xi) \geqslant \int_{q=-\infty}^{\infty} v_{Q_1}(q)f_{Q_2}(\xi-q)(1-F_{\dot{Q}_2|Q_2}(0,\xi-q))dq \tag{10.259}$$

$$+ \int_{q=-\infty}^{\infty} v_{Q_2}(\xi-q)f_{Q_1}(q)(1-F_{\dot{Q}_1|Q_1}(0,q))dq$$

Here $v_{Q_1}(\ )$ is the mean-upcrossing rate function for $Q_1(t)$ and $f_{Q_1}(\ )$ is the marginal or arbitrary-point-in-time probability density function for $Q_1(t)$. The factor $1 - F_{\dot{Q}_1 \mid Q_1}(0, q)$ gives the probability of a positive derivative $\dot{Q}_1(t)$ given that $Q_1(t) = q$. For a normal process this probability is 0.5, independent of q. For the sum of two normal processes the lower bound (10.259) is thus half the value of the upper bound (10.258).

The bounds are easily generalized to cover situations with nonstationary load processes or nonconstant thresholds (Madsen, 1982), and to situations where more than two time-varying loads are acting simultaneously. For the sum of three stationary and independent processes, $Q_1(t)$, $Q_2(t)$, and $Q_3(t)$, the upper bound on $v_Q(\xi)$ is

$$v_Q(\xi) \leqslant \int_{-\infty}^{\infty} v_{Q_1}(q) f_{Q_2 + Q_3}(\xi - q) dq \qquad (10.260)$$

$$+ \int_{-\infty}^{\infty} v_{Q_2}(q) f_{Q_1 + Q_3}(\xi - q) dq + \int_{-\infty}^{\infty} v_{Q_3}(q) f_{Q_1 + Q_2}(\xi - q) dq$$

where, e.g.,

$$f_{Q_1 + Q_2}(q) = \int_{-\infty}^{\infty} f_{Q_1}(x) f_{Q_2}(q - x) dx = f_{Q_1} * f_{Q_2} \quad (10.261)$$

The upper bound can thus be written as

$$v_Q(\xi) \leqslant v_{Q_1} * f_{Q_2} * f_{Q_3} + v_{Q_2} * f_{Q_1} * f_{Q_3} + v_{Q_3} * f_{Q_1} * f_{Q_2} \quad (10.262)$$

The upper bound on $v_Q(\xi)$ is thereby expressed solely in terms of convolution integrals of the mean-upcrossing rate function and the marginal probability density function for each process. In Table 10.2 these two functions are given for various load processes commonly used as load models. Additional results are given in Madsen (1979). A renewal pulse process is defined similar to the Poisson square-wave process defined in Section 7.3.4, except that the pulse shapes need not be rectangular and the Poisson process is replaced by the more general renewal process.

A more general treatment of the point-crossing-term idea for nonlinear load combinations and nonstationary processes is given by Ditlevsen and Madsen (1981) and Ditlevsen (1983). The fact that the results of the point crossing method are in terms of convolution integrals makes a combination of the results with a first-order reliability methods as described in Chapter 5 very simple.

If the upper bound (10.258) on $v(\xi)$ is used in (10.248), a strict upper bound on the failure probability is maintained. This upper bound can still provide a very good approximation to the exact value if the upper bound on $v(\xi)$ is close to the exact value. Table 10.3 presents some exact results taken from Larrabee and Cornell (1981) and Madsen et al. (1979) compared to the upper bound. The ratio of the exact result to the upper bound is in all

TABLE 10.2 Mean-Upcrossing Rate Functions and Marginal Probability Density Functions for Poisson Pulse Processes (Intensity ν)		
Process	$\nu_Q(q)$	$f_Q(q)$
Spike process	$\nu(1 - F_S(q))$	$\delta(q)$
Square-wave process	$\nu F_S(q)(1 - F_S(q))$	$f_S(q)$
Triangular pulse process	$\nu(1 - F_S(q))$	$\int_{s=q}^{\infty} \dfrac{1}{s}\, dF_S(s)$
Parabolic pulse process	$\nu(1 - F_S(q))$	$\int_{s=q}^{\infty} \dfrac{1}{2\sqrt{1 - \dfrac{q}{s}}}\, \dfrac{1}{s}\, dF_S(s)$

cases close to unity, indicating that the upper bound is indeed a good approximation.

The analysis of linear load combinations of stationary and independent load processes can be summarized. It follows that if the extreme-value distribution of the combined load is well approximated in terms of the mean-upcrossing rate function, see (10.248), sufficiently accurate approximations to the extreme-value distribution can be computed from the mean-upcrossing rate function $\nu_Q(\xi)$ and the marginal probability density function $f_Q(q)$ for each process. It can thus be stated that the pair of functions (ν_Q, f_Q) provides sufficient information about each load process in a linear load combination. Several extensions of the bounding technique

TABLE 10.3 Bounds on the Mean-Upcrossing Rate for Sums of Two Stationary Processes	
$\Lambda = \int_{q=-\infty}^{\infty} \nu_{Q_1}(q) f_{Q_2}(\xi - q) dq \; + \; \int_{q=-\infty}^{\infty} \nu_{Q_2}(q) f_{Q_1}(\xi - q) dq$	
Processes combined	Bounds
Normal + normal	$\Lambda / \sqrt{2} \leqslant \nu(\xi) \leqslant \Lambda$
Normal + unimodal renewal pulse process	$0.5\Lambda \leqslant \nu(\xi) \leqslant \Lambda$
Renewal spike process + arbitrary process	$\nu(\xi) = \Lambda$
Renewal rectangular pulse process + arbitrary process	$\nu(\xi) = \Lambda$
Unimodal Poisson pulse process + same	$0.75\Lambda \leqslant \nu(\xi) \leqslant \Lambda$
Filtered renewal rectangular pulse process + arbitrary process	$\nu(\xi) = \Lambda$

explained above to linear combinations of dependent processes are demonstrated by Winterstein (1980). An analysis of the clustering effect on failure rates of combined loads is presented in Winterstein and Cornell (1984).

The method using the point crossing terms in approximation of the failure probability is called the *point crossing method.* Another method which is equally well developed is the *load coincidence method* (see, e.g., Wen, 1977, 1981; Wen and Pearce, 1981, 1983). This method is based on an identification of load coincidences leading to a level crossing by the combined load. The mean rates of various types of load coincidences are computed together with the probability of a level crossing given that a load coincidence occurs. The mean-upcrossing rate of the level is expressed as a sum of products of mean occurrence rates and the conditional probabilities.

10.6.3 Load Combinations in Codified Structural Design

An essential feature of structural design procedures is a set of requirements for load combination. Such formats provide a list of the combinations to be considered and a set of appropriate load factors to be applied to specified or characteristic values of the individual loads. To provide for the many design situations that can arise, most codes have found it useful to categorize loads as either permanent (e.g., self-weight and prestressing forces) or variable. Variable loads can be further decomposed into long term (e.g., furniture loads, snow loads in some regions) and short term (e.g., wind and earthquake). For each type of load, codes specify characteristic or representative values, normally defined to have a specified probability of being exceeded in some specified period. As an example, the characteristic 50-year wind speed described in Example 7.1 is the wind speed that is exceeded with a probability of 2% in one year.

To describe the basic design formats, permanent loads are denoted D and variable loads L, further decomposed into long-term components LL and short-term components LS. With this basic notation, a first subscript k is used to denote a characteristic value, and a second subscript, $j = 1, 2, \ldots$, to denote a particular load type, such as wind or earthquake.

The format proposed by the CEB (1976) and JCSS (1978) is a family of total loads of the general form

$$\gamma_D D_k + \gamma_i L_{ki} + \sum_{j \ne i} \gamma_j \psi_{ij} L_{kj} \qquad (10.263)$$

for the ultimate state, in which γ-values are load factors. The products $\gamma_j \psi_{ij} L_{kj}$ may be called companion values of the loads. The format involves the factored or design value of one load plus factored companion values of the others. There are at least as many such equations as there are load types. This combination method is called the *companion action format*.

Another basic design format in use by the National Building Code of Canada is of the form

$$\gamma_D D_k + \varphi \left(\sum_j \gamma_j L_{kj} \right) \qquad (10.264)$$

in which φ is a probability factor to account for the fact that extreme values of different loads are unlikely to occur together. When only one load acts, $\varphi = 1$. The American Concrete Institute uses a similar format. This combination format is called the *combination factor format*.

The Soviet Union has adopted a slightly different ultimate state format, which can be written (see Allen, 1977)

$$\gamma_D D_k + \gamma_{LL} LL_k + \varphi \left(\sum_j \gamma_j LS_{kj} \right) \qquad (10.265)$$

The long-term loads are considered at their full design values and short-term actions are considered at reduced companion values by means of a common probability factor. Another closely related format is that in the proposed load and resistance factor design for steel structures (Ravindra and Galambos, 1978).

The essential difference between the basic code formats is whether they multiply design loads $\gamma_i L_{ki}$ by combination factors after summation or before. In all cases, serviceability loads are obtained directly from the characteristic values.

Within a geographic region and a specific class of intended use, the physical effects of loads vary from structure to structure and between elements in a structure. The total variable load effect $S(t)$ in a linear or quasi-linear analysis can be written in the form

$$S(t) = c_1\gamma_1 L_1(t) + c_2\gamma_2 L_2(t) + c_3\gamma_3 L_3(t) \qquad (10.266)$$

in which $L_i(t)$ are the random time dependent variable loads, c_i are deterministic influence coefficients, and γ_i are deterministic load factors. Within a single structure, c_i may be zero for one load type at one element and dominant at another element. The relative magnitudes of the random loads $L_i(t)$ depend on geography and intended use.

The fact that any load $L_i(t)$ in a combination can appear alone if $c_j = 0$, $j \neq i$ suggests the following statement of the design load combination problem: Establish a set of companion action factors ψ_{ij} in (10.263) or design probability factors φ in (10.264) such that the probability of exceeding design loads is approximately constant for all situations involving one or more loads, the spectrum of influence coefficients c_i, all geographic areas and intended structural uses, and all materials and types of structural form covered by a code. Given the scope of the problem definition and practical limitations on the number of factors permissible in any design procedure, it is evident that great precision cannot be expected.

Before proceeding to the determination of the load combination factors it is of some interest to view the various load combination formats in the light of the results obtained for linear load combinations. In this context *Turkstra's rule* (Turkstra, 1970), plays a central role. The rule states that the maximum value of the sum of two independent random processes occurs when one of the processes has its maximum value. The rule is an approximation and corresponds to the assumption that the distribution functions of the two random variables

$$Z_1 = \max_{0 \leq t \leq T} (Q_1(t) + Q_2(t)) \qquad (10.267)$$

and

$$Z_2 = \max \begin{cases} \max_{0 \leq t \leq T} Q_1(t) + Q_2(t) \\ Q_1(t) + \max_{0 \leq t \leq T} Q_2(t) \end{cases} \qquad (10.268)$$

are the same.

For Z_2 the complementary cumulative distribution function is

$$P(Z_2 > \xi) = P(\max_{0 \leqslant t \leqslant T} Q_1(t) + Q_2(t) > \xi) \qquad (10.269)$$

$$+ P(Q_1(t) + \max_{0 \leqslant t \leqslant T} Q_2(t) > \xi)$$

$$- P(\max_{0 \leqslant t \leqslant T} Q_1(t) + Q_2(t) > \xi \text{ and } Q_1 + \max_{0 \leqslant t \leqslant T} Q_2(t) > \xi)$$

Neglect of the negative term and use of the bound (10.248) without P_0 leads to

$$P(Z_2 > \xi) \leqslant T \int_{q=-\infty}^{\infty} \nu_{Q_1}(q) f_{Q_2}(\xi - q) dq \qquad (10.270)$$

$$+ T \int_{q=-\infty}^{\infty} \nu_{Q_2}(q) f_{Q_1}(\xi - q) dq$$

Use of the bound (10.248) without P_0 together with the bound (10.258) leads to the same upper bound for $P(Z_1 > \xi)$. Based on these results, it can be concluded that when the upper bound (10.269) is a good approximation for both $P(Z_1 > \xi)$ and $P(Z_2 > \xi)$, Turkstra's rule is also a good approximation. The conditions for the applicability of the upper bound given here are, however, not necessary conditions for good accuracy of Turkstra's rule.

Turkstra's rule indicates that a natural code format for a combination of two loads is

$$\max \begin{cases} \gamma_1 q_{1k} + \gamma_2 \psi_2 q_{2k} \\ \gamma_1 \psi_1 q_{1k} + \gamma_2 q_{2k} \end{cases} \qquad (10.271)$$

where the ψ-factors express the ratio between fractiles in the extreme-value distributions and the marginal distributions. There is thus a logic rationale for the format in (10.263) and studies such as those of Turkstra and Madsen (1980) also show that this format is superior to the others.

A comprehensive study aimed at determining the load combination factors ψ_{ij} in (10.263) has been presented in Turkstra and Madsen (1980). The analysis is restricted to cases where loads do not act in opposite senses, leading to stress reversal. The load combination factors are aimed at being the same for all materials and the criteria of probability levels of 0.01, 0.001, and 0.0001 for individual design loads and combinations are therefore adopted as objective. A linear combination as in (10.266) is used with the complete range of influence factors being covered. The major conclusions of the study are as follows:

The uncertainty in load models is of major importance in the study of individual loads. However, results for the combination of loads are relatively insensitive to the load models used.

Design combination rules depend on the probability level at which comparisons are made. In general, the more likely the exceedence of the design values of individual loads, the less important the combination problem.

Simple addition of design loads can lead to very conservative results. Ignoring load superposition can lead to extremely nonconservative results.

No combinations of transient live, earthquake, and wind loads need be made at the fractile levels used in conventional structural design.

The combination factor format lead to significant errors in a number of cases.

The companion action format, coupled with a simple model for the ψ-factors, leads to design values almost always within 10% and normally within 5% of theoretical values.

10.7 SUMMARY

In this chapter the stochastic methods developed in the previous chapters are used to describe some specific load models, and the combination of loads is also discussed.

The first three sections are devoted to dynamic loads from wind, waves, and earthquakes. The wind load is described in terms of "frozen turbulence," translated by the mean wind velocity. Elements of the theory of homogeneous turbulence are presented leading to appropriate spectra and coherence functions, and the role of aerodynamic damping and admittance is explained by an example. The section on wave loads contains a brief introduction to wave spectra and linear wave theory. The Morison load equation is presented, and various effects of the nonlinear drag force term are discussed, e.g., the extreme force distribution and the expected increase in fatigue damage accumulation. Also earthquake loads are conveniently described in terms of a spectrum, and the mechanical interpretation of the Kanai-Tajimi spectrum and its relations to response analysis are explained. Nonlinear response analysis of a simple hysteretic structure by the equivalent linearization procedure is also discussed.

The following two sections are devoted to models for traffic load and live load in buildings. The stream of traffic is considered as a pattern of loaded and unloaded areas moving with constant speed. The traffic pattern is modeled by a double renewal process, and the covariance function is derived. The live load model for buildings included sustained and transient load. The variability of the sustained load is separated into variations between buildings, floors, and rooms, as well as a spatial distribution within the rooms. The time variation is in the form of a step function. The transient load is concentrated in load cells and acts only in brief time intervals. An equivalent uniformly distributed load (*EUDL*) is introduced, and the problem of finding the distribution of the time maximum of sustained and transient load is discussed.

The chapter concludes with a discussion of problems related to the combination of different load processes. The concept of multidimensional outcrossings is introduced, and bounding techniques by the point crossing terms are discussed.

REFERENCES

ABRAMOWITZ, M. and I. STEGUN, *Handbook of Mathematical Functions,* Dover, New York, 1965.

ALLEN, D. E., "Load Combinations," Presented to the December 9, 1977, meeting of the Limit States Design Committee of the Canadian Society for Civil Engineering at Ottawa, Ottawa, Canada, 1977.

AMIN, M. and A. H. S. ANG, "Nonstationary Stochastic Model of Earthquake Motions," *Journal of the Engineering Mechanics Division,* ASCE, Vol. 94, 1969, pp. 559-583.

AWS (American Welding Society), "Structural Welding Code - Steel," AWS D1.1-83, 1983.

BABER, T. T. and Y.-K. WEN, "Random Vibration of Hysteretic, Degrading Systems," *Journal of the Engineering Mechanics Division,* ASCE, Vol. 107, 1981, pp. 1069-1087.

BATCHELOR, G. K., *The Theory of Homogeneous Turbulence,* Cambridge University Press, London, 1959.

BORGMAN, L. E., "Wave Forces on Piling for Narrow-Band Spectra," *Journal of the Waterways and Harbors Division,* ASCE, Vol. 91, 1965, pp. 65-90.

BORGMAN, L. E., "Spectral Analysis of Ocean Wave Forces on Pilings," *Journal of the Waterways and Harbors Division,* ASCE, Vol. 93, 1967, pp. 129-156.

BORGMAN, L. E., "Random Hydrodynamic Forces on Objects," *Annals of Mathematical Statistics,* Vol. 38, 1967, pp. 37-51.

BREITUNG, K. and R. RACKWITZ, "Nonlinear Combination of Load Processes," *Journal of Structural Mechanics,* Vol. 10, No. 2, 1982, pp. 145-166.

BROUWERS, J. J. H. and P. H. J. VERBEEK, "Expected Fatigue Damage and Expected Extreme Response for Morison-type Wave Loading," *Applied Ocean Research,* Vol. 5, 1983, pp. 129-133.

CANAVIE, A., M. ARHAN and R. EZRATY, "A Statistical Relationship Between Individual Heights and Periods of Storm Waves," in Vol. 2, *Proceedings,* BOSS'76, Trondheim, Norway, 1976, pp. 354-360.

CEB (Comité Europeen du Béton), Joint Committee on Structural Safety CEB-CECM-FIP-IABSE-IASS-RILEM, "First Order Concepts for Design Codes," *CEB Bulletin No. 112,* July 1976.

CLOUGH, R. W. and J. PENZIEN, *Dynamics of Structures,* McGraw-Hill, New York, 1975.

COROTIS, R. B. and V. A. DOSHI, "Probability Models for Live Load Survey Results," *Journal of the Structural Division,* ASCE, Vol. 103, 1977, pp. 1257-1274.

DAVENPORT, A. G., "Gust Loading Factors," *Journal of the Structural Division,* ASCE, Vol. 93, 1967, pp. 11-33.

DAVENPORT, A. G., "The Prediction of the Response of Structures to Gusty Wind," in *Safety of Structures under Dynamic Loading,* ed. I. Holand et al., Tapir, Trondheim, 1977, pp. 257-284.

DITLEVSEN, O., *Extremes and First Passage Times,* Dissertation, Copenhagen, Denmark, 1971.

DITLEVSEN, O., *Uncertainty Modeling with Applications to Multidimensional Civil Engineering Systems,* McGraw Hill, New York, 1981.

DITLEVSEN, O., "Gaussian Outcrossing from Safe Convex Polyhedrons," *Journal of the Engineering Mechanics Division,* ASCE, Vol. 109, 1983, pp. 127-148.

DITLEVSEN, O., "Level Crossings of Random Processes," *Reliability Theory and Its Applications in Structural and Soil Mechanics,* ed. P. Thoft-Christensen, NATO ASI Series E, Martinus Nijhoff, The Hague, 1983, pp. 57-83.

DITLEVSEN, O. and H. O. MADSEN, "Probabilistic Modeling of Man-Made Load Processes and Their Individual and Combined Effects," in *Structural Safety and Reliability,* ed. T. Moan and M.

Shinozuka, Proceedings ICOSSAR'81, Trondheim, Norway, June 1981, pp. 103-134.

DnV (Det norske Veritas), "Rules for the Design, Construction and Inspection of Offshore Structures, Appendix C Steel Structures," Reprint with Corrections, 1982.

FULLER, J. R., "Boundary Excursions for Combined Random Loads," *AIAA Journal,* Vol. 20, 1982, pp. 1300-1305.

GAVER, D. P. and P. A. JACOBS, "On Combination of Random Loads," *SIAM Journal of Applied Mathematics,* Vol. 40, 1981, pp. 454-466.

HARMAN, D. J. and A. G. DAVENPORT, "A Statistical Approach to Traffic Loads on Bridges," in *Proceedings,* Specialty Conference on Probabilistic Mechanics and Structural Reliability, ASCE, Tucson, Ariz., January 1979, pp. 170-175.

HARRIS, I. R., "The Nature of the Wind," in *The Modern Design of Wind Sensitive Structures,* CIRIA, London, 1971.

HASOFER, A. M., "Time Dependent Maximum of Floor Live Loads," *Journal of the Engineering Mechanics Division,* ASCE, Vol. 100, 1974, pp. 1086-1091.

HASSELMANN, K. et al., "Measurements of the Wind Wave Growth and Swell Decay during the Joint North Sea Wave Project (JONSWAP)," *Deutchen Hydrographischen Zeitschrift,* Ergazungsheft, Reihe A, Nr. 12, Hamburg, 1973.

HOLMES, P and R. G. TICKELL, "Full Scale Wave Loading on Cylinders," in *Proceedings, BOSS'79,* Imperial College, London, August 28-31 1979, Paper 79.

HOUSNER, G. W. and P. C. JENNINGS, "Generation of Artificial Earthquakes," *Journal of the Engineering Mechanics Division,* ASCE, Vol. 90, 1964, pp. 113-150.

ISSC (International Ship Structures Congress), Report of Committee 1, Proceedings of the Second International Ship Structures Congress, Delft, The Netherlands, July 20-24, 1964.

IWAN, W. D. and L. D. LUTES, "Response of the Bilinear Hysteretic System to Stationary Random Excitation," *Journal of the Acoustical Society of America,* Vol. 13, 1968, pp. 545-552.

IYENGAR, R. N. and K. T. S. R. IYENGAR, "A Nonstationary Random Process Model for Earthquake Accelerograms," *Bulletin of the Seismological Society of America,* Vol. 59, 1969, pp. 1163-1188.

JCSS (Joint Committee on Structural Safety), "General Principles of Safety and Serviceability Regulations for Structural Design," ed. L. Ostlund, Lund Institute of Technology, Lund, Sweden, 1978.

KAIMAL, J. C., J. C. WYNGAARD, Y. IZUMI and O. R. COTE, "Spectral Characteristics of Surface Layer Turbulence," *Quarterly Journal of the Royal Meteorological Society,* Vol. 98, 1972, pp. 563-589.

KANAI, K., "Some Empirical Formulas for the Seismic Characteristics of the Ground," *Bulletin of the Earthquake Research Institute,* University of Tokyo, Vol. 35, 1957, pp. 309-325.

KRISTENSEN, L. and N. O. JENSEN, "Lateral Coherence in Isotropic Turbulence and in the Natural Wind," *Boundary Layer Meteorology,* Vol. 17, 1979, pp. 353-373.

LARRABEE, R. D., "Modeling Extreme Vehicle Loads on Highway Bridges," in *Proceedings,* Specialty Conference on Probabilistic Mechanics and Structural Reliability, ASCE, Tucson, Ariz., January 1979, pp. 176-180.

LARRABEE, R. D. and C. A. CORNELL, "Combination of Various Load Processes," *Journal of the Structural Division,* ASCE, Vol. 107, 1981, pp. 223-239.

LIGHTHILL, J., *Waves in Fluids,* Cambridge University Press, Cambridge, 1978.

LIN, Y. K., *Probabilistic Theory of Structural Dynamics,* Krieger Publishing, Huntington, N. Y., 1976.

LINDGREN, G. and I. RYCHLIK, "Wave Characteristic Distributions for Gaussian Waves — Wavelength, Amplitude and Steepness," *Ocean Engineering,* Vol. 9, 1982, pp. 411-432.

LOMNITZ, C. and E. ROSENBLUETH eds., *Seismic Risk and Engineering Decisions,* Elsevier, Amsterdam, 1976.

LONGUET-HIGGINS, M. S., "On the Joint Distribution of the Period and Amplitude of Sea Waves," *Journal of Geophysical Research,* Vol. 80, 1975, pp. 2688-2694.

LONGUET-HIGGINS, M. S., "On the Joint Distribution of Wave Periods and Amplitudes in a Random Wave Field," *Proceedings of the Royal Society of London,* Vol. A 389, 1983, pp. 241-258.

MACCAMY, R. S. and R. A. FUCHS, "Wave Forces on Piles: A Diffraction Theory," U.S. Army Corps of Engineers, Beach Erosion Board, *Technical Memo No. 69,* Washington, D.C., 1954.

MADSEN, H. O., "Load Models and Load Combinations," Report No. R113, Department of Structural Engineering, Technical University of Denmark, February 1979.

MADSEN, H. O., "A Line Model for Furniture Loads," Report ST.79-8, Department of Civil Engineering and Applied Mechanics, McGill University, Montreal, August, 1979.

MADSEN, H. O., "Reliability under Combination of Nonstationary Load Processes," in *DIALOG 1-82,* Danish Engineering Academy, Lyngby, Denmark, 1982, pp. 45-58.

MADSEN, H. O. and O. DITLEVSEN, "Transient Load Modeling: Markov Rectangular Pulse Processes," The Danish Center for Applied Mathematics and Mechanics, Report No. 220, 1981.

MADSEN, H. O., R. KILCUP and C. A. CORNELL, "Mean Upcrossing Rate for Sums of Pulse-Type Stochastic Load Processes," in *Proceedings,* Specialty Conference on Probabilistic Mechanics and Structural Reliability, ASCE, Tucson, Ariz., January 1979, pp. 54-58.

MADSEN, H. O and C. J. TURKSTRA, "Residental Floor Loads, Theoretical and Field Study," Report ST.79-9, Department of Civil Engineering and Applied Mechanics, McGill University, Montreal, October 1979.

MALHOTRA, A. K. and J. PENZIEN, "Nondeterministic Analysis of Offshore Structures," *Journal of the Engineering Mechanics Division,* ASCE, Vol. 96, 1970, pp. 985-1003.

McGUIRE, R. K. and C. A. CORNELL, "Live Load Effects in Office Buildings," *Journal of the Structural Division, ASCE, Vol. 100, 1974,* pp. 1351-1366

MOE, G. and S. H. CRANDALL, "Extremes of Morison-Type Wave Loading on a Single Pile," *Journal of Mechanical Design, ASME,* Vol. 100, 1978, pp. 100-104.

MORISON, J. R., M. P. O'BRIEN, J. W. JOHNSON and S. A. SHAAF, "The Force Exerted by Surface Waves on Piles," *Petroleum Transactions,* AIME, Vol. 189, 1950, pp. 149-154.

PEIR, J. C. and C. A. CORNELL, "Spatial and Temporal Variability of Live Loads," *Journal of the Structural Division,* ASCE, Vol. 99, 1973, pp. 903-922.

PHILLIPS, O. M., "The Equilibrium Range in the Spectrum of Wind Generated Waves," *Journal of Fluid Mechanics,* Vol. 4, No. 4, 1958, pp. 426-434.

PIERSON, W. J. and P. HOLMES, "Irregular Wave Forces on Piles,"

Journal of the Waterways and Harbors Division, ASCE, Vol. 91, 1965, pp. 1-10.

PIERSON, W. J. and L. MOSKOWITZ, "A Proposed Spectral Form for Fully Developed Wind Seas Based on the Similarity Theory of S. A. Kitaigorodskii," *Journal of Geophysical Research,* Vol. 69, 1964, pp. 5181-5190.

RAVINDRA, M. K. and T. V. GALAMBOS, "Load and Resistance Factor Design for Steel," *Journal of the Structural Division,* ASCE, Vol. 104, 1978, pp. 1337-1353.

RUIZ, P. and J. PENZIEN, "Stochastic Seismic Response of Structures," *Journal of the Engineering Mechanics Division,* ASCE, Vol. 97, 1971, pp. 441-456.

SARPKAYA, T. and M. ISAACSON, *Mechanics of Wave Forces on Offshore Structures,* Van Nostrand Reinhold, New York, 1981.

SENTLER, L., "A Stochastic Model for Live Loads on Floors in Buildings," Report 60, Division of Building Technology, Lund Institute of Technology, Lund, Sweden, 1975.

SHINOZUKA, M. and Y. SATO, "Simulation of Nonstationary Random Processes," *Journal of the Engineering Mechanics Division,* ASCE, Vol. 93, 1967, pp. 11-40.

SIGBJØRNSSON, R., "Stochastic Theory of Wave Loading Processes," *Engineering Structures,* Vol. 1, 1979, pp. 58-64.

SIMIU, E. and R. U. SCANLAN, *Wind Effects on Structures; An Introduction to Wind Engineering,* John Wiley, New York, 1978.

TAJIMI, H., "A Statistical Method of Determining the Maximum Response of a Building Structure During an Earthquake," *Proceedings,* Second World Conference on Earthquake Engineering, Tokyo and Kyoto, Vol. II, 1960, pp. 781-796.

TAYLOR, R. E. and A. RAJAGOPALAN, "Dynamics of Offshore Structures, Part I: Perturbation Analysis," *Journal of Sound and Vibration,* Vol. 83, 1982, pp. 401-416.

TICKELL, R. G., "Continuous Random Wave Loading on Structural Members," *The Structural Engineer,* Vol. 55, 1977, pp. 209-222.

TURKSTRA, C. J., *Theory of Structural Safety,* SM Study No. 2, Solid Mechanics Division, University of Waterloo, Waterloo, Ontario, 1970.

TURKSTRA, C. J. and H. O. MADSEN, "Load Combinations in Codified Structural Design," *Journal of the Structural Division,* ASCE, Vol. 106, 1980, pp. 2527-2543.

VANMARCKE, E. H., *Random Fields,* MIT Press, Cambridge, Mass., 1983.

VENEZIANO, D., M. GRIGORIU and C. A. CORNELL, "Vector-Process Models for System Reliability," *Journal of the Engineering Mechanics Division,* ASCE, Vol. 103, 1977, pp. 441-460.

WAUGH, C. B., "Approximate Models for Stochastic Load Combinations," Report R77-1, Department of Civil Engineering, Massachusetts Institute of Technology, Cambridge, Mass., January 1977.

WEN, Y.-K., "Methods for Random Vibration of Hysteretic Systems," *Journal of the Engineering Mechanics Division,* ASCE, Vol. 102, 1976, pp. 249-263.

WEN, Y.-K., "Statistical Combination of Extreme Loads," *Journal of the Structural Division,* ASCE, Vol. 103, 1977, pp. 1079-1093.

WEN, Y.-K., "Equivalent Linearization for Hysteretic Systems under Random Excitation," *Journal of Applied Mechanics,* Vol. 47, 1980, pp. 150-154.

WEN, Y.-K., "A Clustering Model for Correlated Load Processes," *Journal of the Structural Division,* ASCE, Vol. 107, 1981, pp. 965-983.

WEN, Y.-K. and H. T. PEARCE, "Stochastic Models for Dependent Load Processes," Structural Research Series No. 489, *UILU-ENG-81-2002,* Civil Engineering Studies, University of Illinois, Urbana, Ill., March 1981.

WEN, Y.-K. and H. T. PEARCE, "Combined Dynamic Effects of Correlated Load Processes," *Nuclear Engineering and Design,* 1983.

WINTERSTEIN, S. R., "Combined Dynamic Response : Extremes and Fatigue Damage," Report R80-46, Department of Civil Engineering, Massachusetts Institute of Technology, Cambridge, Mass., December 1980.

WINTERSTEIN, S. R. and C. A. CORNELL, "Load Combinations and Clustering Effects," *Journal of the Structural Division,* ASCE, Vol. 110, 1984, pp. 2690-2708.

REFERENCES

ABRAMOWITZ, M. and I. STEGUN, *Handbook of Mathematical Functions,* Dover, New York, 1965.

ALLEN, D. E., "Load Combinations," Presented to the December 9, 1977, meeting of the Limit States Design Committee of the Canadian Society for Civil Engineering at Ottawa, Ottawa, Canada, 1977.

ALLEN, D. E., "Criteria for Design Safety Factors and Quality Assurance Expenditure," in *Structural Safety and Reliability,* ed. T. Moan and M. Shinozuka, Proceedings ICOSSAR'81, Trondheim, Norway, June 1981, pp. 667-678.

AMIN, M. and A. H. S. ANG, "Nonstationary Stochastic Model of Earthquake Motions," *Journal of the Engineering Mechanics Division,* ASCE, Vol. 94, 1969, pp. 559-583.

API (American Petroleum Institute), "Recommended Practice for Planning, Designing and Constructing Fixed Offshore Platforms," API RP2A, 13th ed., January 1982.

ASCE (American Society of Civil Engineers), Committee on Fatigue and Fracture Reliability of the Committee on Structural Safety and Reliability of the Structural Division, "Fatigue Reliability 1-4," *Journal of the Structural Division,* ASCE, Vol. 108, 1982, pp. 3-88.

ATALIK, T. S. and S. UTKU, "Stochastic Linearization of Multi-Degree-of-Freedom Nonlinear Systems," *Earthquake Engineering and Structural Dynamics,* Vol. 4, 1976, pp. 411-420.

AWS (American Welding Society), "Structural Welding Code — Steel," AWS D1.1-83, 1983.

BABER, T. T. and Y.-K. WEN, "Random Vibration of Hysteretic, Degrading Systems," *Journal of the Engineering Mechanics Division,* ASCE, Vol. 107, 1981, pp. 1069-1087.

BARLOW, R. E. and F. PROSCHAN, *Statistical Theory of Reliability and Life Testing,* Holt, Rinehart and Winston, New York, 1975.

BARLOW, R. E., J. B. FUSSEL and N. D. SINGPURWALLA, eds., *Theoretical and Applied Aspects of System Reliability and Safety Assessment,* Society for Industrial and Applied Mathematics, Philadelphia, 1975.

BASLER, E., "Analysis of Structural Safety," paper presented to the ASCE Annual Convention, Boston, Mass., June 1960.

BATCHELOR, G. K., *The Theory of Homogeneous Turbulence,* Cambridge University Press, London, 1959.

BENDAT, J. S. and A. G. PIERSOL, *Measurement and Analysis of Random Data,* John Wiley, New York, 1966.

BERNARD, M. C. and J. W. SHIPLEY, "The First-Passage Problem for Stationary Random Structural Vibration," *Journal of Sound and Vibration,* Vol. 24, 1972, pp. 121-132.

BOGDANOFF, J. L., "A New Cumulative Damage Model, Part 1," *Journal of Applied Mechanics,* Vol. 45, 1978, pp. 246-250.

BOGDANOFF, J. L., "A New Cumulative Damage Model, Part 3," *Journal of Applied Mechanics,* Vol. 45, 1978, pp. 733-739.

BOGDANOFF, J. L. and F. KOZIN, "A New Cumulative Damage Model, Part 4," *Journal of Applied Mechanics,* Vol. 47, 1980, pp. 40-44.

BOGDANOFF, J. L. and W. KRIEGER, "A New Cumulative Damage Model, Part 2," *Journal of Applied Mechanics,* Vol. 45, 1978, pp. 251-257.

BOLOTIN, V. V., *Statistical Methods in Structural Mechanics,* Holden-Day, San Francisco, 1969.

BOLOTIN, V. V., *Wahrscheinlichkeitsmethoden zur Berechnung von Konstruktionen,* VEB Verlag für Bauwesen, Berlin, 1981.

BORGMAN, L. E., "Wave Forces on Piling for Narrow-Band Spectra," *Journal of the Waterways and Harbors Division,* ASCE, Vol. 91, 1965, pp. 65-90.

BORGMAN, L. E., "Spectral Analysis of Ocean Wave Forces on Pilings," *Journal of the Waterways and Harbors Division,* ASCE, Vol. 93, 1967, pp. 129-156.

BORGMAN, L. E., "Random Hydrodynamic Forces on Objects," *Annals of Mathematical Statistics,* Vol. 38, 1967, pp. 37-51.

BREITUNG, K., "Asymptotic Approximations for Multinormal Integrals," *Journal of the Engineering Mechanics Division,* ASCE, Vol. 110, 1984, pp. 357-366.

BREITUNG, K. and R. RACKWITZ, "Nonlinear Combination of Load Processes," *Journal of Structural Mechanics,* Vol. 10, No. 2, 1982, pp. 145-166.

BROUWERS, J. J. H. and P. H. J. VERBEEK, "Expected Fatigue Damage and Expected Extreme Response for Morison-type Wave Loading," *Applied Ocean Research,* Vol. 5, 1983, pp. 129-133.

BROWN, C. B., "A Fuzzy Safety Measure," *Journal of the Engineering Mechanics Division,* ASCE, Vol. 105, 1979, pp. 855-872.

CANAVIE, A., M. ARHAN and R. EZRATY, "A Statistical Relationship Between Individual Heights and Periods of Storm Waves," in Vol. 2, *Proceedings,* BOSS'76, Trondheim, Norway, 1976, pp. 354-360.

CARTWRIGHT, D. E. and M. S. LONGUET-HIGGINS, "The Statistical Distribution of the Maxima of a Random Function," *Proceedings of the Royal Society of London,* Vol. A237, 1956, pp. 212-232.

CAUGHEY, T. K., "Equivalent Linearization Techniques," *Journal of the Acoustical Society of America,* Vol. 35, 1963, pp. 1706-1711.

CEB (Comité Europeen du Béton), Joint Committee on Structural Safety CEB-CECM-FIP-IABSE-IASS-RILEM, "First Order Concepts for Design Codes," *CEB Bulletin No. 112,* 1976.

CEB (Comité Europeen du Béton), Joint Committee on Structural Safety CEB-CECM-FIP-IABSE-IASS-RILEM, "Common Unified Rules for Different Types of Construction and Materials," *CEB Bulletin No. 116E,* 1976.

CHEN, X. and N. C. LIND, "Fast Probability Integration by Three Parameter Normal Tail Approximation," *Structural Safety,* Vol. 1, 1983, pp. 269-276.

CHRISTENSEN, J. K. and J. D. SØRENSEN, "Simulering ·af Gaussiske Processer på Datamaskine med Henblik på Fastlæggelse af Brudsandsynligheder," Report 8007 (in Danish), Institute of Building Technology and Structural Engineering, AUC Aalborg, 1980.

CINLAR, E., "Markov Renewal Theory: A Survey," *Management Science,* Vol. 21, 1975, pp. 727-752.

CIRIA (Construction Industry Research and Information Association), "Rationalisation of Safety and Serviceability Factors in Structural Codes," *CIRIA Report No. 63,* London, 1977.

CLOUGH, R. W. and J. PENZIEN, *Dynamics of Structures,* McGraw-Hill, New York, 1975.

CORNELL, C. A., "Bounds on the Reliability of Structural Systems," *Journal of the Structural Division,* ASCE, Vol. 93, 1967, pp. 171-200.

CORNELL, C. A., "Some Thoughts on 'Maximum Probable Loads' and 'Structural Safety Insurance'," *Memorandum,* Department of Civil Engineering, Massachusetts Institute of Technology, to Members of ASCE Structural Safety Committee, March, 1967.

CORNELL, C. A., "A Probability-Based Structural Code," *Journal of the American Concrete Institute,* Vol. 66, No. 12, 1969, pp. 974-985.

CORNELL, C. A. and R. D. LARRABEE, "Representation of Loads for Code Purposes," in *Proceedings,* ICOSSAR'77, ed. H. Kupfer et al., Werner Verlag, Düsseldorf, 1977, pp. 135-148.

COROTIS, R. B. and V. A. DOSHI, "Probability Models for Live Load Survey Results," *Journal of the Structural Division,* ASCE, Vol. 103, 1977, pp. 1257-1274.

COX, D. R. and H. D. MILLER, *The Theory of Stochastic Processes,* Chapman & Hall, London, 1977.

CRANDALL, S. H., "Perturbation Techniques for Random Vibration of Nonlinear Systems," *Journal of the Acoustical Society of America,* Vol. 35, 1963, pp. 1700-1705.

CRANDALL, S. H., "Non-Gaussian Closure for Random Vibration of Nonlinear Oscillators," *International Journal of Non-Linear Mechanics,* Vol. 15, 1980, pp. 303-313.

CRANDALL, S. H., K. L. CHANDIRAMANI and R. G. COOK, "Some First-Passage Problems in Random Vibration," *Journal of Applied Mechanics,* ASME, Vol. 33, 1966, pp. 532-538.

CSA (Canadian Standards Association), "Standards for the Design of Cold-Formed Steel Members in Buildings," *CSA S-136,* 1974, 1981.

DANIELS, H. E., "The Statistical Theory of the Strength of Bundles of Threads," *Proceedings of the Royal Society,* Vol. A183, 1945, pp. 405-435.

DAVENPORT, A. G., "Note on the Distribution of the Largest Value of a Random Function with Application in Gust Loading," *Proceedings of the Institution of Civil Engineers London,* Vol. 28, 1964, pp. 187-196.

DAVENPORT, A. G., "Gust Loading Factors," *Journal of the Structural Division,* ASCE, Vol. 93, 1967, pp. 11-33.

DAVENPORT, A. G., "The Prediction of the Response of Structures to Gusty Wind," in *Safety of Structures under Dynamic Loading,* ed. I. Holand et al., Tapir, Trondheim, 1977, pp. 257-284.

DAVENPORT, A. G., "On the Assessment of the Reliability of Wind Loading on Low Buildings," in *Proceedings,* 5th Colloquium on Industrial Aerodynamics, Aachen, West Germany, June 14-16, 1982.

DEÁK, I., "Three Digit Accurate Multiple Normal Probabilities," *Numerische Mathematik,* Vol. 35, Springer Verlag, Berlin, 1980, pp. 369-380.

DEPARTMENT OF ENERGY, "New Fatigue Design Guidance for Steel Welded Joints in Offshore Structures," Recommendation, UK, 1982.

DER KIUREGHIAN, A., "Structural Response to Stationary Excitation," *Journal of the Engineering Mechanics Division*, ASCE, Vol. 106, 1980, pp. 1195-1213.

DITLEVSEN, O., *Extremes and First Passage Times*, Dissertation, Copenhagen, Denmark, 1971.

DITLEVSEN, O., "Structural Reliability and the Invariance Problem," Research Report No. 22, Solid Mechanics Division, University of Waterloo, Waterloo, Canada, 1973.

DITLEVSEN, O., "Generalized Second Moment Reliability Index," *Journal of Structural Mechanics*, Vol. 7, 1979, pp. 435-451.

DITLEVSEN, O., "Narrow Reliability Bounds for Structural Systems," *Journal of Structural Mechanics*, Vol. 7, 1979, pp. 453-472.

DITLEVSEN, O., *Uncertainty Modeling with Applications to Multidimensional Civil Engineering Systems*, McGraw-Hill, New York, 1981.

DITLEVSEN, O., "Principle of Normal Tail Approximation," *Journal of the Engineering Mechanics Division*, ASCE, Vol. 107, 1981, pp. 1191-1208.

DITLEVSEN, O., "Model Uncertainty in Structural Reliability," *Structural Safety*, Vol. 1, 1982, pp. 73-86.

DITLEVSEN, O., "Gaussian Outcrossing from Safe Convex Polyhedrons," *Journal of the Engineering Mechanics Division*, ASCE, Vol. 109, 1983, pp. 127-148.

DITLEVSEN, O., "Gaussian Safety Margins," in *Proceedings*, Fourth International Conference on Application of Statistics and Probability in Soil and Structural Engineering, ICASP4, University of Firenze, Italy, June 1983, pp. 785-824.

DITLEVSEN, O., "Level Crossings of Random Processes," *Reliability Theory and Its Applications in Structural and Soil Mechanics*, ed. P. Thoft-Christensen, NATO ASI Series E, Martinus Nijhoff, The Hague, 1983, pp. 57-83.

DITLEVSEN, O., "Taylor Expansion of Series System Reliability," *Journal of the Engineering Mechanics Division*, ASCE, Vol. 110, 1984, pp. 293-307.

DITLEVSEN, O. and H. O. MADSEN, "Probabilistic Modeling of Man-Made Load Processes and Their Individual and Combined Effects," in *Structural Safety and Reliability*, ed. T. Moan and M. Shinozuka, Proceedings ICOS-SAR'81, Trondheim, Norway, June 1981, pp. 103-134.

DnV (Det norske Veritas), "Rules for the Design, Construction and Inspection of Offshore Structures, Appendix C Steel Structures," Reprint with Corrections, 1982.

DOWLING, N. E., "Fatigue Failure Prediction Methods for Complicated Stress-Strain Histories," *Journal of Materials*, Vol. 7, 1972, pp. 71-84.

DUDDECK, H., "The Role of Research Models and Technical Models in Engineering Science," in *Proceedings, ICOSSAR'77*, ed. H. Kupfer et al., Werner Verlag, Düsseldorf, 1977, pp. 115-118.

ELLINGWOOD, B. et al., "Development of a Probability Based Load Criterion for American National Standard A58," National Bureau of Standards Publication 577, Washington, D. C., 1980.

FERRY-BORGES, J., "Implementation of Probabilistic Safety Concepts in International Codes," in *Proceedings, ICOSSAR'77*, ed. H. Kupfer et al., Werner Verlag, Düsseldorf, 1977, pp. 121-133.

FERRY-BORGES, J. and M. CASTANHETA, *Structural Safety*, Laboratorio Nacional de Engenharia Civil, Lisbon, 1971.

FIESSLER, B., H.-J. NEUMANN and R. RACKWITZ, "Quadratic Limit States in Structural Reliability," *Journal of the Engineering Mechanics Division*, ASCE, Vol. 105, 1979, pp. 661-676.

FISHER, R. A. and L. H. C. TIPPETT, "Limiting Forms of the Frequency Distribution of the Largest or Smallest Member of a Sample," *Proceedings of the Cambridge Philosophical Society*, Vol. 24, 1928, pp. 180-190.

FORSSELL, C., "Economics and Buildings," *Sunt Förnuft*, 4 (in Swedish), 1924, pp. 74-77. Translated in excerpts in *Structural Reliability and Codified Design*, ed. N. C. Lind, SM Study No. 3, Solid Mechanics Division, University of Waterloo, Waterloo, Canada, 1970.

FOSS, K. A., "Co-ordinates Which Uncouple the Equations of Motion of Damped Linear Dynamic Systems," *Journal of Applied Mechanics*, ASME, Vol. 25, 1958, pp. 361-364.

FREUDENTHAL, A. M., "The Safety of Structures," *Trans. ASCE*, Vol. 112, 1947.

FULLER, J. R., "Boundary Excursions for Combined Random Loads," *AIAA Journal*, Vol. 20, 1982, pp. 1300-1305.

GASPARINI, D. A., "Response of MDOF Systems to Nonstationary Random Excitation," *Journal of the Engineering Mechanics Division*, ASCE, Vol. 105, 1979, pp. 13-27.

GAVER, D. P. and P. A. JACOBS, "On Combination of Random Loads," *SIAM Journal of Applied Mathematics*, Vol. 40, 1981, pp. 454-466.

GEORGE, K. P. and A. A. BASMA, "An Extreme-Value Model for Strength of Stiff Clay," in *Probabilistic Characterization of Soil Properties: Bridge Between Theory and Practice*, eds. D. S. Bowles and H.-Y. Ko, ASCE, 1984, pp. 157-169.

GOODING, E. J., "Investigation of the Tensile Strength of Glass," *Journal of the Society of Glass Technology*, 16, 1932.

GRIFFITH, A. A., "The Phenomenon of Rupture and Flow in Solids," *Philosophical Transactions of the Royal Society*, A221, 1920, pp. 163-168.

GUMBEL, E. J., *Statistics of Extremes,* Columbia University Press, New York, 1958.

HALD, A., *Statistical Theory with Engineering Applications,* John Wiley, New York, 1952.

HAPGOOD, F., "Risk Benefit Analysis: Putting a Price on Life," *Atlantic,* Vol. 243, 1979, pp. 33-38.

HARLOW, D. G. and S. L. PHOENIX, "Probability Distributions for the Strength of Composite Materials I: Two-Level Bounds," *International Journal of Fracture,* Vol. 17, No. 4, 1981, pp. 347-371.

HARMAN, D. J. and A. G. DAVENPORT, "A Statistical Approach to Traffic Loads on Bridges," in *Proceedings,* Specialty Conference on Probabilistic Mechanics and Structural Reliability, ASCE, Tucson, Ariz., January 1979, pp. 170-175.

HARRIS, I. R., "The Nature of the Wind," in *The Modern Design of Wind Sensitive Structures,* CIRIA, London, 1971.

HASOFER, A. M., "Time Dependent Maximum of Floor Live Loads," *Journal of the Engineering Mechanics Division,* ASCE, Vol. 100, 1974, pp. 1086-1091.

HASOFER, A. M. and N. C. LIND, "Exact and Invariant Second Moment Code Format," *Journal of the Engineering Mechanics Division,* ASCE, Vol. 100, 1974, pp. 111-121.

HASSELMANN, K. et al., "Measurements of the Wind Wave Growth and Swell Decay during the Joint North Sea Wave Project (JONSWAP)," *Deutchen Hydrographischen Zeitschrift,* Ergazungsheft, Reihe A, Nr. 12, Hamburg, 1973.

HOGG, R. V. and A. T. CRAIG, *Introduction to Mathematical Statistics,* Collier-Macmillan, New York, 1966.

HOHENBICHLER, M., "An Approximation to the Multivariate Normal Distribution, in *Proceedings,* EUROMECH 155 Reliability Theory of Structural Engineering Systems, DIALOG 6-82, Danish Engineering Academy, Lyngby, Denmark, 1982, pp. 79-100.

HOHENBICHLER, M., "Resistance of Large Brittle Parallel Systems," in *Proceedings,* Fourth International Conference on Application of Statistics and Probability in Soil and Structural Engineering, ICASP4, University of Firenze, Italy, June 1983, pp. 1301-1312.

HOHENBICHLER, M., "An Asymptotic Formula for the Probability of Intersections," *Berichte zur Zuverlässigkeitstheorie der Bauwerke,* Heft 69, LKI, Technische Universität München, 1984, pp. 21-48.

HOHENBICHLER, M., "Mathematische Grundlagen der Zuverlässigkeitsmethode Erste Ordnung und Einige Erweiterungen," Doctoral Thesis at the Technical University of Munich, Munich, West Germany, 1984.

HOHENBICHLER, M. and R. RACKWITZ, "Nonnormal Dependent Vectors in Structural Reliability, *Journal of the Engineering Mechanics Division,* ASCE, Vol. 107, 1981, pp. 1127-1238.

HOHENBICHLER, M. and R. RACKWITZ, "First-Order Concepts in System Reliability," *Structural Safety,* Vol. 1, 1983, pp. 177-188.

HOHENBICHLER, M. and R. RACKWITZ, "Reliability of Parallel Systems under Imposed Uniform Strain," *Journal of the Engineering Mechanics Division,* ASCE, Vol. 109, 1983, pp. 896-907.

HOLMES, P. and R. G. TICKELL, "Full Scale Wave Loading on Cylinders," in *Proceedings,* BOSS'79, Imperial College, London, August 28-31 1979, Paper 79.

HOUSNER, G. W. and P. C. JENNINGS, "Generation of Artificial Earthquakes," *Journal of the Engineering Mechanics Division,* ASCE, Vol. 90, 1964, pp. 113-150.

IRVING, P. E. and L. N. MCCARTNEY, "Prediction of Fatigue Crack Growth Rates: Theory, Mechanisms, and Experimental Results," Fatigue 77 Conference, Cambridge University, *Metal Science,* August/September 1977, pp. 351-361.

ISSC (International Ship Structures Congress), Report of Committee 1, Proceedings of the Second International Ship Structures Congress, Delft, The Netherlands, July 20-24, 1964.

IWAN, W. D. and L. D. LUTES, "Response of the Bilinear Hysteretic System to Stationary Random Excitation," *Journal of the Acoustical Society of America,* Vol. 13, 1968, pp. 545-552.

IYENGAR, R. N., "Random Vibration of a Second-Order Nonlinear Elastic System," *Journal of Sound and Vibration,* Vol. 40, 1975, pp. 155-165.

IYENGAR, R. N. and K. T. S. R. IYENGAR, "A Nonstationary Random Process Model for Earthquake Accelerograms," *Bulletin of the Seismological Society of America,* Vol. 59, 1969, pp. 1163-1188.

JOHNSON, A. I., *Strength, Safety and Economical Dimensions of Structures,* Statens Kommitte för Byggnadsforskning, Meddelanden No. 22, Stockholm, 1953.

JOHNSON, N. L. and S. KOTZ, *Distributions in Statistics: Continuous Multivariate Distributions,* John Wiley, New York, 1972.

JOHNSON, N. L. and B. L. WELCH, "Applications of the Non-Central *t*-Distribution," *Biometrika,* Vol. 31, 1940, pp. 362-389.

JCSS (Joint Committee on Structural Safety), "General Principles of Safety and Serviceability Regulations for Structural Design," ed. L. Ostlund, Lund Institute of Technology, Lund, Sweden, 1978.

KAIMAL, J. C., J. C. WYNGAARD, Y. IZUMI and O. R. COTE, "Spectral Characteristics of Surface Layer Turbulence," *Quarterly Journal of the Royal Meteorological Society,* Vol. 98, 1972, pp. 563-589.

KANAI, K., "Some Empirical Formulas for the Seismic Characteristics of the Ground," *Bulletin of the Earthquake Research Institute,* University of Tokyo, Vol. 35, 1957, pp. 309-325.

KAUFMANN, A., D. GROUCHKO and R. CRUON, *Mathematical Models for the Study of the Reliability of Systems,* Academic Press, New York, 1977.

KRENK, S., "A Double Envelope for Stochastic Processes," The Danish Center for Applied Mathematics and Mechanics, Report No. 134, Lyngby, Denmark, 1978.

KRENK, S., "Nonstationary Narrow-Band Response and First-Passage Probability," *Journal of Applied Mechanics,* ASME, Vol. 46, 1979, pp. 919-924.

KRENK, S., "Generalized Hermite Polynomials and the Spectrum of Non-linear Random Vibration," in *Probabilistic Methods in the Mechanics of Solids and Structures,* Springer-Verlag, Berlin, 1985.

KRENK, S., H. O. MADSEN and P. H. MADSEN, "Stationary and Transient Response Envelopes," *Journal of Engineering Mechanics,* ASCE, Vol. 109, 1983, pp. 263-278.

KRENK, S. and P. H. MADSEN, "Stochastic Response Analysis," in *Reliability Theory and Its Applications in Structural and Soil Mechanics,* ed. P. Thoft-Christensen, NATO ASI Series E, Martinus Nijhoff, The Hague, 1983, pp. 103-172.

KRISTENSEN, L. and N. O. JENSEN, "Lateral Coherence in Isotropic Turbulence and in the Natural Wind," *Boundary Layer Meteorology,* Vol. 17, 1979, pp. 353-373.

LAIRD, C., "The Influence of Metallurgical Structure on the Mechanism of Fatigue Crack Propagation," in *Fatigue Crack Propagation,* ASTM STP 415, 1967, pp. 131-181.

LARRABEE, R. D., "Modeling Extreme Vehicle Loads on Highway Bridges," in *Proceedings,* Specialty Conference on Probabilistic Mechanics and Structural Reliability, ASCE, Tucson, Ariz., January 1979, pp. 176-180.

LARRABEE, R. D. and C. A. CORNELL, "Combination of Various Load Processes," *Journal of the Structural Division,* ASCE, Vol. 107, 1981, pp. 223-239.

LEVE, H. L., "Cumulative Damage Theories," in *Metal Fatigue Theory and Design,* ed. A. F. Madayag, John Wiley, New York, 1969.

LIGHTHILL, J., *Waves in Fluids,* Cambridge University Press, Cambridge, 1978.

LIN, Y. K., *Probabilistic Theory of Structural Dynamics,* Krieger Publishing, Huntington, N. Y., 1976.

LIND, N. C., "The Design of Structural Design Norms," *Journal of Structural Mechanics,* Vol. 1, No. 3, 1973, pp. 357-370.

LIND, N. C., "Approximate Analysis and Economics of Structures," *Journal of the Structural Division,* ASCE, Vol. 102, 1976, pp. 1177-1196.

LIND, N. C., "Management of Gross Errors," in *Proceedings,* Fourth International Conference on Application of Statistics and Probability in Soil and Structural Engineering, ICASP4, University of Firenze, Italy, June 1983, pp. 669-682.

LIND, N. C., "Models of Human Error in Structural Reliability," *Structural Safety,* Vol. 1, 1983, pp. 167-175.

LIND, N. C., "Structural Quality and Human Error," in *Reliability Theory and Its Applications in Structural and Soil Mechanics,* ed. P. Thoft-Christensen, NATO ASI Series E, Martinus Nijhoff, The Hague, 1983, pp. 225-236.

LIND, N. C., C. J. TURKSTRA and D. T. WRIGHT, "Safety, Economy and Rationality of Structural Design," in *Proceedings,* IABSE 7th Congress, Rio de Janeiro, Preliminary Publication, 1964, pp. 185-192.

LINDGREN, G. and I. RYCHLIK, "Wave Characteristic Distributions for Gaussian Waves — Wave Length, Amplitude and Steepness," *Ocean Engineering,* Vol. 9, 1982, pp. 411-432.

LINDLEY, D. V., *Probability and Statistics, Vol. 2: Inference,* Cambridge University Press, Cambridge, 1965.

LOMNITZ, C. and E. ROSENBLUETH eds., *Seismic Risk and Engineering Decisions,* Elsevier, Amsterdam, 1976.

LONGUET-HIGGINS, M. S., "The Distribution Intervals between Zeros of a Stationary Random Function," *Philosophical Transactions of the Royal Society of London,* Vol. A254, 1962, pp. 557-599.

LONGUET-HIGGINS, M. S., "On the Joint Distribution of the Period and Amplitude of Sea Waves," *Journal of Geophysical Research,* Vol. 80, 1975, pp. 2688-2694.

LONGUET-HIGGINS, M. S., "On the Joint Distribution of Wave Periods and Amplitudes in a Random Wave Field," *Proceedings of the Royal Society of London,* Vol. A389, 1983, pp. 241-258.

LUCE, R. D. and H. RAIFFA, *Games and Decisions,* John Wiley, New York, 1957.

MACCAMY, R. S. and R. A. FUCHS, "Wave Forces on Piles: A Diffraction Theory," U.S. Army Corps of Engineers, Beach Erosion Board, *Technical Memo No. 69,* Washington, D.C., 1954.

MADSEN, H. O., "Some Experience with the Rackwitz and Fiessler Algorithm for the Calculation of Structural Reliability under Combined Loading," in *DIALOG 77,* Danish Engineering Academy, Lyngby, Denmark, 1978, pp. 73-98.

MADSEN, H. O., "Load Models and Load Combinations," Report No. R113, Department of Structural Engineering, Technical University of Denmark, February 1979.

MADSEN, H. O., "A Line Model for Furniture Loads," Report ST.79-8, Department of Civil Engineering and Applied Mechanics, McGill University, Montreal, August, 1979.

MADSEN, H. O., "Reliability under Combination of Nonstationary Load Processes," in *DIALOG 1-82,* Danish Engineering Academy, Lyngby, Denmark, 1982, pp. 45-58.

MADSEN, H. O., "Deterministic and Probabilistic Models for Damage Accumulation Due to Time Varying Loading," DIALOG 5-82, Danish Engineering Academy, Lyngby, Denmark, 1982.

MADSEN, H. O., "Random Fatigue Crack Growth and Inspection," in *Proceedings ICOSSAR'85,* Kobe, Japan, May 1985.

MADSEN, H. O. and O. BACH-GANSMO, "Design Wave Determination by Fast Integration Technique," in *System Modelling and Optimization,* Proceedings of the 11th FIP Conference, Copenhagen, July 25-29, 1983, Springer-Verlag, Berlin, 1984, pp. 471-477.

MADSEN, H. O. and O. DITLEVSEN, "Transient Load Modeling: Markov Rectangular Pulse Processes," The Danish Center for Applied Mathematics and Mechanics, Report No. 220, Lyngby, Denmark, 1981.

MADSEN, H. O., R. KILCUP and C. A. CORNELL, "Mean Upcrossing Rate for Sums of Pulse-Type Stochastic Load Processes," in *Proceedings,* Specialty Conference on Probabilistic Mechanics and Structural Reliability, ASCE, Tucson, Ariz., January 1979, pp. 54-58.

MADSEN, H. O. and C. J. TURKSTRA, "Residental Floor Loads, Theoretical and Field Study," Report ST.79-9, Department of Civil Engineering and Applied Mechanics, McGill University, Montreal, October 1979.

MADSEN, P. H. and S. KRENK, "Stationary and Transient Response Statistics," *Journal of the Engineering Mechanics Division,* ASCE, Vol. 108, 1982, pp. 622-635.

MADSEN, P. H. and S. KRENK, "An Integral Equation for the First-passage Problem in Random Vibration," *Journal of Applied Mechanics,* Vol. 51, 1984, pp. 674-679.

MALHOTRA, A. K. and J. PENZIEN, "Nondeterministic Analysis of Offshore Structures," *Journal of the Engineering Mechanics Division,* ASCE, Vol. 96, 1970, pp. 985-1003.

MATOUSEK, M., "Outcome of a Survey on 800 Construction Failures," in *Proceedings,* IABSE Colloquium on Inspection and Quality Control Institute of Structural Engineering, Swiss Federal Institute of Technology, Zurich, 1977.

MAYER, M., *Die Sicherheit der Bauwerke,* Springer-Verlag, Berlin 1926.

MCGUIRE, R. K. and C. A. CORNELL, "Live Load Effects in Office Buildings," *Journal of the Structural Division,* ASCE, Vol. 100, 1974, pp. 1351-1366.

MELCHERS, R. E. and M. V. HARRINGTON, "Structural Reliability as Affected by Human Error," in *Proceedings, Fourth International Conference on Application of Statistics and Probability in Soil and Structural Engineering,* ICASP4, University of Firenze, Italy, June 1983, pp. 683-694.

MILLER, K. S., *Complex Stochastic Processes,* Addison-Wesley, Reading, Mass., 1974.

MILTON, R. C., "Computer Evaluation of the Multivariate Normal Integral," *Technometrics,* Vol. 14, 1972, pp. 881-889.

MINER, M. A., "Cumulative Damage in Fatigue," *Journal of Applied Mechanics,* ASME, Vol. 12, 1945, pp. A159-A164.

MOE, G. and S. H. CRANDALL, "Extremes of Morison-Type Wave Loading on a Single Pile," *Journal of Mechanical Design,* ASME, Vol. 100, 1978, pp. 100-104.

MORISON, J. R., M. P. O'BRIEN, J. W. JOHNSON and S. A. SHAAF, "The Force Exerted by Surface Waves on Piles," *Petroleum Transactions,* AIME, Vol. 189, 1950, pp. 149-154.

NBCC (National Building Code of Canada), National Research Council of Canada, Ottawa, Ontario, 1975, 1977, 1980.

NIELSEN, S. K., "Probability of Failure of Structural Systems under Random Vibration," Report No. 8001, Institute of Building Technology and Structural Engineering, AUC Aalborg, 1980.

NKB (The Nordic Committee on Building Regulations), "Recommendations for Loading and Safety Regulations for Structural Design," *NKB-Report,* No. 36, Copenhagen, November 1978.

NOWAK, A. S. and N. C. LIND, "Practical Bridge Code Calibration," *Journal of the Structural Division,* ASCE, Vol. 105, 1979, pp. 2497-2510.

OFVERBECK, P., "Small Sample Control and Structural Safety," Report TVBK-3009, Department of Structural Engineering, Lund Institute of Technology, Lund, Sweden, 1980.

OHBDC (Ontario Highway Bridge Design Code), Ontario Ministry of Transportation and Communication, Downsview, Ontario, 1983.

ORTIZ, K., "Stochastic Modeling of Fatigue Crack Growth," Ph.D. dissertation, Stanford University, Stanford, California, 1984.

PALMGREN, A., "Die Lebensdauer von Kugellagern," *Zeitschrift der Vereines Deutches Ingenieure,* Vol. 68, No. 14, 1924, pp. 339-341.

PAPOULIS, A., *Probability, Random Variables, and Stochastic Processes,* McGraw-Hill, Tokyo, 1965.

PARIS, P. C. and F. ERDOGAN, "A Critical Analysis of Crack Propagation Laws," *Journal of Basic Engineering,* ASME, Vol. 85, 1963, pp. 528-534.

PARIS, P. C. and G. C. SIH, "Stress Analysis of Cracks," in *Fracture Toughness Testing and Its Applications,* ASTM STP 381, 1965, pp. 30-82.

PARZEN, E., *Stochastic Processes,* Holden-Day, San Francisco, 1962.

PEIR, J. C. and C. A. CORNELL, "Spatial and Temporal Variability of Live Loads," *Journal of the Structural Division,* ASCE, Vol. 99, 1973, pp. 903-922.

PHILLIPS, O. M., "The Equilibrium Range in the Spectrum of Wind Generated Waves," *Journal of Fluid Mechanics,* Vol. 4, No. 4, 1958, pp. 426-434.

PIERSON, W. J. and P. HOLMES, "Irregular Wave Forces on Piles," *Journal of the Waterways and Harbors Division,* ASCE, Vol. 91, 1965, pp. 1-10.

PIERSON, W. J. and L. MOSKOWITZ, "A Proposed Spectral Form for Fully Developed Wind Seas Based on the Similarity Theory of S. A. Kitaigorodskii," *Journal of Geophysical Research,* Vol. 69, 1964, pp. 5181-5190.

PLUM, N. M., "Is the Design of Our Houses Rational When Initial Cost, Maintenance, and Repair Are Taken into Account?" *Ingeniøren,* 50, 1950, p. 454.

PRIESTLY, M. B., "Evolutionary Spectra and Nonstationary Processes," *Journal of the Royal Statistical Society,* Vol. B27, 1965, pp. 204-237.

PRIESTLY, M. B., "Power Spectral Analysis of Nonstationary Random Processes," *Journal of Sound and Vibration,* Vol. 6, 1967, pp. 86-97.

PUGSLEY, A., *The Safety of Structures,* Edward Arnold, London, 1966.

RACKWITZ, R., "Close Bounds for the Reliability of Structural Systems," *Berichte zur Zuverlassigkeitstheorie der Bauwerke,* Heft 29, LKI, Technische Universität München, 1978, pp. 67-78.

RACKWITZ, R. and B. FIESSLER, "Structural Reliability under Combined Random Load Sequences," *Computers & Structures,* Vol. 9, 1978, pp. 489-494.

RAVINDRA, M. K. and T. V. GALAMBOS, "Load and Resistance Factor Design for Steel," *Journal of the Structural Division,* ASCE, Vol. 104, 1978, pp. 1337-1353.

RAVINDRA, M. K. and N. C. LIND, "Theory of Structural Code Optimization," *Journal of the Structural Division,* ASCE, Vol. 99, 1973, pp. 541-553.

RAVINDRA, M. K. and N. C. LIND, "Trends in Safety Factor Optimization," *Beams and Beam Columns,* ed. R. Narayanan, Applied Science Publishers, Barking, Essex, England, 1983, pp. 207-236.

RAVINDRA, M. K., N. C. LIND and W. W. SIU, "Illustrations of Reliability-Based Design," *Journal of the Structural Division,* ASCE, Vol. 100, 1974, pp. 1789-1811.

RICE, J. R. and F. P. BEER, "On the Distribution of Rises and Falls in a Continuous Random Process," *Journal of Basic Engineering,* Vol. 87, 1965, pp. 398-404.

RICE, J. R. and F. P. BEER, "First-Occurrence Time of High-Level Crossings in a Continuous Random Process," *Journal of the Acoustical Society of America,* Vol. 39, 1966, pp. 323-335.

RICE, S. O., "Mathematical Analysis of Random Noise," *Bell System Technical Journal,* Vol. 23, 1944, pp. 282-332 and Vol. 24, 1944, pp. 46-156. Reprinted in *Selected Papers in Noise and Stochastic Processes,* ed. N. Wax, Dover, New York, 1954.

RICE, S. O., "Distributions of Quadratic Forms in Normal Random Variables − Evaluation by Numerical Integration," SIAM, *J. Sci. Stat. Comp.,* Vol. 1, 1980, pp. 438-448.

RICHARDS, C. W., "Size Effect in the Tension Test of Mild Steel," *Proceedings,* ASTM, Vol. 54, 1954, pp. 995-1002.

ROBERTS, J. B., "Probability of First-Passage for Nonstationary Random Vibration," *Journal of Applied Mechanics,* ASME, Vol. 42, 1975, pp. 716-720.

ROSENBLATT, M., "Remarks on a Multivariate Transformation, *The Annals of Mathematical Statistics,* Vol. 23, 1952, pp. 470-472.

ROSENBLUETH, E. and L. ESTEVA, "Reliability Basis for Some Mexican Codes," *ACI Publication SP-31,* 1972, pp. 1-41,

RUBEN, H., "An Asymptotic Expansion for the Multivariate Normal Distribution and Mill's Ratio," *Journal of Research NBS,* Vol. 68B (Mathematics and Mathematical Physics), No. 1, 1964, pp. 3-11.

RUBINSTEIN, Y., *Simulation and the Monte Carlo Method,* John Wiley, New York, 1981.

RUIZ, P. and J. PENZIEN, "Stochastic Seismic Response of Structures," *Journal of the Engineering Mechanics Division,* ASCE, Vol. 97, 1971, pp. 441-456.

SARPKAYA, T. and M. ISAACSON, *Mechanics of Wave Forces on Offshore Structures,* Van Nostrand Reinhold, New York, 1981.

SAVAGE, I. R., "Mill's Ratio for Multivariate Normal Distributions," *Journal of Research NBS,* Vol. 66B (Mathematics and Mathematical Physics), No. 3, 1962, pp. 93-96.

SCHIJVE, J., "The Accumulation of Fatigue Damage in Aircraft Materials and Structures," *AGARD Conference Proceedings No. 118,* Symposium on Random Load Fatigue, Lyngby, Denmark, April 1972, pp. 3.1-3.120.

SCHIJVE, J., "Four Lectures on Fatigue Crack Growth," *Engineering Fracture Mechanics,* Vol. 11, 1979, pp. 167-221.

SCHWARTZ, M. W., et al., "Load Combination Methodology Development. Load Combination Program: Project II Final Report," NUREG/CR-2087, UCRL-53025, Lawrence Livermore Laboratory, Pasadena, Cal., July 1981.

SENTLER, L., "A Stochastic Model for Live Loads on Floors in Buildings," Report 60, Division of Building Technology, Lund Institute of Technology, Lund, Sweden, 1975.

SHINOZUKA, M. and Y. SATO, "Simulation of Nonstationary Random Processes," *Journal of the Engineering Mechanics Division,* ASCE, Vol. 93, 1967, pp. 11-40.

SHIPLEY, J. W. and M. C. BERNARD, "The First Passage Time Problem for Simple Structural Systems," *Journal of Applied Mechanics,* ASME, Vol. 39, 1972, pp. 911-917.

SIGBJØRNSSON, R., "Stochastic Theory of Wave Loading Processes," *Engineering Structures,* Vol. 1, 1979, pp. 58-64.

SIMIU, E. and R. U. SCANLAN, *Wind Effects on Structures; An Introduction to Wind Engineering,* John Wiley, New York, 1978.

SIU, W. W., S. R. PARIMI and N. C. LIND, "Practical Approach to Code Calibration," *Journal of the Structural Division,* ASCE, Vol. 101, 1975, pp. 1469-1480.

SMITH, D. W., "Bridge Failures," *Proceedings of the Civil Engineers,* Vol. 60, 1976, pp. 367-382.

SMITH, E., "On Nonlinear Random Vibrations," Report No. 78-3, Division of Structural Mechanics, The Norwegian Institute of Technology, Trondheim, Norway, 1978.

SPANOS, P.-T. D., "Spectral Moments Calculation of Linear System Output," *Journal of Applied Mechanics,* ASME, Vol. 50, 1983, pp. 901-903.

STALLMEYER, J. E. and W. H. WALKER, "Cumulative Damage Theories and Application," *Journal of the Structural Division,* ASCE, Vol. 94, 1968, pp. 2739-2750.

STOKER, J. J., *Nonlinear Vibration,* Interscience, New York, 1950.

STRATONOVICH, R. L., *Topics in the Theory of Random Noise,* Gordon and Breach, New York, 1963.

SU, H. L., "Statistical Approach to Structural Design," *Proceedings of the Institute of Civil Engineers,* Vol. 13, 1959, pp. 353-362.

TAJIMI, H., "A Statistical Method of Determining the Maximum Response of a Building Structure During an Earthquake," *Proceedings,* Second World

Conference on Earthquake Engineering, Tokyo and Kyoto, Vol. II, 1960, pp. 781-796.

TAYLOR, R. E. and A. RAJAGOPALAN, "Dynamics of Offshore Structures, Part I: Perturbation Analysis," *Journal of Sound and Vibration,* Vol. 83, 1982, pp. 401-416.

THOFT-CHRISTENSEN, P. and M. BAKER, *Structural Reliability Theory and Its Applications,* Springer-Verlag, Berlin, 1982.

TICKELL, R. G., "Continuous Random Wave Loading on Structural Members," *The Structural Engineer,* Vol. 55, 1977, pp. 209-222.

TURKSTRA, C. J., *Theory of Structural Safety,* SM Study No. 2, Solid Mechanics Division, University of Waterloo, Waterloo, Ontario, 1970.

TURKSTRA, C. J. and H. O. MADSEN, "Load Combinations in Codified Structural Design," *Journal of the Structural Division,* ASCE, Vol. 106, 1980, pp. 2527-2543.

TVEDT, L., "Two Second Order Approximations to the Failure Probability," Veritas Report RDIV/20-004-83, Det norske Veritas, Oslo, Norway, 1983.

VANMARCKE, E. H., "On the Distribution of the First-Passage Time for Normal Stationary Random Processes," *Journal of Applied Mechanics,* ASME, Vol. 42, 1975, pp. 215-220.

VANMARCKE, E. H., *Random Fields,* MIT Press, Cambridge, Mass., 1983.

VENEZIANO, D., "Contributions to Second Moment Reliability Theory," Research Report R74-33, Department of Civil Engineering, Massachusetts Institute of Technology, Cambridge, Mass., April 1974.

VENEZIANO, D., "Probabilistic Seismic Resistance of Reinforced Concrete Frames," in *Structural Safety and Reliability,* ed. T. Moan and M. Shinozuka, Proceedings ICOSSAR'81, Trondheim, Norway, June 1981, pp. 241-258.

VENEZIANO, D., M. GRIGORIU and C. A. CORNELL, "Vector-Process Models for System Reliability," *Journal of the Engineering Mechanics Division,* ASCE, Vol. 103, 1977, pp. 441-460.

VIRKLER, D. A., B. M. HILBERRY and P. K. GOEL, "The Statistical Nature of Fatigue Crack Propagation," *Journal of Engineering Materials and Technology,* ASME, Vol. 101, 1979, pp. 148-153.

WANG, M. C. and G. E. UHLENBECK, "On the Theory of the Brownian Motion II," *Review of Modern Physics,* Vol. 17, 1945, pp. 323-342. Reprinted in *Selected Papers on Noise and Stochastic Processes,* ed. N. Wax, Dover, New York, 1954.

WAUGH, C. B., "Approximate Models for Stochastic Load Combinations," Report R77-1, Department of Civil Engineering, Massachusetts Institute of Technology, Cambridge, Mass., January 1977.

WEIBULL, W., "A Statistical Theory of the Strength of Materials," *Proceed-

ings, Royal Swedish Institute of Engineering Research, No. 151, Stockholm, Sweden, 1939.

WEIBULL, W., "The Phenomenon of Rupture in Solids," *Proceedings, Royal Swedish Institute of Engineering Research,* No. 153, Stockholm, Sweden, 1939.

WEIBULL, W., "A Statistical Distribution Function of Wide Applicability," *Journal of Applied Mechanics,* ASME, Vol. 18, 1951.

WEN, Y.-K., "Methods for Random Vibration of Hysteretic Systems," *Journal of the Engineering Mechanics Division,* ASCE, Vol. 102, 1976, pp. 249-263.

WEN, Y.-K., "Statistical Combination of Extreme Loads," *Journal of the Structural Division,* ASCE, Vol. 103, 1977, pp. 1079-1093.

WEN, Y.-K., "Equivalent Linearization for Hysteretic Systems under Random Excitation," *Journal of Applied Mechanics,* Vol. 47, 1980, pp. 150-154.

WEN, Y.-K., "A Clustering Model for Correlated Load Processes," *Journal of the Structural Division,* ASCE, Vol. 107, 1981, pp. 965-983.

WEN, Y.-K. and H. T. PEARCE, "Stochastic Models for Dependent Load Processes," Structural Research Series No. 489, *UILU-ENG-81-2002,* Civil Engineering Studies, University of Illinois, Urbana, Ill., March 1981.

WEN, Y.-K. and H. T. PEARCE, "Combined Dynamic Effects of Correlated Load Processes," *Nuclear Engineering and Design,* 1983.

WINTERSTEIN, S. R., "Combined Dynamic Response: Extremes and Fatigue Damage," Report R80-46, Department of Civil Engineering, Massachusetts Institute of Technology, Cambridge, Mass., December 1980.

WINTERSTEIN, S. R. and C. A. CORNELL, "Load Combinations and Clustering Effects," *Journal of the Structural Division,* ASCE, Vol. 110, 1984, pp. 2690-2708.

WIRSCHING, P. H. and A. M. SHEHATA, "Fatigue under Wide Band Random Stresses Using the Rain-Flow Method," *Journal of Engineering Materials and Technology,* ASME, July 1977, pp. 205-211.

YANG, J.-N., "First Excursion Probability in Nonstationary Random Vibration," *Journal of Sound and Vibration,* Vol. 27, 1973, pp. 165-182.

YANG, J.-N., "Approximation to First Passage Probability," *Journal of the Engineering Mechanics Division,* ASCE, Vol. 101, 1975, pp. 361-372.

YANG, J.-N. and M. SHINOZUKA, "On the First Excursion Probability in Stationary Narrow-Band Random Vibration," *Journal of Applied Mechanics,* ASME, Vol. 38, 1971, pp. 1017-1022.

YANG, J.-N. and M. SHINOZUKA, "On the First Excursion Probability in Stationary Narrow-Band Random Vibration, II," *Journal of Applied Mechanics,* ASME, Vol. 39, 1972, pp. 733-730.

ZECH, B. and F. H. WITTMAN, "Probabilistic Approach to Describe the Behavior of Materials," *Transactions of SMIRT 4,* 1977, pp. 575-584.

INDEX